Images of Materials

IMAGES OF MATERIALS

Edited by

DAVID B. WILLIAMS

Department of Materials Science and Engineering
Lehigh University
Bethlehem, Pennsylvania

ALAN R. PELTON

Metals Division
Raychem Corporation
Menlo Park, California

RONALD GRONSKY

Department of Materials Science
and Mineral Engineering
University of California
Berkeley, California

With a Foreword by
Sir Peter B. Hirsch
Department of Metallurgy
and Science of Materials
University of Oxford

New York   Oxford
OXFORD UNIVERSITY PRESS
1991

Oxford University Press

Oxford   New York   Toronto
Delhi   Bombay   Calcutta   Madras   Karachi
Petaling Jaya   Singapore   Hong Kong   Tokyo
Nairobi   Dar es Salaam   Cape Town
Melbourne   Auckland

and associated companies in
Berlin   Ibadan

Copyright © 1991 by Oxford University Press, Inc.

Published by Oxford University Press, Inc.,
200 Madison Avenue, New York, New York 10016

Library of Congress Cataloging-in-Publication Data
Images of materials /
edited by David B. Williams, Alan R. Pelton, Ronald Gronsky.
p.   cm.
Includes bibliographical references and index.
ISBN 0-19-505856-9
1. Materials—Microscopy.   I. Williams, David B. (David Bernard), 1949–
II. Pelton, Alan R.   III. Gronsky, R. (Ronald)
TA418.7.I44   1991   620.1′127—dc20   90-14181

9 8 7 6 5 4 3 2 1

Printed in the United States of America
on acid-free paper

Foreword

The relationship between structure and property is central to materials science. The properties of materials are controlled by composition and structure ranging from atomic to macroscopic dimensions, and the materials scientist needs to have techniques available that provide this compositional and structural information both for the interior and the surface of materials. Optical metallography, pioneered by Sorby in the middle of the last century, was the first technique that revealed the microstructure of metals from images of the surface at moderate magnification. This technique, vastly improved over the years, is firmly established as an essential tool in materials science. But also over the last fifty years, developments mainly in electron and ion optical techniques have led to an impressive range of new methods for imaging structures inside and at the surface of materials down to atomic dimensions, and for identifying phases and measuring composition changes on a scale of a few atoms. Electron microscopy and electron diffraction were the techniques that spearheaded many of these new developments; electron probe microanalysis and field-ion microscopy followed. These techniques have undergone spectacular developments over the years, many of them pioneered by the materials scientists themselves. They now provide a powerful armory and wide variety of tools for the characterization of materials. In the early days of electron microscopy the general approach paralleled that of optical metallography, obtaining microstructural information from images of surface structure, and that remains an important, well-established method. But the need for observing the interior of materials led to the thin foil technique and eventually to the high-resolution transmission electron microscopes of today that are capable of atomic resolution. And the requirement for structural and compositional information with high resolution has resulted in the current sophisticated microanalytical transmission electron microscopes, which combine imaging with facilities such as selected area diffraction, convergent beam electron diffraction, energy dispersive x-ray analysis, and electron energy-loss spectroscopy. The scanning electron microscope has proved equally versatile in providing various modes for imaging with secondary or backscattered electrons for topographic and elemental composition contrast, voltage and magnetic field contrast, and with electron-beam-induced conductivity for semiconductors. Crystallographic information can be obtained from channeling patterns, composition from x-ray microanalysis, spectroscopic data from cathodoluminescence, etc. Field-ion microscopy has developed similarly into a technique providing both structural and compositional information, with the atom probe and the position-sensitive atom probe providing data from which three-dimensional composition maps with near atomic resolution can be reconstructed. Advances in microelectronics, computers, and vacuum technology have played an important role in all these developments, and the availability of powerful computers in particular has led to equally important developments in image processing and interpretation.

As if this were not already sufficient to satisfy the materials scientist, the scanning tunneling and atomic force microscopes mark a revolutionary advance for imaging surfaces, providing information on atomic positions and surface electronic structure in unprecedented detail, and much potential for further development, e.g., on mechanical response of materials on a nanometer scale. At the micron level of resolution the acoustic microscope provides a powerful means of imaging differences of elastic response in a material.

The availability of this wide range of different techniques makes it essential to have a reference work assembled that gives a simple, but state-of-the-art, introduction to the various techniques, their applications, and the use of image-processing methods. *Images of Materials,* which consists of contributions from experts in the various fields, is such a resource book. This volume should give the materials scientist a valuable overview of the powerful tools now available for the characterization of materials.

18 December 1990   P. B. Hirsch

Preface

Images of Materials celebrates the beauty of the structures of the materials that we use in our everyday life—the metals that we drive and fly in; the polymers that we wear; the ceramics that may one day provide us with resistance-free electrical transmission through the marvellous new superconductors; and of course the ubiquitous silicon semiconductors, without which society today would fall into disarray and confusion. The pictures in this book are also a tribute to the engineers whose technical achievements have given materials scientists, chemists, and physicists an ever-more versatile array of instrumentation to probe the structure and chemistry of materials down to the atomic level.

It is almost two hundred years since it was first realized that metals had structure. In 1805 Howard in England and, independently, Count Alois von Widmanstätten in Germany a few years later observed the coarse structure of iron meteorites that is visible to the naked eye. Amazingly, the Widmanstätten structure still remains a topic of active research, and examples of this classic structure are to be found in this book. What Widmanstätten and Howard could see were the boundaries between different phases in the meteorite, and these boundaries are one of many different kinds of structural defects in materials. It wasn't until nearly sixty years after these initial observations that the then well-established technique of light microscopy was applied to the study of steels by Sorby in Sheffield. From Sorby's initial observations it became apparent that the properties of steels could be related to their defect structures, and this discovery laid the foundations of metallurgical science and the technique of "Metallography" was born. From many of the pictures in this book it will become apparent that it is the imaging of the defects in materials that remains the most important aspect of metallography. This fact has resulted in the often-quoted statement that "Materials are like people: it is their defects that make them interesting." This book is replete with pictures of defects, and from such pictures we attempt to show the achievements of what is still known as metallography, despite the fact that other materials are competing strongly with metals to play the major role in the materials world.

Light microscopy remained the only technique for observing microstructures for 70 years or so, and it is appropriate therefore that this book begins with a review of the current state of the art of light microscopy. It wasn't until the 1930s that Ruska and others developed the electron microscope, which has grown since then to become the most widespread and versatile technique for the imaging and analysis of materials. Therefore several of the chapters deal with the information to be gained when a beam of electrons strikes a piece of material. The light microscope is restricted to a few thousand times magnification by the wavelength of light, but the understanding that electrons also had wave characteristics immediately provided the means to overcome that limitation. Modern transmission electron microscopes image the atomic structure of materials with relative ease. Before this could be achieved, however, the art of specimen preparation had to be developed to provide thin foils of materials, representative of the bulk sample, but transparent to the high energy electron beams. Such processes were only mastered in the 1950s, but after that time transmission electron microscopy techniques expanded rapidly to the point where it is now possible to identify every known type of crystalline defect in a routine manner. The concept of electron diffraction is inextricably linked to the image-forming process in a transmission electron microscope, and a separate chapter dealing with this phenomenon is included. It can be argued that since the diffraction patterns are Fourier transforms of the electron distribution emerging from the sample, they are not images in the true sense of the word (whatever that is!). However, it cannot be doubted that these patterns exhibit the most startling details and symmetries, especially when formed with a fine converging beam of electrons. It is therefore on aesthetic grounds that we include the topic of "diffraction images."

From transmission electron microscopy came the related, complementary scanning electron microscope, which was long thought a useless technique until revived by Oatley in Cambridge in the 1950s. The scanning electron microscope bridges the magnification gap between the light microscope and the transmission electron microscope, and also overcomes the other major limit of the light microscope, namely its inability to provide in-focus images of anything other than a very flat surface. Through one of Nature's more favorable whims, the angular distribution of secondary electrons emanating from the surface of a sample obeys the same law as the reflection of light, and so secondary electron images look like the familiar reflected light images that we always see, but secondary electron images can be obtained at much higher magnifications and from rough, unpolished surfaces. As a result we can create marvellous images of microscopic samples that show the full three-dimensional shape of the sample, all in focus at the same time. The scanning electron microscope also introduced the idea of lens-less magnification, a technique that has been used to produce images from any of the signals that can be emitted when materials are bombarded with some form of radiation. A plethora of techniques now exist, from which we have chosen, somewhat arbitrarily, scanning ion microscopy and scanning acoustic microscopy. Developments in scanning technology have recently come full circle with the arrival of scanning light microscopy, either in the form of "confocal microscopy," or near-field microscopy, which has broken the wavelength limit that constrained classical light microscopy. However, we have not included a chapter on this topic since applications in materials science are not yet common.

The field ion microscope, pioneered by Muller in Chicago in the early 1950s, produced the first images of atoms twenty years before they were seen in the transmission electron microscope. The field ion microscope has now reached the stage where not only can we see individual atoms, but it is feasible to pull the atoms off the sample one at a time and determine their elemental nature. This may seem to represent the ultimate in metallography, but it remains to be seen what can be achieved with the latest microscopy technique that emerged with much publicity in the early 1980s, the scanning tunneling microscope. The development of this instrument earned the Nobel prize for the inventors, Binnig and Rohrer in Zurich, and prompted the tardy award of the same prize to Ruska, the father of the transmission electron microscope. However, in contrast to the slow progress in the development of the transmission electron microscope, the scanning tunneling microscope has generated an almost unprecedented wave of research since it combines the high resolution of the best electron and field ion microscopes, with the simplicity, compactness and relatively low price of a light microscope. Since the images in a tunneling microscope are representations of variations in the quantum-mechanical tunneling current as a fine needle is moved across the sample surface, we have reached the stage where an 'image' is far removed from the familiar concept of what we see when we peer down a light microscope. While the scanning tunneling microscope is best known for its ability to show the atom distribution on the surface of a sample, there is much more information present in terms of spectroscopic techniques that can be applied to the tunneling signal. It may be that these aspects prove more useful in the long run, given that we have been able to image atoms for over 30 years with other microscopes. Similarly, spectroscopic techniques coupled with electron microscopes such as x-ray spectrometry and electron energy-loss spectroscopy in the transmission electron microscope, cathodoluminescence in the scanning electron microscope, and Raman spectroscopy on a laser-light microscope all make modern metallographic techniques much more versatile and powerful than their predecessors. However, spectroscopic outputs tend to be graphical, and as such have little aesthetic value. This book therefore ignores these aspects, and endeavors only to emphasize the beauty in images of the structures of engineering materials.

Finally, it must be noted that the modern microscope, in any form, is not complete without its attendant computer. The ability to digitize any signal and feed it into a computer has opened the door to image analysis and image processing of all microscope outputs, and the final few chapters introduce some of the recent developments in this area. Seeing is no longer a prerequisite for believing, since the computer often has to be consulted before the image can be understood. But the computer can also quantify what were hitherto only qualitative, analogue images. As a result it can be argued that we are about to see another major advance in our ability to comprehend the structure of materials, once the power of modern computers is fully utilized for image storage and analysis.

ACKNOWLEDGMENTS

This volume contains selected manuscripts from a symposium of the same name that the editors organized at the World Materials Congress in Chicago in September

1988. The symposium was organized under the auspices of the Structures Activity Committee of ASM International, and was presented in the TMS program at the Congress. Thanks are due to Marlene Karl of TMS for her help in organizational aspects of the symposium, and to TMS for financial help in getting many of the speakers to the Congress. In addition to the professional materials societies, the symposium would not have been possible without substantial financial help from many of the commercial companies involved in the area of microscopy. We wish to acknowledge with grateful thanks the support of the following companies: Cameca Instruments Inc., EDAX International Inc., E.A. Fischione Instrument Manufacturing, JEOL USA Inc., Kevex Corporation, Leco Inc., Link Analytical Inc., NSI Hitachi Scientific Instruments, Philips Electronic Instruments, Ted Pella Inc., and Carl Zeiss Inc.

Bethlehem, Pa.   D.B.W.
Menlo Park, Calif.   A.R.P.
Berkeley, Calif.   R.G.

Contents

Contributors

DR. G. A. D. BRIGGS

Department of Metallurgy and
   Science of Materials
University of Oxford
Oxford, England

DR. D. S. BRIGHT

National Institute of Standards and
   Technology
Center for Analytical Chemistry
Building 222
Gaithersburg, Maryland

DR. M. G. BURKE

Westinghouse Research and
   Development Center
Pittsburgh, Pennsylvania

DR. J. M. CHABALA

University of Chicago
Enrico Fermi Institute
Chicago, Illinois

DR. S. CHIANG

IBM Research Division
Almaden Research Center
San Jose, California

DR. C. GIROD-HALLÉGOT

Department of Biology
University of Chicago
Chicago, Illinois

PROFESSOR R. GRONSKY

Department of Materials Science
   and Mineral Engineering
University of California
Berkeley, California

PROFESSOR J. I. GOLDSTEIN

Department of Materials Science
   and Engineering
Lehigh University
Bethlehem, Pennsylvania

DR. P. HALLÉGOT

Department of Biology
University of Chicago
Chicago, Illinois

DR. M. HOPPE

Ernst Leitz Wetzlar GmbH
Wetzlar, West Germany

PROFESSOR D. C. JOY

E. M. Facility
University of Tennessee
Knoxville, Tennessee

PROFESSOR R. LEVI-SETTI

University of Chicago
Enrico Fermi Institute
Chicago, Illinois

DR. R. B. MARINENKO

National Institute of Standards and
   Technology
Center for Analytical Chemistry
Gaithersburg, Maryland

DR. M. K. MILLER

Metals and Ceramics Division
Oak Ridge National Laboratory
Oak Ridge, Tennessee

DR. R. L. MYKLEBUST

National Institute of Standards and
   Technology
Center for Analytical Chemistry
Gaithersburg, Maryland

DR. D. E. NEWBURY

National Institute of Standards and
   Technology
Center for Analytical Chemistry
Gaithersburg, Maryland

PROFESSOR J. C. RUSS

Department of Materials Science
   and Engineering
North Carolina State University
Raleigh, North Carolina

DR. E. B. STEEL

National Institute of Standards and
   Technology
Center for Analytical Chemistry
Gaithersburg, Maryland

PROFESSOR G. THOMAS

Department of Materials Science
   and Mineral Engineering
University of California
Berkeley, California

MR. G. F. VANDER VOORT

Carpenter Technology Corporation
Carpenter Steel Division
Reading, Pennsylvania

PROFESSOR K. S. VECCHIO

Department of AMES
University of California, San
   Diego
La Jolla, California

DR. Y. WANG

AT&T Bell Laboratories
Murray Hill, New Jersey

PROFESSOR D. B. WILLIAMS

Department of Materials Science
   and Engineering
Lehigh University
Bethlehem, Pennsylvania

DR. R. J. WILSON

IBM Research Division
Almaden Research Center
San Jose, California

Images of Materials

1

Imaging by Light Optical Microscopy

G. F. VANDER VOORT

The light optical microscope, a relatively simple instrument compared with most of the other instruments discussed in this book, is still the single most important, most widely used instrument for the study of microstructure. Nearly all studies of microstructure should begin with examination by light microscopy. Such examinations are relatively rapid, compared with electron microscopy techniques, and efficient. Such an examination will permit categorization of the basic microstructural constituents present; definition of their relative size, shape, and quantity; assessment of the normalcy or abnormality of the microstructure relative to composition, processing history, and service conditions; and an indication of the need for further examination by more specialized techniques.

All studies of microstructure should begin at low magnification. Indeed, it is often useful to precede such studies with macrostructural evaluations (Vander Voort, 1978, 1984) using techniques such as macroetching or sulfur printing to reveal gross conditions and areas to be examined by light microscopy. Thus, macroexamination is an aid to sampling for microstructural studies. The specimens selected in this manner will reveal either abnormal or worse conditions, or typical conditions. Without such guidance, materials evaluation will be "hit or miss." Such procedures are often critical for successful failure analysis and quality control studies.

After the specimens have been selected, they must be properly prepared in order to reveal the true microstructure fully. Proper preparation techniques for metallographic examination have been discussed in specialized monographs such as Vander Voort (1984) and in *Metallography and Microstructures,* Volume 9 of the

Metals Handbook (1985). Sloppy preparation techniques lead to development of false structures, or poorly revealed true structures, which will either produce erroneous or incomplete interpretations. If quantitative methods are to be used to define microstructural features, good specimen preparation is even more critical. The author has witnessed numerous cases of inaccurate or incorrect microstructural interpretations, even in litigation work, due to improper specimen preparation. These techniques are relatively easy to learn, do require more effort and attention, but are absolutely essential.

After the specimen has been properly sectioned and polished, it is ready for examination. In many cases, it is important to begin the examination with the as-polished specimen, particularly for quality control or failure analysis work. Certain microstructural constituents, such as inclusions, intermetallics, cracks, or porosity, are best evaluated in the as-polished condition, because etching (to bring up other microstructural features) may obscure such details. In the as-polished condition there is a minimum of extra information for the observer to deal with, which makes examination of these features most efficient. Some materials do not require etching, and, in a few cases, examination is less satisfactory after etching. Relevant examples of this problem will be illustrated in this chapter.

For many metals, some technique must be used to reveal the microstructure fully. A wide range of processes is used, the most common being etching with either acidic or basic solutions. For most metals and alloys, there are a number of general purpose etchants that should be used first to reveal the microstructure. Subsequent examination may indicate that other more specialized "etching" techniques may be useful. For example, the examination may reveal certain constituents the metallographer may wish to measure. Measurements, particularly those performed with automatic image analyzers, will be simpler to perform and more reliable if the desired constituent can be revealed selectively, and numerous procedures are available to produce selective phase contrasting (Vander Voort, 1981, 1984, 1986). Some examples will be shown in the text.

Since most metallographic specimens are opaque, the vast majority of optical microscopy work is performed with incident bright-field illumination. Therefore, this discussion will be limited to the use of the reflected (incident) light metallurgical microscope (Vander Voort, 1984, 1985b). While most studies of the microstructure of materials should begin with bright-field illumination, other illumination methods should be tried, and they will often be found to be helpful, sometimes even essential, for such work. Obviously, for the examination of certain optically anisotropic metals (those with noncubic crystallographic structures), examination with cross-polarized light is required for best results. Indeed, in some cases, these metals are difficult to etch, and, if they can be etched, the bright-field image may often be much less satisfactory than the cross-polarized light image of the as-polished specimen.

Examination with cross-polarized light, dark-field illumination, or differential interference contrast does require a higher-quality polish than for bright field because fine scratches that may not be visible with bright-field illumination will be observed using these other methods. These fine scratches may not be objectionable for routine work but are aesthetically unacceptable for quality photomicroscopy.

Examples comparing bright-field images of microstructures with those produced by these other illumination modes will be presented to illustrate their value.

After the microstructure has been examined, it may be necessary to document the images so that others can see the results. At present, there are three basic approaches that can be employed—traditional wet-processed films, instant films, or electronic procedures. The latter technique is improving at a rapid pace thanks to the revolutionary progress in personal computers, digitizing boards, and hard-copy devices.

Currently, the highest-quality images, either black and white or color, are produced with the use of traditional wet-processed films. This approach also has the advantage of use of a negative image, which permits generation of any number of prints. Image contrast can also be improved by choice of films, development technique, and choice of printing papers. While the cost of the materials for such work is relatively low, the processing is labor intensive. For those laboratories that use photography heavily, automation in both negative development and printing can be justified and greatly speeds up the process.

The instant films have become very popular with metallographers because of the elimination of dark room processing. However, the cost per print, ignoring labor, is rather high, and the higher-quality instant films are significantly more expensive. The image quality of the better instant films approaches that of wet-processed films, but the lower-cost instant films are noticeably inferior. Most instant films do not provide the metallographer with a negative, which makes production of multiple prints less efficient. Some of the instant films must be coated to prevent them from fading, and, if not properly dried, prints can stick together. The latitude of instant films is rather limited. Hence, it is usually imperative to employ an exposure meter to obtain optimum prints and avoid film wastage. For those instant films that produce a negative, the restricted latitude generally results in a rather thin negative, which restricts the ability to utilize printing techniques fully to enhance print contrast further.

The electronic techniques have the potential to become a very important tool for the metallographer. High-resolution cameras are now available for both black and white or color work, although resolution is greatest for monochrome cameras. Several systems are now available, using the NTSC (National Television Systems Committee) video standard, for storing digitized images and producing hard copies. The quality of these prints, either black and white or color, is comparable with most instant films. The multiple copy problem with instant films is not present, and the prints do not require coating. Additionally, information can be added to the print electronically to document the specimen identity, etch, magnification, and so forth. One can also employ digital video signal processing devices to enhance the image digitally. Examples of such enhancement will be given in this chapter. Higher-resolution storage and printing devices are also available currently at greater cost. Further advances should make this technology very attractive. Other contributors to this volume, for example, Newbury et al. (Chapter 10), Bright et al. (Chapter 11), and Russ (Chapter 12), also describe various aspects of the use of video technology and image analysis in materials science.

As can be seen from these introductory comments, the process for obtaining high-quality microstructural images by light microscopy requires the systematic correct execution of each step from sampling to image documentation. Failure to implement one step in the chain properly will impair the final results. Thus, we must take a systems approach to our work.

THE LIGHT OPTICAL MICROSCOPE

When Henry Clifton Sorby made his first examination of the microstructure of iron on July 28, 1863, he was using a transmitted light petrographic microscope made by Smith, Beck and Beck of London that he had purchased in 1861 (Samuels, 1963; Moore, 1964; Humphries, 1967). Although capable of magnifications up to 400×, most of his examinations were conducted at 30, 60, or 100× and his first micrographs were produced at only 9×. The objectives of this microscope were equipped with Lieberkühns silvered concave reflectors for focusing light on opaque specimens.

Sorby quickly realized that reflected light produced in this way was inadequate, and in 1863 he started work on alternative illumination methods, two of which he developed. Subsequently, others developed vertical illuminators using prisms or plane glass reflectors, and Sorby's systems were not further developed. In 1886 Sorby reported on the use of "very high power" (650×) for the study of pearlite, a fine-scale, two-phase structure present in many steels. This was accomplished using a 45° cover glass vertical illuminator made for him by Beck.

Developments in both metallography and reflected light microscopy have been rapid since the days of Sorby (Campbell, 1926; Vilella, 1951; Quarrell, 1963) and metallography is now considered to be a mature science. Although the metallurgical microscope is considered to be a mature instrument, one should not assume that it cannot be further improved. On the contrary, the past ten years have witnessed substantial improvements (Vander Voort, 1988).

For many years photomicroscopy was conducted using specially built reflected microscopes known as metallographs. These devices represented the "top of the line" in metallurgical microscopes and were essential for quality work. In the late 1960s and early 1970s, manufacturers developed metallographs that were easier to use from the eyepiece viewing position. The temperamental carbon–arc light sources were replaced by xenon arc lamps. The unwieldy bellows systems for altering magnification were replaced by zoom systems. Vertical illuminators using single-position objectives were equipped with four-to-six–position rotatable nosepiece turrets to minimize objective handling. The light path was deflected so that the film plane was near at hand. Universal-type vertical illuminators and objectives were introduced so that the illumination mode could be readily switched from bright field to dark field, polarized light or differential interference contrast. Such systems were very attractive in that separate vertical illuminators and objectives were no longer needed for each illumination mode, and handling was eliminated.

Exposure meters were also added at this time, chiefly as a result of the rising popularity of instant films, for which such devices are needed to minimize film wastage.

While these metallographs were certainly more convenient to use, they were not without their own problems. The early xenon arc lamps produced substantial ozone, which required installation of an exhaust hood above the unit. Fortunately, it was not long before ozone-free xenon lamps were introduced. Early zoom systems often introduced a small degree of uncertainty into magnifications that could only be controlled by calibration with a stage micrometer before critical measurements were made. The extra lenses, prisms, mirrors, or reflectors placed in the light path by the zoom system and the new film plane position resulted in longer exposure times, which led to the use of much higher-wattage light sources and poorer images at very high magnifications. Besides these problems, strong inflation through the 1960s and 1970s and a dramatic increase in the value of the West German Deutschmark relative to the U.S. dollar in the 1970s occurred, which caused the price of these metallographs, particularly those made by Leitz and Zeiss, to approach $100,000 by the late 1970s. Few laboratories, even the larger ones, could justify purchase of a light microscope at this price, although even greater sums could usually be obtained for electron metallographic instruments.

Microscope manufacturers recognized these problems, and, starting in 1979, they introduced a number of very high-quality, reasonably priced, compact metallographs. All of these microscopes can be obtained with a wide variety of objective lens types and auxiliary accessories to meet the metallographer's needs. They are available with universal vertical illuminators that permit easy switching from one illumination mode to another using the same set of objective lenses. Furthermore, the manufacturers have introduced new glass compositions and lens formulations, developed generally by computer-aided design, for improved image brightness and contrast. Tungsten halogen filament lamps have largely replaced xenon arc lamps as the preferred light source. Figure 1.1 shows a cutaway view of the light path in a modern reflected light microscope.

ILLUMINATION MODES

The vast majority of metallographic examination and photomicroscopy is performed using bright-field illumination. Examination of as-polished specimens for inclusions and intermetallics is most efficient with bright field. Metallographers working with noncubic, optically anisotropic metals, such as beryllium, α-titanium, hafnium, uranium, zirconium, and some of the precious metals, utilize cross-polarized light extensively. Many of these alloys are difficult to etch, and these microstructures are best revealed using polarized light. Aside from this, the use of other illumination modes, such as dark field and differential interference contrast, is infrequent. This is unfortunate, since these techniques are very useful.

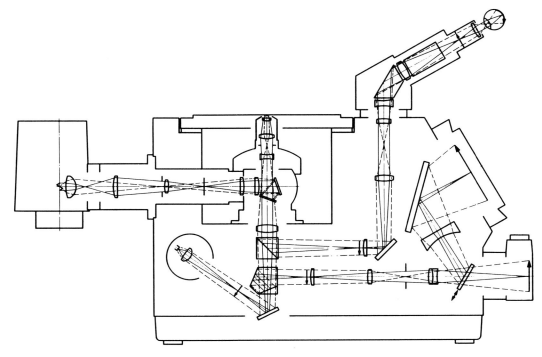

Figure 1.1 Cutaway view of a modern inverted reflected light microscope. Courtesy of C. Zeiss, Inc.

Bright-Field Illumination

In reflected bright-field illumination, light that strikes flat features in the image (perpendicular to the optic axis) is reflected back through the objective to the eyepiece, where it appears bright. Light that strikes inclined features is scattered and appears dark in the image with the degree of darkening depending upon the angle of the feature relative to the optic axis.

Dark-Field Illumination

Dark-field illumination produces a reversal of the image contrast compared with bright field. The light that is scattered is collected and appears bright in the image, while the light that reflects directly back to the eyepiece is blocked and is dark in the image. This technique is very sensitive to surface irregularities and features such as grain or twin boundaries, slip lines, etch pits, and so forth. Because of the "self-luminous" nature of the image, detail may often be observed in dark field that cannot be observed in bright field. The chief disadvantage of dark field is the rather long exposures that are required to obtain good micrographs.

Cross-Polarized Light

To obtain polarized light, one places a polarizer and analyzer in the light path with the analyzer rotated 90° to the polarizer. This is the crossed position. When an isotropic metal surface is examined with cross-polarized light, the image is dark (extinction). If we now place a noncubic, optically anisotropic polished metal in the light path, an image of the structure is produced. If the stage can be rotated through 360°, we will observe four positions of maximum and minimum light intensity in each grain when using monochromatic light. With white light, we may observe color variations in each grain upon rotation of the stage.

Most metallurgical microscopes use Polaroid® filters for the polarizer and analyzer. These are satisfactory for most work. Some manufacturers also have optional prism polarizers that produce brighter images under crossed conditions, which reduces exposure time. Color effects can be enhanced by inserting a sensitive tint plate in the light path. The most commonly used compensator is the sensitive tint plate, also known as a λ plate, a full-wave plate, or a gypsum plate.

Differential Interference Contrast

Differential interference contrast (DIC) is produced by adding a double quartz prism, such as a Wollaston prism, between the vertical illuminator and objective lens while employing cross-polarized light. This produces interference effects, which display height differences in the image much like those obtained by oblique illumination. Color is produced by adding a sensitive tint plate. Adjustment of the prism produces a change in color.

Oblique Illumination

Prior to the development of DIC, many metallographs employed oblique illumination to emphasize height differences in a microstructure. Oblique illumination is obtained by decentering either the condenser assembly or the mirror; however, only a minor amount can be tolerated as the numerical aperture is reduced, which impairs resolution, and nonuniform lighting across the image occurs. Oblique illumination is still a useful technique but is now offered on only a few microscopes.

BRIGHT-FIELD IMAGES OF MATERIALS

Metallic Materials

A properly polished specimen usually must be etched to reveal the microstructure. The etchant must be compatible with the specimen, and the degree of etching must be carefully controlled. The metallurgical literature describes a vast number of etchants, particularly for common metals and alloys, and the beginner is faced with a real problem in choosing the optimum etch for a particular alloy and heat treatment. Standard compilations of etchants (Vander Voort, 1984; *Metals Handbook,* 1985) are very helpful in solving this problem.

Generally, one starts with a simple chemical etchant that will reveal the overall microstructure. Such etchants are applied by either immersion with agitation or by swabbing. The latter technique is usually required with those metals and alloys that are more corrosion resistant due to the presence of thin oxide films on the surface. It is best to etch these alloys as soon as possible after polishing.

Figure 1.2 shows an example of control of the degree of etching. In this example, a specimen of annealed 270 grade nickel was etched with Kalling's No. 2 ("waterless" Kalling's). The first micrograph is underetched slightly. The grain and twin boundaries are crisp, but some have not been revealed. The second micrograph reveals all of the boundaries present, but the thickness of the boundaries is greater.

There are numerous types of etchants that can be used to reveal grain structures. In many cases, the standard chemical etchants are grain-orientation sensitive so that some boundaries between grains appear heavily etched while others are not etched at all. This is a common problem with the use of nital (a few percent of nitric acid in ethanol) for etching ferrite grains in steel. Ferrite is almost pure iron with a small amount of carbon in solution. One excellent way to overcome this problem is to use tint etchants that will color the grains according to their crystallographic orientation (Beraha and Shipgher, 1977; Vander Voort, 1984, 1985a). Figure 1.3 and Plate 1.1 show a low-carbon steel specimen that has been etched with 2% nital, 4% picral, (a few percent of picric acid in ethanol) and with a tint etch (aqueous 10% sodium thiosulfate and 3% potassium metabisulfite). The use of nital produces a "flat" etch where only the ferrite grain boundaries are visible (plus a few carbides and inclusions). Not all of the grain boundaries are visible and some are etched more heavily than others. Actually, the grain boundaries are not etched *per se,* as the etchant dissolves the ferrite grains at different rates according to their orientation. Picral attacks the ferrite uniformly, revealing the outline of the carbides present, but does not reveal the ferrite grain structure (if the degree of etching is heavy, ferrite grain boundaries may be revealed). Tint etching (see Plate 1.1) reveals all of the grains quite clearly in color contrast (enhanced with polarized light nearly crossed), while the cementite (iron carbide, Fe_3C) is unaffected.

As another example, Figure 1.4 and Plate 1.2 show the microstructure of annealed phosphor bronze. One of the most popular etchants for copper-based alloys is a mixture of equal parts of ammonium hydroxide and 3% hydrogen peroxide used by swabbing. On many copper-based specimens, this etchant produces a grain

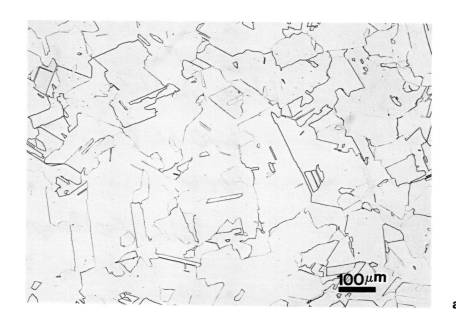

a

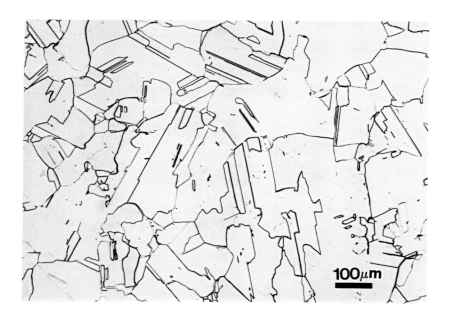

b

Figure 1.2 Example of the control of the degree of etching showing 270 grade nickel etched with Kalling's No. 2 reagent: (a) underetched, and (b) properly etched (bright field).

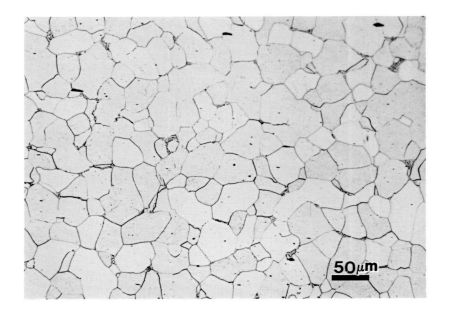

Figure 1.3 Different etchants can reveal microstructures differently, as shown by a specimen of low-carbon sheet steel etched with (a) 2% nital, and (b) 4% picral (bright field).

12

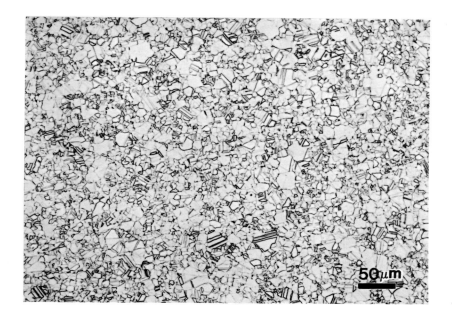

a

b

Figure 1.4 Grain structures of CDA 510 phosphor bronze revealed by (a) equal parts of NH$_4$OH and 3% H$_2$O$_2$, and (b) Klemm's I tint etch (bright field).

contrast etch. However, on this specimen a "flat" etch was obtained (a). Tint etching, (b) using Klemm's I (50 ml water saturated with sodium thiosulfate plus 1 g potassium metabisulfite) produced coloration (Plate 1.2) of the grains and twins. Note that the tint etch also reveals segregation present. Both black and white and color images are shown for comparison.

Once we have defined the overall microstructural constituents present, we may want to utilize an etchant that is specific to certain constituents in the structure. Many are available (Vander Voort, 1981, 1984, 1985b, 1986) for such work. To illustrate selective etching, Figure 1.5 and Plate 1.3 show the microstructure of annealed AISI W2 carbon tool steel. The specimen was first etched using the two most commonly used chemical etchants, (a) 2% nital and (b) 4% picral. Nital reveals the spheroidized (almost spherical) cementite plus the ferrite grain boundaries, although the structure within some grains is not clearly revealed. Picral, however, reveals only the cementite, which is an advantage if the metallographer is evaluating the degree of spheroidization. Note that etching is uniform, unlike with nital. If we wish to measure this structure using automatic image analysis, neither nital nor picral is useful. Instead, we need to darken either the ferrite matrix or the cementite preferentially and leave the other phase unaffected. This can be readily accomplished since etching with boiling alkaline sodium picrate (c) will darken the cementite, while tint etching with Klemm's I (d) will color the ferrite. We can also tint etch the cementite (Plate 1.3) with a cathodic tint etch, Beraha's sodium molybdate reagent (e).

Figure 1.6 and Plate 1.4 show the microstructure of α-β brass, a two-phase alloy of Cu–40% Zn, that was heated to 940°F (504°C), held 1 hour, and water quenched. Etching with aqueous 10% ammonium persulfate, a common etch for copper-based alloys, produced different effects depending on the degree of etching. Light etching outlined the β phase, while heavy etching brought out the α-matrix phase grain structure. Swab etching with equal parts of ammonium hydroxide, 3% hydrogen peroxide, and water contrast etched the α-matrix phase. Etching with aqueous 10% cupric ammonium chloride (add ammonium hydroxide until the pH is slightly basic) darkened the β phase preferentially. Tint etching with Klemm's I reagent colored (Plate 1.4) the β phase selectively.

Figure 1.7 and Plate 1.5 show the microstructure of a duplex (two-phase) stainless steel, 7-Mo PLUS®, using bright-field illumination. Most of the standard chemical etchants that would normally be used to reveal the microstructure of stainless steels are somewhat sensitive to crystal orientation. These etchants reveal only a portion of the phase boundaries between ferrite (body-centered cubic iron-rich phase) and austenite (face-centered cubic iron-rich phase) in this alloy. One chemical etchant that does reveal the grain structure of this alloy properly is ethanolic 15% HCl, although a 30 minute immersion is required. However, the results are good, as shown in Figure 1.7, although one cannot tell which phase is which, except by familiarity with the grade. Many of the electrolytic etchants used for stainless steels do produce excellent results. Most of these reveal only the phase boundaries, as illustrated by the use of aqueous 60% HNO_3, while a few produce phase coloration as well, as illustrated by the use of aqueous 20% NaOH, which colors the ferrite light orange/tan. More vivid coloration can be obtained by tint etching, as demonstrated by Plate 1.5. The tint etch was an aqueous solution containing 10% HCl plus 1% potassium metabisulfite. Grain coloration is a function of the deposited film thickness, which depends upon the etching time, grain orientation, and grain composition.

Geological Materials

The microstructure of metallic meteorites (which are iron-nickel alloys) can be revealed using bright-field illumination after suitable etching. While standard chemical etchants, such as nital, are widely used for this purpose, much better results can be obtained using tint etchants. Kamacite (ferrite) in meteorites is a low-nickel phase with a body-centered cubic structure and can be selectively colored using an aqueous solution of 10% sodium thiosulfate and 3% potassium metabisulfite. Plate 1.6 shows the microstructure of Coahuila, a hexahedrite meteorite (Fe–5.59% Ni) found in Mexico. Because the specimen was a single crystal (i.e., no internal grain boundaries), the ferrite matrix was uniformly colored blue. The long diagonal lines are mechanical twins (Neumann's bands) due to an extraterrestrial shock event. The rod-shaped and prismatic-shaped particles are rhabdite, an iron-nickel phosphide.

Plate 1.7 shows the microstructure of Gibeon, an Fe–7.93% Ni fine octahedrite found in Southwest Africa. Octahedrites are noted for their network of criss-crossing Widmanstätten kamacite (ferrite) grains. These elongated grains are colored according to their crystal orientation. Note the short parallel lines within these elongated ferrite grains. These lines are mechanical twins, like those observed in Coahuila. Between the elongated kamacite grains is a fine two-phase mixture of kamacite and taenite (a more nickel-rich phase with a face-centered cubic structure) called *plessite*. This particular form of plessite would be referred to as "finger" plessite. In between some of the elongated kamacite grains, we can find some taenite films that appear white (unetched).

Plate 1.8 shows the microstructural appearance of another form of plessite, "pearlitic" plessite, observed in Odessa, an Fe–7.35% Ni coarse octahedrite found in Texas. At the top of the field of view is a white/light green band of taenite that has accumulated at the interface between the plessite and a kamacite grain above the taenite, outside the field of view. The top edge of the taenite has a darker appearance; this is the so-called "cloudy" zone, which is an unresolved two-phase structure in this region, highly enriched in nickel (compared to the bulk Ni content).

Plate 1.9 shows the microstructure of Washington County, an Fe–9.9% Ni ataxite meteorite found in Colorado. Note the absence of mechanical twins in the colored kamacite grains. Also, the kamacite grains are equiaxed rather than Widmanstätten. The variety of colors within the kamacite grains indicates a random crystal orientation. The white (unetched) particles are taenite.

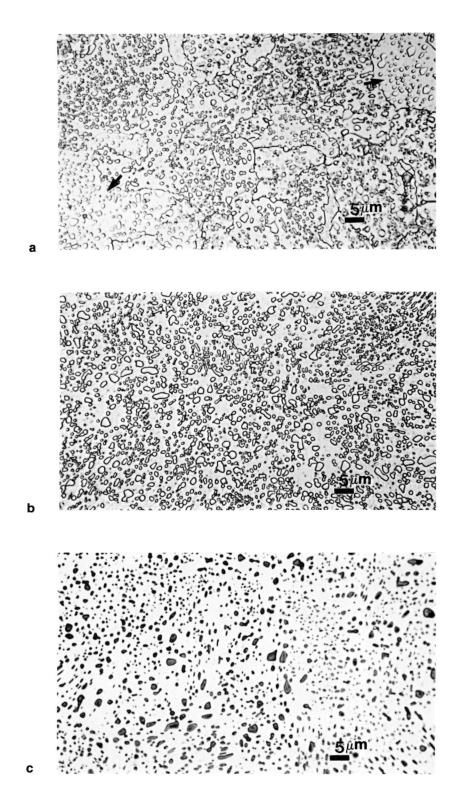

a

b

c

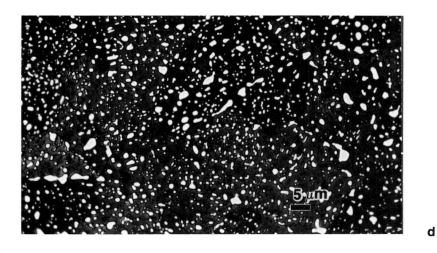

d

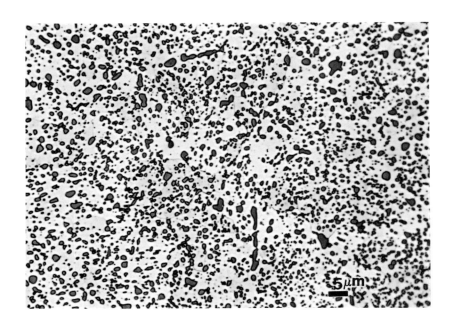

e

Figure 1.5 Comparison of etchants for revealing spheroidized cementite in AISI W2 carbon tool steel: (a) 2% nital, (b) 4% picral, (c) boiling alkaline sodium picrate, (d) Klemm's I tint etch, and (e) Beraha's sodium molybdate cathodic tint etch (bright field).

17

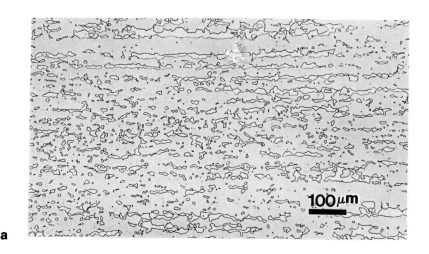

c

d

Figure 1.6 Comparison of etchants for revealing the microstructure of α-β brass (Cu−40% Zn): (a) and (b) aqueous 10% ammonium persulfate, light versus heavy etching, (c) equal parts NH₄OH, 3% H₂O₂, and H₂O; and (d) buffered 10% cupric ammonium chloride (bright field).

19

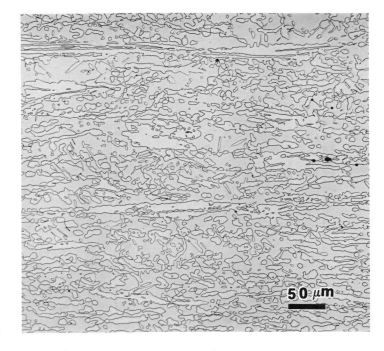

a

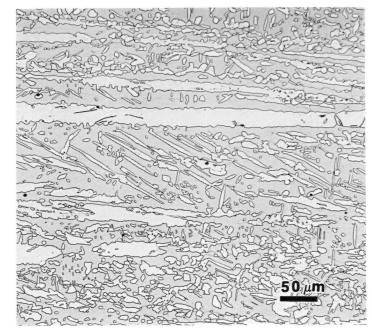

b

20

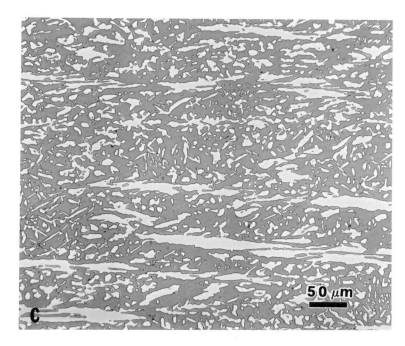

c

Figure 1.7 Comparison of etchants for revealing the microstructure of a duplex (austenite and ferrite) stainless steel, 7-Mo PLUS®: (a) 30 minute immersion in ethanolic 15% HCl, (b) electrolytic etch with aqueous 60% HNO_3 (Pt cathode, 1 V dc, 40 s), and (c) electrolytic etch with aqueous 20% NaOH (Pt cathode, 4 V dc, 10 s), which colors the ferrite.

CROSSED-POLARIZED LIGHT IMAGES OF NONCUBIC MATERIALS

The previous examples demonstrated the use of different etchants to produce good contrast images with bright-field illumination. As mentioned previously, the microstructure of certain noncubic metals can be clearly revealed using cross-polarized light. To illustrate the use of cross-polarized light, Figure 1.8 and Plate 1.10 show the microstructure of beryllium using both a Polaroid filter polarizer and an Ahrens prism polarizer, with or without a Berek compensator prism. Use of a Polaroid filter polarizer produced an acceptable structure that was, however, devoid of color. The addition of the Berek prism increased the exposure time and produced a minor degree of coloration. Use of the Ahrens prism polarizer produced an improved image with fair colors. The addition of the Berek prism decreased the exposure time, contrary to its use with the Polaroid polarizer, and the best color image (shown in both black and white and color for comparison).

Plate 1.11 shows four polarized light images of a $YBa_2Cu_3O_{7-x}$ superconducting oxide. For these four images, Polaroid filters were used for both the polarizer and analyzer. The first micrograph shows the polarizer and analyzer crossed. Some of

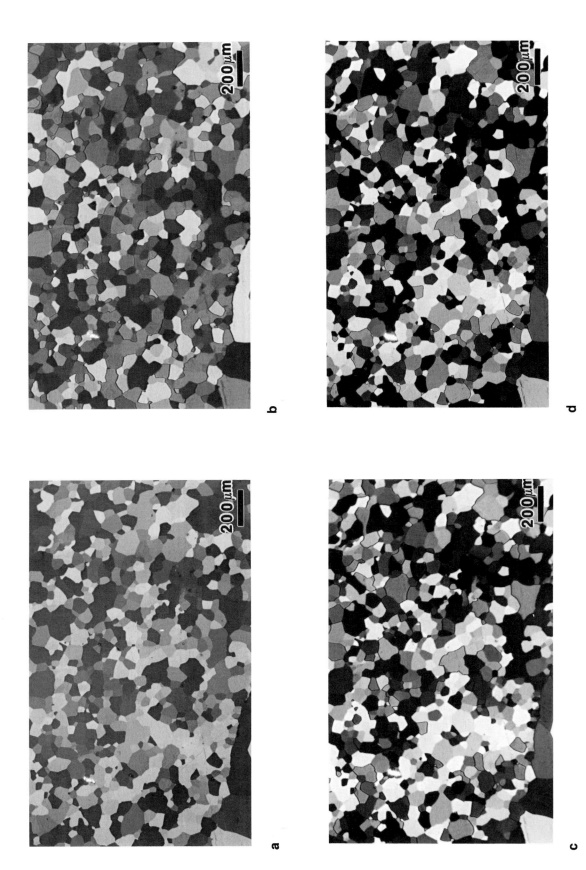

Figure 1.8 Microstructure of polycrystalline beryllium examined with cross-polarized light: (a) and (b) using crossed Polaroid filters, (c) and (d) using an Ahrens prism polarizer. In (b) and (d), a Berek compensator has been added. Micrographs taken with Kodak Tri-X Orthochromatic film (320 ISO); exposure times (a) 32 s, (b) 108 s, (c) 82 s, and (d) 25 s.

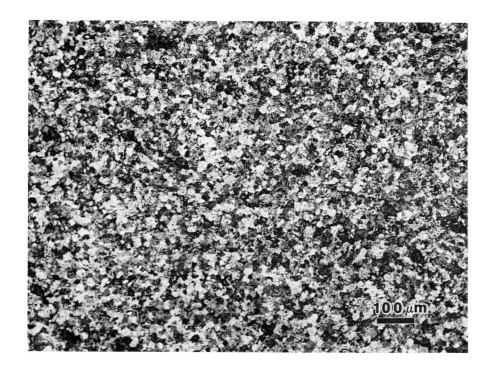

Figure 1.9 Microstructure of α-phase Ti–5% Al–2.5% Sn etched with aqueous 0.5% HF (bright field).

the elongated grains are light in color and nearly featureless, while others are colored and show internal twinning (fine parallel lines). If the microscope stage can be rotated, the light grains will become colored while the colored grains will become lighter. The same effect can be obtained as the polarizer is rotated slightly off the crossed position either clockwise or counter-clockwise. The second micrograph shows the resulting image produced by insertion of a full-wave compensation plate between the polarizer and analyzer. Note that the image shows more hues, but some of the fine image detail has been lost. The third example shows the resulting image produced by insertion of a Berek prism compensator between the polarizer and analyzer. This improves the quality of the polarized light and produces the best overall image. Compared with the first image, using only crossed polarizers, the addition of the Berek prism produces an image that is brighter and more vivid. Although it is not as effective in introducing color to the image as the full-wave plate (sensitive tint), it does not degrade the fine details. The fourth image in Plate 1.11 shows the use of crossed polarizers plus both the full-wave plate and the Berek prism compensator. This is slight more colorful than the second image and brighter, due to the Berek prism. This example, as with the beryllium specimen, shows the value of the Berek prism compensator for polarized light studies. Plate 1.12 shows a higher-magnification view of this specimen using an Ahrens polarizer, a Polaroid filter analyzer, and the Berek prism compensator. Naturally, the polarizer and analyzer are in the crossed position. The detail in this image is excellent despite the very high magnification, which requires strain-free objective lenses of the highest quality.

Plate 1.13 shows the grain structure of as-polished ruthenium in cross-polarized light plus a full-wave plate. Plate 1.14 shows the microstructure of a gravity casting alloy, Zn-12, after tint etching with Klemm's I reagent and viewing with cross-polarized light. While the microstructure of this alloy is colored somewhat by the tint etch, the coloration is greatly enhanced by the use of polarized light. Nearly as good results can be obtained on this specimen without use of the tint etch. Cross-polarized light, in a few cases, is more effective with noncubic metals after they have been etched. Figure 1.9 and Plate 1.15 show the microstructure of an α-phase Ti–5% Al–2.5% Sn alloy etched with aqueous 0.5% hydrofluoric acid. The bright-field image shows the grain structure but not as clearly as the polarized light images (taken with a calcite prism and full-wave plate). Note how adjustment of the full-wave plate alters the colors in Plate 1.15.

Polarized light can also be used effectively for the study of polymeric materials. Plate 1.16 shows the microstructure of a polished, unetched specimen of a graphite-fiber-reinforced polysulfone composite. The reinforcement particles are in focus, and we can see coloration within the transparent polymer matrix. One can then focus down into the plastic and image defects below the surface. When doing this type of work, one must be careful not to change the focus so much that the front lens of the objective strikes the specimen surface! Dark-field illumination, demonstrated in Plate 1.16 as well is also useful for the study of such materials. In comparison, the use of bright field produces markedly inferior results (not shown); only the opaque reinforcement can be observed.

There are also some cases in which the microstructure can be observed in bright field, but additional detail can be obtained using cross-polarized light. Perhaps the best-known example of this is the examination of graphite nodules in ductile iron, as demonstrated in Figure 1.10 and Plate 1.17. Cross-polarized light reveals the growth pattern of the nodules, which cannot be observed with bright field.

DARK-FIELD AND DIFFERENTIAL INTERFERENCE CONTRAST IMAGES OF MATERIALS

Dark-field illumination has been used as a check for success in using calcium treatment to modify manganese sulfides in steels (Farrell and Hiltz, 1976). In bright field, both CaS and MnS appear similar in color (gray), while in dark field, CaS is bright while MnS is outlined but dark, as shown in Figure 1.11. Polarized light has been commonly used to study and identify inclusions in many metals and alloys (Morrough, 1952; Wilson and Wells, 1973). Dark field is also highly suitable for reflected light examination of polished woods. Plate 1.18 shows the microstructure of walnut in dark field, for example.

Dark field and differential interference contrast (DIC) are both quite useful for examining grain structures. Figures 1.12 and 1.13 and Plate 1.19 show Waspaloy, a nickel-based superalloy, in the solution-annealed and double-aged condition. In Figure 1.12 the Waspaloy is etched with glyceregia. In bright field we can see most of the grain and twin boundaries, but these are further emphasized using dark field and DIC. Figure 1.13 and Plate 1.19 show the microstructure of the Waspaloy after tint etching in both black and white and color for comparison. Tint etching again reveals the total grain structure vividly.

A similar example is provided by Figures 1.14 and 1.15 and Plate 1.20, which show the microstructure of a duplex (ferrite-austenite) stainless steel, AISI 312. Figure 1.14 shows the structure etched with glyceregia in bright field, dark field, and DIC. It is not really obvious which phase is which. However, if we tint etch (aqueous 10% HCl plus 1% potassium metabisulfite), as in Figure 1.15 (black and white) and Plate 1.20 (color), the ferrite phase is colored preferentially, and interpretation is simplified.

These other illumination modes are also very useful for nonrecrystallized (i.e., deformed) microstructures as obtained due to warm working or controlled rolling with a low finishing temperature. To illustrate this, Figure 1.16 shows three micrographs of a high manganese nitrogen-stabilized austenitic stainless steel alloy, where hot working was continued to a temperature below the recrystallization range. Note that, despite the use of a very aggressive etchant, acetic glyceregia, and prolonged etching, the bright-field image is rather devoid of detail. However, with either dark field or DIC, we do obtain much greater detail.

Differential interference contrast illumination can be very effective in certain circumstances, chiefly those in which a certain degree of roughness is present. This roughness, however, must be preferential to the microstructure and must be created, and carefully controlled, by final polishing and/or etching. To illustrate this, Plate 1.21 shows the microstructure of an as-polished, unetched specimen of as-cast Cu–

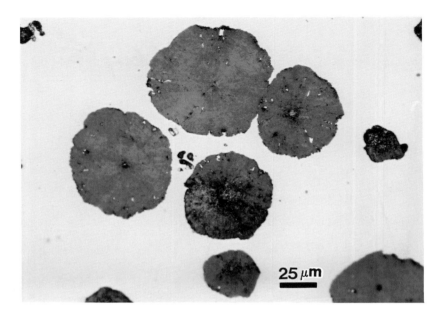

a

b

Figure 1.10 Micrographs of annealed ductile iron containing nodular graphite: (a) bright field at 400×, and (b) cross-polarized light (Polaroid filters) plus a sensitive tint plate.

26

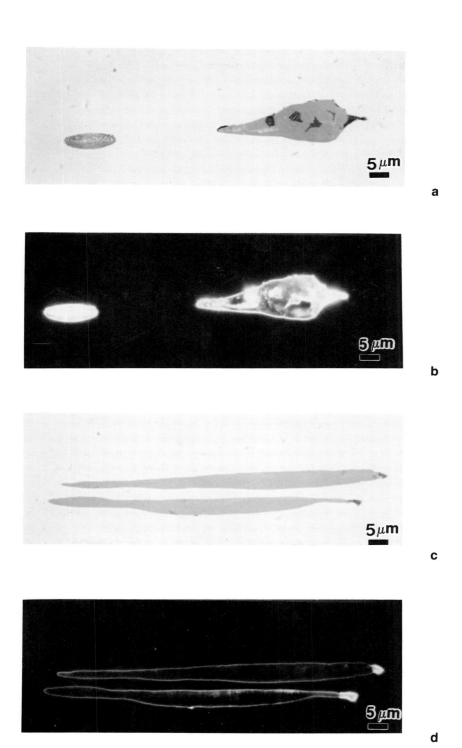

Figure 1.11 Dark-field illumination identifies sulfides modified by calcium treatment: (a) calcium-manganese sulfide, bright field; (b) same field of view as in (a) but dark field; (c) MnS with calcium-manganese sulfide tips, bright field; and (d) same field of view as in (c) but in dark field.

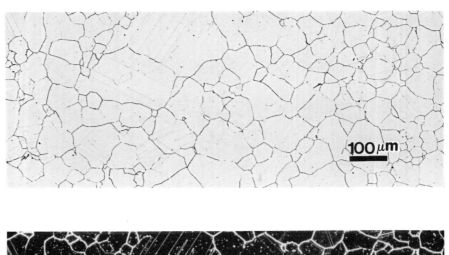

a

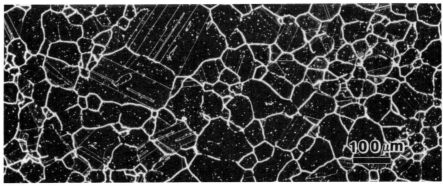

b

c

Figure 1.12 Microstructure of solution-annealed and double-aged Waspaloy (Ni-based superalloy) etched with glyceregia: (a) bright field, (b) dark field, and (c) differential interference contrast.

28

Figure 1.13 Microstructure of Waspaloy specimen tint etched with a modified Beraha's solution (bright field).

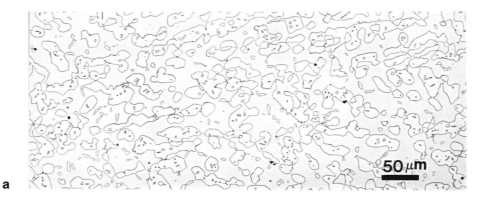

a

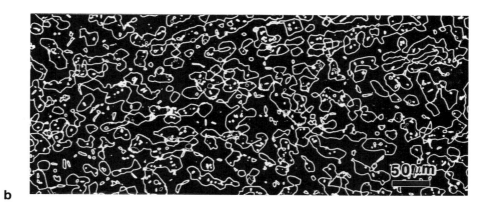

b

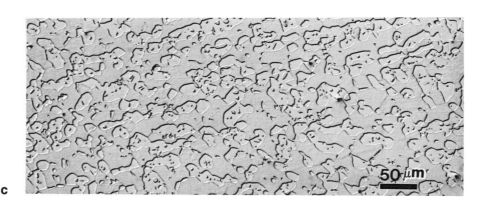

c

Figure 1.14 Microstructure of AISI 312 duplex stainless steel etched with glyceregia: (a) bright field, (b) dark field, and (c) differential interference contrast.

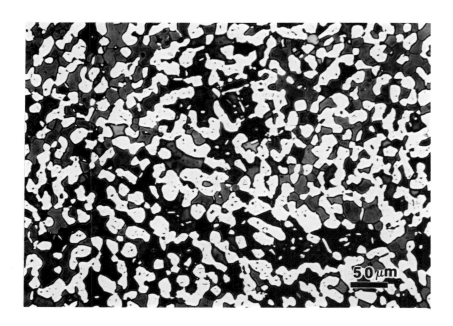

Figure 1.15 Microstructure of AISI 312 (Figure 1.14) specimen tint etched with a modified Beraha's solution to color the ferrite (bright field).

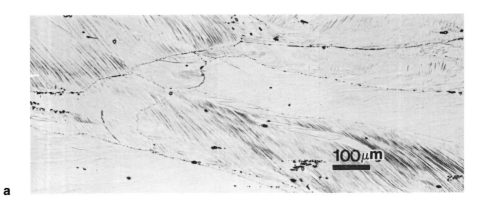

a

b

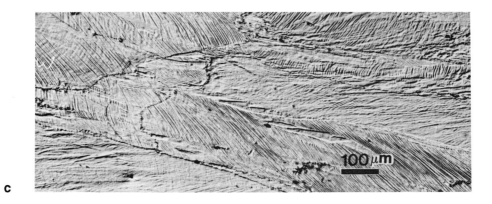

c

Figure 1.16 Microstructure of a warm-worked high-manganese austenitic stainless steel drill collar alloy etched with acetic glyceregia: (a) bright field, (b) dark field, and (c) differential interference contrast.

32

37% Zn viewed with DIC. The dendritic nature of the solidification structure is quite easily observed. Plate 1.22 shows an example of an as-polished fiberglass-reinforced plastic specimen, sectioned perpendicular to the reinforcement direction, where DIC has been used to examine the microstructure. We see that a certain degree of polishing relief has occurred that slightly rounds the outer periphery of the fibers. Bright-field imaging of this specimen produces only a very faint, flat image (not shown). Another excellent example of the use of DIC is given in Plate 1.23, which shows β_1 martensite in a Cu–26% Zn–5% Al shape memory alloy. Martensite is a low-temperature phase produced by shearing the atomic lattice to create a different crystal structure. While cross-polarized light can be used to examine this microstructure, the introduction of some relief in final polishing enables DIC to be used. The DIC images are much more colorful than the cross-polarized light images (not shown) and have much shorter exposure times. As a final example of the use of DIC, Plate 1.24 shows "butterfly" martensite, which was formed on the surface of a polished specimen of an Fe–22% Ni–3% Cr alloy during subzero cooling. The specimen was polished before refrigeration, and the expansion that accompanies the martensitic transformation permits observation using DIC alone without the risk of harmfully altering the surface, which can occur during etching.

These examples demonstrate that good image contrast is dependent on specimen preparation, choice of etchant, and choice of illumination mode. Naturally, one needs a high-quality microscope for optimum results. The microscope must be kept clean, and the light source must be properly aligned. All of these factors must be properly chosen or performed in order to get quality results.

VIDEO MICROSCOPY

Until recently, few metallographers have used video microscopy in any manner except for more convenient group viewing, compared with projection screens, or for classroom instruction. This scenario should soon change, since the value of video microscopy technology has been demonstrated in recent years by biologists, botanists, and medical researchers. The introduction of high-resolution cameras, cameras sensitive to light outside the visual range, and cameras sensitive to very low light levels, plus both hard-wired and software signal processing devices, has spawned new approaches for light microscopy studies. Much of the signal-processing image enhancement methodology comes from satellite reconnaissance technology. A number of textbooks (Pratt, 1978; Hall, 1979; Baxes, 1984; Inoue, 1986) have been written on this subject.

Video microscopy, as the words imply, couples video devices to the light microscope to detect the image (which in some cases cannot be seen in the eyepieces due to the low light level), which is then digitized, manipulated to enhance its quality, stored in memory, and finally documented by a number of printing devices. Image analysis may also be conducted so that information about the structure can be quantified. A video camera is attached to the microscope using a C-mount. The image is picked up by the camera, producing an analog signal. This signal is sent to the "frame grabber," which converts the analog signal into digital information that is stored in the computer memory as an array of numbers. The signal-processing device, which may be either hard-wired or software based, is used to manipulate the digitized image to enhance the regions of interest and sharpen the features.

It is common practice to examine a histogram of the gray-level image (black and white systems) and then employ a number of techniques to alter the histogram and thus change the contrast. A variety of techniques can also be used to clean the image and reduce noise. For example, one can sample a number of successive frames and average the gray level at each point in the image. The average image is then stored in memory and further processed. The processed image is stored, using any one of a number of storage media, and a hard copy of the image may be produced, again using one of several approaches.

The video signal is viewed with a television monitor during this work. The electron beam scans horizontally, left to right, and top to bottom to produce a "field" in $\frac{1}{60}$ second (NTSC). For the next field, the scan begins again in the upper left of the screen, but the scan lines are offset slightly so that the first scan line of the second field falls between the first and second scan lines of the first field. These two fields are "interlaced" to form a "frame" and a complete, nonflickering image every $\frac{1}{30}$ second.

The U.S. NTSC standard uses 525 horizontal scan lines per frame or 262.5 lines per field for a scan rate of 15.75 kHz. However, not all of the 525 (or 262.5) scan lines are visible, as the lower portion is blanked off to permit the electron beam to return to the top of the screen. Thus, for each field, only 240 lines are visible, or 480 per frame. The number of visible scan lines per inch of screen height gives an estimate of the vertical resolution. Thus we can calculate the vertical distance between scan lines per field or per frame. This explains why the image quality of smaller TV screens is better than for larger TV screens.

In the U.S. NTSC standard, there are 330 picture elements ("pixels" or picture points) across the screen. Again, if we know the width of the screen, we can calculate the number of pixels per inch (or per cm). Again, this is a limiting factor in the resolution on the monitor. The NTSC standard has a relatively low resolution, which is a serious limitation to its use for video microscopy. However, there are many higher-resolution systems available that produce quite excellent image resolution.

Video Camera

The video camera used to pick up the image from the microscope is a critical part of the system. Either black and white or color cameras can be used; however, greater resolution can be obtained with black and white (monochrome) cameras, and they are usually preferred. Important camera characteristics are: sensitivity (to both light level and wavelength), resolution (measured in terms of the number of scan lines per field or the number of pixels), contrast (gray levels), lag (persistence of image detail for two or more frames after the excitation is turned off), noise, distortion, anti-blooming ("bleeding" of image between adjacent pixels), gamma (the exponent of the equation relating the output signal to the input signal), video standard (such as NTSC), size and weight, and cost.

Cameras are either of vacuum tube or solid-state construction. Vacuum-tube types have a much longer history and exhibit greater resolution and sensitivity. The solid-state cameras, such as the CCD (charge-coupled device), are smaller in size and lighter and have been improved substantially in recent years.

Vacuum-tube cameras, such as the vidicon, plumbicon, chalnicon, and new-vicon, employ an electron gun to scan electrons across a target where the image is focused. The target material varies with the type of tube; for example, antimony trisulfide is used in the vidicon camera. The target material chosen influences the camera characteristics, such as sensitivity, resolution, lag, and gamma. Mono-chrome vacuum-tube cameras provide the highest number of scan lines currently available.

The solid-state cameras use a semiconductor chip for each pixel in the field. CCD cameras usually provide pixel densities of around 500 (vertical) by 500 (horizontal), although some newer CCD cameras have pixel densities around 1024 × 1024. These, of course, are rather expensive. The aligned pixel array of the solid-state camera is an advantage from the viewpoint of image distortion.

Color cameras of either vacuum-tube type or solid-state construction are available from many manufacturers. Historically, color image resolution has been about half that of black and white cameras. Today, color cameras with approximately 500 × 500 pixels are available and produce excellent images.

Signal Digitizer

The analog signal from the camera is fed into the digital image signal-processing computer. In most cases, the digitizer is calibrated to provide 256 gray levels (8 bits), where black has a value of 0 and white has a value of 255. This, of course, represents a gray-level discrimination four to six times greater than the human eye can distinguish. A histogram of the gray levels present in an image can be produced as an aid to subsequent signal processing, as demonstrated later in the examples.

The digitizer also establishes the pixel density of the image, 512 × 512 (8 bits) being quite popular. Higher pixel density digitizers are available; for example, 1024 × 1024 is becoming more common. Naturally, greater computer memory is required as the pixel density increases and signal-processing time becomes greater. Thus, a more powerful computer is required for such systems. For a 512 × 512 pixel array and 256 gray levels, we need about 262 kbytes of memory per image.

Image Documentation

To provide prints for reports, some type of hard-copy device is required. Systems are available, particularly for the NTSC standard, to produce prints, but their resolution is less than the 512 × 512 pixel density of most digitizers. Nevertheless, the quality of these prints is acceptable for most purposes, although significantly poorer than obtainable by wet-processed films. One can also photograph the screen using a standard photographic camera and obtain reasonably good results. This technique is, of course, somewhat less convenient. Recently introduced color printers exhibit resolutions comparable to that of the better television monitors. Video cassette recorders (VCR) can also be used to document images; however, the resolution of the VHS format is poorer than the NTSC standard. Better resolution can be obtained using $\frac{3}{4}$-and 1-inch formats (Umatic). The recently developed Super VHS (S-VHS) format provides excellent resolution and should become popular for video microscopy.

Image Storage

Image storage can be accomplished in several ways. A certain limited number of images can be stored in the processing computer or in a hard drive. Systems are now available that use 2-inch diskettes that can store either 50 fields or 25 frames per diskette. With some of these systems, it is possible to place a number of these diskettes in a cartridge and access all of them from a computer. In such cases, it is possible to store and access about 1500 images. One benefit of this technology is that database software can be used to catalog the stored images as they are taken and retrieve them later when needed. The hard drive in the computer can be used for temporary storage of images, or to store images used for comparison purposes. It is also possible to transmit the image file by telephone to a colleague or customer, as long as both parties have the required equipment—which is readily available.

Digital laser disks provide another approach for storing images. At present, these are write once, read many times (WORM), devices but erasable systems are currently being developed. These disks are available in several sizes and have vast memory capacity permitting storage of about 1600 to several thousand 512×512 pixel by 256 gray-level images, depending on the capacity of the disk. The cost for storing an image is about \$0.10, which is very attractive. Also, images can be recalled in a fraction of a second.

Ransick (1988) has described construction of a system for storing microstructural images using an optical laser disk. His system uses a 1000-line vacuum-tube camera, a frame grabber and image-processing software, a $5\frac{1}{4}$-inch optical disk drive, a video-to-film replay unit for hard copies, a black and white monitor, and an 80386 computer (about \$20,000 for the system). The $5\frac{1}{4}$-inch optical disk has a 400 Mbyte capacity and can store about 1600 micrographs. The replay unit has a high-resolution CRT (1300 lines), and film holders for a 4×5 inch, 8×10 inch, or 35 mm films can be used to photograph the image from the CRT, as is done with scanning electron microscopes.

Image Processing

One of the most fascinating aspects of video microscopy technology is the ability to enhance or modify the nature of the digitized image to produce a better image, either simply for viewing or for subsequent feature analysis. A wide range of procedures is available for contrast enhancement, edge enhancement, line intensification, pseudo-three-dimensional contrast, pseudo color assignment, shading correction, and image cleaning. Such operations require a substantial amount of computer memory for storage of the initial image and manipulation. The reader is referred to Chapter 12 by Russ for more in-depth coverage of the specific operations involved in general image processing.

Video results are best when the image covers the entire 256 gray-level range. An excessively white image will oversaturate the camera; if the image is too black, the camera is undersaturated. The microscopist should have the light source properly centered and the illumination level set correctly. The black-level setting on the camera must also be set correctly. Then the gray-level histogram can be adjusted to cover the entire 256 range.

There are a number of commercially available image-processing systems. One can assemble a video microscopy system using these devices plus a camera, monitor, computer, and hard-copy devices. The basic textbooks (Pratt, 1978; Hall, 1979; Baxes, 1984; Inoue, 1986) describe system construction, as do a number of other authors (Inoue, 1981; Agar, 1983; Rigney, 1985; Maranda, 1987; Jarvis, 1987). Systems can be assembled with a wide range of sophistication and cost.

Many of the better-quality video cameras come with a control box that provides some of the simple image-processing features. Typical features would include contrast enhancement, shading correction, video boosting, and inversion (i.e., reversal). These are hard-wired devices. Several companies offer more sophisticated hard-wired devices for image manipulation. Most of the available image-processing systems are software-based systems. Such systems offer greater versatility at a lower cost, but they are also slower than hard-wired devices. The two approaches can be combined, and some systems do so, to provide both speed and flexibility. Those who are more adventurous can assemble their own systems by purchasing components and software from various vendors.

VIDEO IMAGES OF MATERIALS

To illustrate the nature of some of these signal enhancement techniques, a number of examples will be shown using several types of systems and different approaches for producing images. Figure 1.17 shows two micrographs of Pyromet® 718, a nickel-based superalloy, which was heat treated to form the δ phase. These micrographs were taken using color print film and a 35-mm attachment on the Polaroid Freeze Frame®. Figure 1.17(a) shows the live image, and Figure 1.17(b) shows the image after averaging and edge enhancement using a hard-wired device. Figure 1.18 shows the use of this same hard-wired device to selectively enhance the edges of the δ phase that are (a) vertical, (b) horizontal, and (c) diagonal in the image. Figure 1.19 shows the microstructure of a W–7%Ni–3%Fe powder metallurgy alloy enhanced with this device. Figure 1.19(a) shows the live image, large globular tungsten particles in a Ni-Fe matrix. In Figure 1.19(b) the image has been subjected to pseudo-three-dimensional enhancement followed by edge enhancement. Figure 1.19(c) shows the microstructure after image averaging and pseudo-three-dimensional enhancement.

Figures 1.20 and 1.21 illustrate digital image processing using a software program added to an automatic image analyzer. In this work the monitor was photographed using color print 35-mm film. An upright microscope was used to produce the initial images. Figure 1.20 shows two micrographs of an improperly heat-treated, cold-work tool steel. Figure 1.20(a) shows the live image, while Figure 1.20(b) shows the same image after edge sharpening. Figure 1.21 shows four views of another W–7% Ni–3% Fe specimen, this one at a lower magnification than in Figure 1.19. Figure 1.21(a) shows the live image, while Figure 1.21(b) shows the image after edge sharpening. Figure 1.21(c) shows the image after application of a Sobel edge detection kernel. Note the contrast reversal. Figure 1.21(d) shows the image after Sobel edge detection, image averaging, and contrast inversion.

Figures 1.22 to 1.25 were produced using a commercially available frame grabber board and software installed in a PC clone, a monochrome CCD camera (565 lines) and a monitor. The images were produced by photographing the monitor using black and white 35-mm film. An inverted microscope was used to produce the initial images. Figure 1.22 shows the microstructure of a modified AISI H13 hot work tool steel austenitized at 1950°F, isothermally held to precipitate pearlite on the prior-austenite grain boundaries, and quenched. Figure 1.22(a) shows the live image while Figure 1.22(b) shows the image after edge sharpening. In Figures 1.22(c) and (d) the contrast was adjusted by thresholding to emphasize the grain boundaries and wash out the matrix structure [Figure 1.22(c)] and to enhance both the grain boundaries and the martensitic matrix [Figure 1.22(d)]. Figure 1.23 shows an example of histogram equilization to enhance the microstructure of heat-treated Pyromet® 350, a precipitation hardenable stainless steel. Figure 1.23(a) shows the live microstructure and a histogram of the gray levels present. In Figure 1.23(b), the gray-level histogram has been stretched and equalized to alter the contrast.

Figure 1.24 shows two views of René 95, a powder metallurgy nickel-based superalloy, in the hot isostatically pressed condition, where both carbides and coarse γ' (Ni_3Al) are present in an austenitic matrix. Figure 1.24(a) shows the live image. This image was processed using an unsharp masking procedure (the image is blurred and then subtracted from the original image) followed by contrast enhancement, as in Figure 1.24(b). Figure 1.25 shows the W–7% Ni–3% Fe alloy. Figure 1.25(a) shows the live image. The midtones of the tungsten grains were selected (see box in upper left corner), and then these tones were subtracted from the image to produce inverted contrast, as shown in Figure 1.25(b).

These examples illustrate only a few of the many techniques that can be applied to enhance images of metallographic specimens. They demonstrate the potential of video enhancement for improving images, either for documentation or for subsequent measurements.

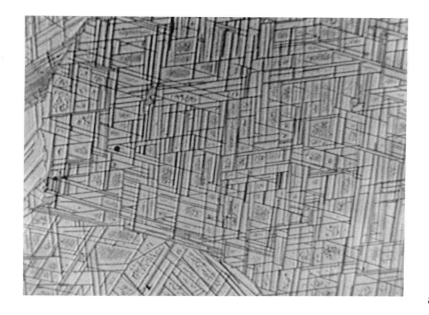

a

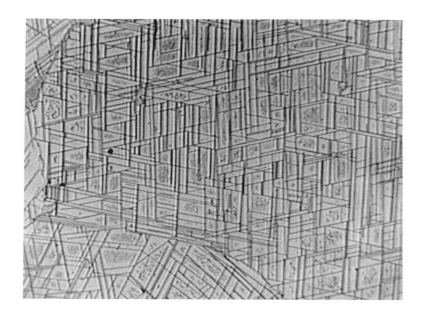

b

Figure 1.17 Microstructure of Pyromet® 718 superalloy heat treated to form the δ phase, etched with glyceregia: (a) live video image, and (b) image after averaging and edge sharpening (~500×).

39

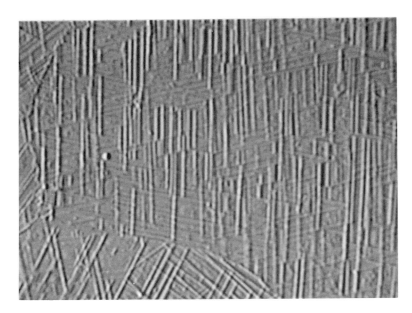

a

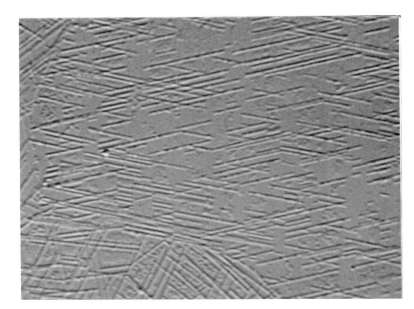

b

40

c

Figure 1.18 Microstructure of Pyromet® 718, as in Figure 1.18, after enhancement, all images averaged and then: (a) vertical edges enhanced, (b) horizontal edges enhanced, and (c) diagonal edges enhanced (~500×).

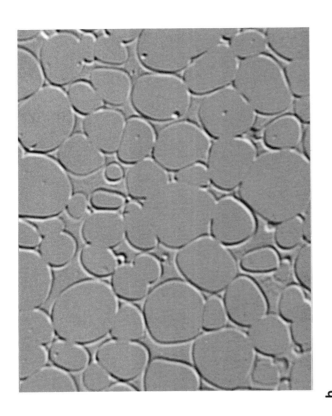

b

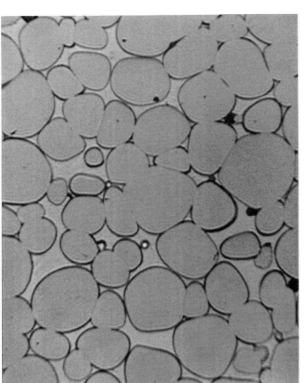

a

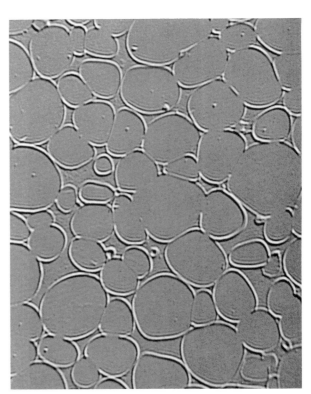

c

Figure 1.19 Microstructure of powder metallurgy W–7% Ni–3% Fe, etched with Kalling's No. 2 reagent: (a) live video image, (b) pseudo-three-dimensional image, edges enhanced, and (c) image averaged and pseudo-three-dimensional enhancement (~500×).

a

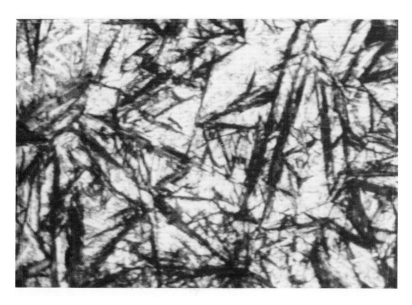

b

Figure 1.20 Microstructure of an improperly heated-treated, cold-work tool steel etched with Vilella's reagent: (a) live video image, and (b) edges sharpened (~400×).

43

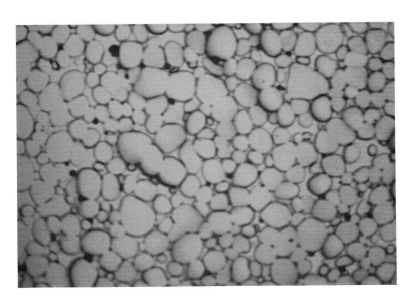

a

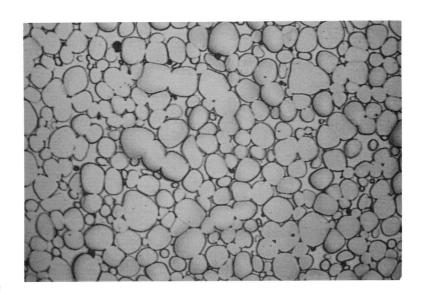

b

44

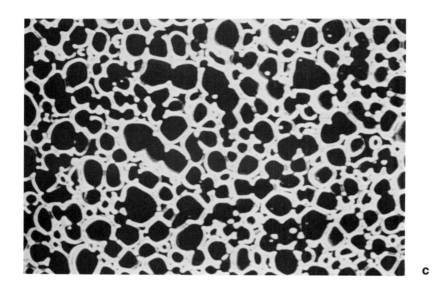

c

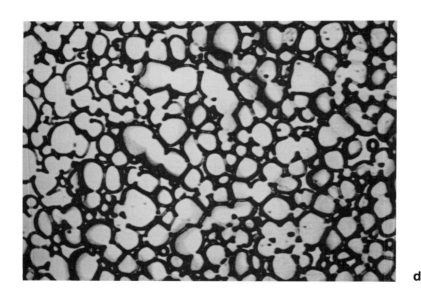

d

Figure 1.21 Microstructure of PM W–7% Ni–3% Fe, etched with Kalling's No. 2 reagent: (a) live video image, (b) edges sharpened, (c) Sobel edge detection, and (d) Sobel edge detection, image averaged and inverted (~400×).

45

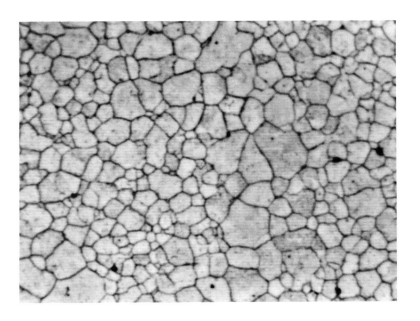

a

b

46

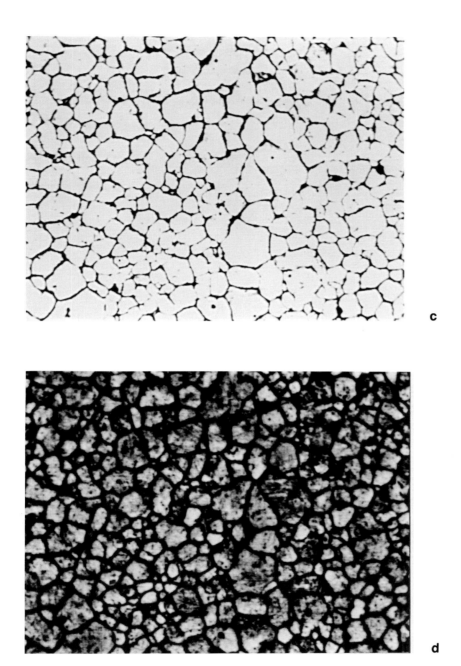

c

d

Figure 1.22 Microstructure of Modified AISI H13 tool steel heat-treated to reveal the prior-austenite grain boundaries, etched with Vilella's reagent: (a) live video image, (b) sharpened, (c) contrast adjusted by thresholding to suppress the matrix and enhance the boundaries, and (d) contrast adjusted by thresholding to enhance the matrix and boundaries (~200×).

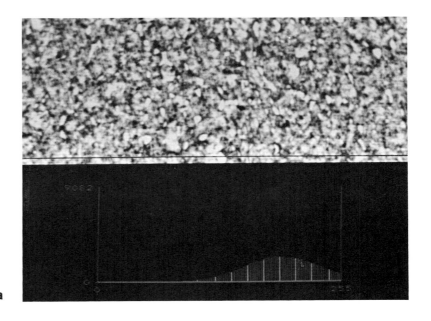

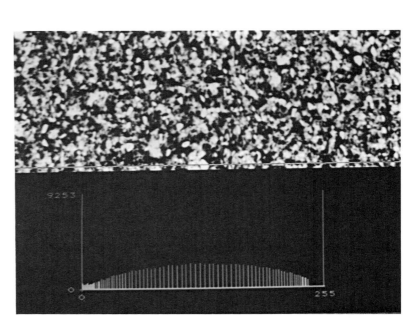

Figure 1.23 Microstructure of heat-treated Pyromet® 350 PH stainless steel etched with Vilella's reagent: (a) live video image and histogram, and (b) contrast adjusted by histogram equilization (~200×).

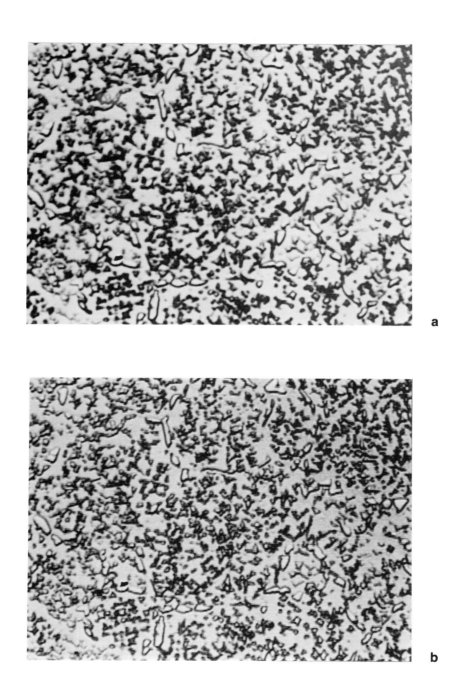

Figure 1.24 Microstructure of as-hot isostatically pressed René 95 Ni-based superalloy etched with glyceregia: (a) live video image, and (b) unsharp mask followed by contrast readjustment (~1000×).

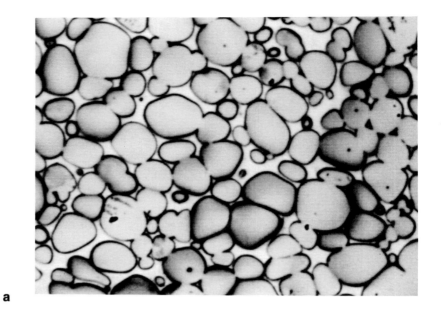

a

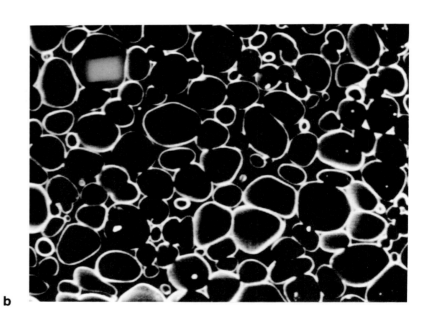

b

Figure 1.25 Microstructure of PM W–7% Ni–3% Fe etched with Kalling's No. 2. reagent: (a) live video image, and (b) midtones of tungsten grains (box in upper left corner) subtracted from image (~200×).

50

PHOTOMICROSCOPY

In the initial work of Sorby, the only way that the microstructure could be documented was by drawing. While still a useful technique for learning about microstructures, drawing of microstructures is a tedious process, depending on the nature of the microstructure, and subject to the artistic limitations of the metallographer. Sorby quickly realized that the new science of photography would be a much better procedure for documenting images, and he enlisted the aid of a commercial photographer, Charles H. Hoole, to make the first photomicrographs at 9× in 1864. Since that time, great progress has been made, both in photomicrography and in photographic materials.

The introduction of instant films by the Polaroid Corporation freed the metallographer from the laborious dark room work required by wet-processed films. Instant films are adequate for the majority of work, but the quality of the images does vary with the film type. For the most critical work, many metallographers prefer to use wet-processed films. Considerable automation of the wet-processing film development and printing processes has been accomplished, but only the larger laboratories can justify purchase of such equipment.

In the past few years, digital technology has spawned several systems based on the NTSC standard. The images produced by these systems are reasonably good but would be better if they were compatible with higher-resolution video formats. This ability probably will be available on future systems. There are a number of systems that can produce slides directly from the monitor, but slides are not as convenient as prints. A number of other approaches, such as color printers, can also be used to generate hard copy. Considerable progress has been made, and advances are sure to follow.

SUMMARY

To obtain the best possible images using the light optical microscope, each step in the process from sampling to printing must be carefully executed. The importance of proper metallographic specimen preparation cannot be overstated. Short cutting the preparation sequence will result in poorer-quality images, at best, and artifact structures, at worst. These techniques are not difficult to learn. The development of automated specimen preparation equipment makes high-quality specimen preparation attainable with a minimum amount of frustration and effort.

There are many different techniques that can be used to reveal microstructure, only a few of these have been shown in this chapter. Chemical etchants are most commonly employed. The optimum etchant for each particular alloy and heat treatment must be selected and applied to the proper extent. Tint etchants, although much more challenging to use, do provide excellent microstructural rendition and are generally selective to certain constituents.

The microscope must be of high quality. Fortunately, very high-quality versatile microscopes, at reasonable cost, are now readily available from a number of manufacturers. However, the best microscope is impaired if it is not kept clean and properly aligned. A regular maintenance program is essential.

Metallographers should make greater use of illumination modes other than bright field. In general, studies should begin with bright-field illumination, and in many cases it is adequate. However, some structures can be revealed better using other illumination modes, such as dark field, polarized light, and differential interference contrast. Cross-polarized light, of course, is absolutely essential when working with noncubic, optically anisotropic metals.

Video microscopy, a newly emerging technique of great promise for the metallographer, has the potential for producing a revolution in metallography. The ability to work with images of any light level, produce enhanced images, and store them on electronic devices, is very attractive. Improvements need to be made in producing higher-quality hard copy at a reasonable price, particularly for those methods that copy directly from the stored image. The intense competition in this field is sure to achieve this goal in the near future.

The last step in the process, photomicroscopy, can be accomplished in a variety of ways to meet any need. The highest-quality images are obtained using wet-processed films and printing, a mature technology. Instant films provide a range of image quality, depending on film cost, and meet most needs. Digital technology now provides a third alternative for photomicroscopy, and the quality of prints made by these methods is improving at a rapid pace.

ACKNOWLEDGMENTS

The author acknowledges the support and encouragement of the management of Carpenter Technology Corporation and his co-workers. The cooperation of personnel from E. Leitz, Reichert-Jung, Optomax, Universal Imaging Corporation, and Cadmet, Inc., in producing illustrations using their equipment is appreciated. 7-Mo Plus® and Pyromet® are registered trademarks of Carpenter Technology Corporation, and Polaroid® is a registered trademark of the Polaroid Corporation.

REFERENCES

Agar, A. W. (1983). *J. Microsc.* **130,** 339.
Baxes, G. A. (1984). *Digital Image Processing.* Prentice-Hall, Inc., Englewood Cliffs, N.J.
Beraha, E., and Shipgher, B. (1977). *Color Metallography.* ASM, Metals Park, Ohio.
Campbell, W. (1926). *Trans. AIME* **73,** 1135.
Farrell, J. W., and Hiltz, D. C. (1976). *Iron & Steelmaker* **3** (8), 17.
Hall, E. L. (1979). *Computer Image Processing and Recognition.* Academic Press, N.Y.
Humphries, D. W. (1967). *The Microscope and Crystal Front* **15,** 351.
Inoue, S. (1981). *J. Cell Biol.* **89,** 346.

———— (1986). *Video Microscopy*. Plenum Press, N.Y.

Jarvis, L. R. (1987). *Acta Stereologica* **6**, 299.

Maranda, B. (1987). *Amer. Lab.* **19** (4), 41, 42, 44, 46, 48, 50–52.

Metallography and Microstructures (1985). Vol. 9, 9th ed. Metals Handbook, ASM, Metals Park, Ohio.

Moore, H. (1964). *Metallography 1963*. The Iron and Steel Institute, Margate, U.K., Special Report 80.37.

Morrough, H. (1952). *Polarized Light in Metallography 88*. Butterworth & Co., Ltd., London.

Pratt, W. K. (1978). *Digital Image Processing*. J. Wiley, N.Y.

Quarrell, A. G. (1963). *JISI* **201,** 563.

———— (1964). *Metallography 1963,* The Iron and Steel Institute, Margate, U.K., 1.

Ransick, M. H. (1988). *Metallography of Advanced Materials, Microstructural Science* **16,** 511. ASM, Metals Park, Ohio.

Rigney, D. R. (1985). *Micron and Microscopica Acta* **16,** 125.

Samuels, L. E. (1963). *J. Australian Inst. Metals* **8** (3), 264.

Vander Voort, G. F. (1978). *Metallography in Failure Analysis,* 33. Plenum Press, N.Y.

———— (1981). *Microstructural Science* **9,** 135. Elsevier North-Holland, N.Y.

———— (1984). *Metallography: Principles and Practice,* McGraw-Hill, N.Y.

———— (1985a). *Metal Progress* **127** (4), 31–33, 36–38, 41.

———— (1985b). In *Metallography and Microstructures,* Vol. 9, 9th ed. 71 Metals Handbook, ASM, Metals Park, Ohio.

———— (1986). *Applied Metallography,* 1. Van Nostrand Reinhold, N.Y.

———— (1988). *Standardization News* **16** (11), 50.

Vilella, J. R. (1951). *Trans. AIME* **191,** 605.

Wilson, W. G., and Wells, R. G. (1973). *Metal Progress* **104** (7), 75.

2

Scanning Electron Microscopy

D. C. JOY AND J. I. GOLDSTEIN

The scanning electron microscope (SEM) is one of the most versatile instruments available for the examination and analysis of the microstructural characteristics of solid objects. The primary reason for the SEM's usefulness is the high resolution that can be obtained when bulk objects are examined; values as good as 1 nm (10 Å) are now quoted for commercial instruments. This resolution approaches the bonding distance of atoms and results in photographs with useful detail at magnifications in excess of 100,000×. Such magnifications lie in the range normally achieved with the transmission electron microscope (TEM) (see Chapter 5). For example, Figure 2.1 shows gold-palladium particles evaporated on carbon taken with a field-emission instrument at an original magnification of 300,000× (300k×) and an electron beam voltage of 30,000 V (30 kV).

Another important feature of the SEM is the three-dimensional appearance of the specimen image. This three-dimensional appearance is a direct result of the large depth of field, where features of the specimen are in focus despite the fact that they lie in different planes. Two examples of the application of the large depth of field of the SEM are given in Figure 2.2. Figure 2.2(a) shows a molybdenum crystal formed in the exit port of a hydrogen furnace. This photograph was taken at an original magnification of 34×. Figure 2.2(b) shows zinc chloride crystalline sediments from a copper water pipe. This photograph was taken at an original magnification of 500×. The greater depth of field of the SEM provides much more information about the specimen than can be achieved with the light optical microscope (LOM) (see Chapter 1) at comparable magnifications. In fact, a casual observation of published SEM photographs indicates that enhanced depth of field is the feature that is of most value to the SEM user.

The SEM spans the magnification range between the light optical microscope, in

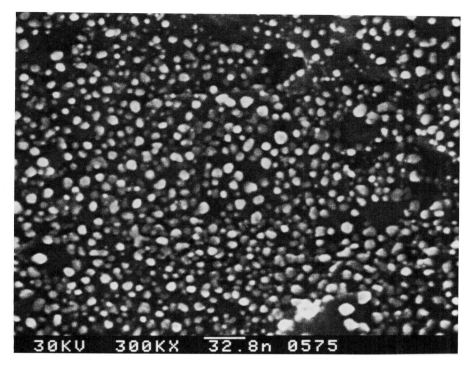

Figure 2.1 Gold-palladium particles evaporated on carbon. Photomicrograph taken with a field-emission SEM (30 kV, 300,000× original magnification).

which the maximum magnification is less than 1500× and the transmission electron microscope, in which magnifications above 10,000× are more common. In addition, most specimens require virtually no specimen preparation, other than a thin coating to draw electrons from the sample surface to an electronic ground. Finally, in most cases image interpretation is relatively simple, which allows the user easily to take advantage of the imaging features of the SEM.

SEM INSTRUMENT AND ITS OPERATION

Figure 2.3 shows a schematic of a modern SEM, a field-emission instrument. The SEM consists of an electron optical column, on the left, which generates and focuses the electron beam and scans the beam over the specimen surface, and an operating desk, on the right, where the operator controls the instrument, selects the focus and magnification, and takes a photomicrograph from the TV screen display. The electron column operates under vacuum because, unlike light, electrons are

a

absorbed in air. The electron gun is the source of electrons. The more conventional electron guns are tungsten (W) or lanthanum hexaboride (LaB$_6$) filaments. The brightest electron source, which ultimately produces the highest-resolution pictures, operates using a field-emission electron gun and requires an ultrahigh vacuum in the gun chamber.

The light source in the LOM or the electron beam source in the TEM illuminates the entire specimen area of interest. In the SEM, however, the electron beam is focused to a fine probe and is then scanned across the specimen to obtain an image. The condenser lenses (Figure 2.3) and the stigmators are used to produce a fine electron beam spot size of less than 10^{-5} cm (1000 Å, 100 nm). The lenses are capable of focusing the electron beam to finer and finer probe sizes. Unfortunately, the current in the electron beam, which ultimately determines the signal strength and image quality, decreases with decreasing probe size. In practice, probe sizes of the

b

Figure 2.2 (a) Molybdenum crystal formed in the exit port of a hydrogen furnace. Photograph taken at 35× original magnification. Courtesy of Polaroid and K. Tremaine and A. Miller (GTE Products Corp.). (b) Zinc chloride crystalline sediments from a copper water pipe. Photomicrograph taken at 500× original magnification. Courtesy of Polaroid and A. Schliessus and J. Heinz (MPA, government materials testing institute).

order of 7 nm (70 Å) are available with conventional W or LaB$_6$ filament guns, while probe sizes of 1 to 2 nm (10–20 Å) are formed with field-emission guns. The ultimate resolution of the instrument is dependent on how small the electron probe can be focused on the specimen with sufficient current in the probe to extract a measurable signal. The voltage of the electron gun is increased to maximize the resolution. Low gun voltages, with subsequent loss of resolution, are often used to minimize specimen damage and electron beam charging.

The specimen is mounted on the goniometer stage, which allows vertical (Z),

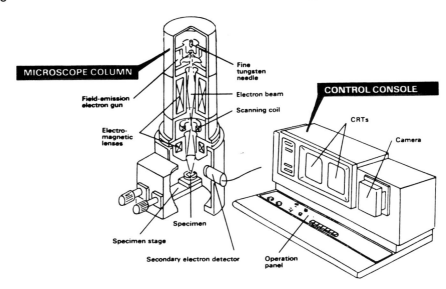

Figure 2.3 Schematic of the SEM microscope column and control console. Courtesy of *The Morning Call* newspaper, Allentown, Pa.

horizontal (X,Y), tilting, and rotating of the specimen in the specimen chamber vacuum. In order to obtain an image, the focused electron probe is moved across the specimen surface (X,Y) with the scanning coils. The measured signals are the emitted electrons generated by the focused electron probe. Two types of electrons are generated: (1) low-energy secondary electrons (SE) of less than 10 eV, which come from the near surface layers ($\approx$ 50 Å) of the specimen and (2) high-energy electrons, which are backscattered (B) from some depth into the specimen ($\approx$ 0.5 μm) and are close in voltage to the impinging electron beam. The high-energy backscattered electrons are typically generated much deeper in the specimen than the secondary electrons. The amount of secondary electrons generated under the focused beam is typically more sensitive to surface topography, while the number of backscattered electrons generated under the beam is proportional to the atomic number of the host specimen.

The two types of electrons (SE and B) are measured by electron detectors in the specimen chamber. Figure 2.4 shows a schematic drawing of the Everhart-Thornley electron detector, which is optimized for measuring SE. A small positive bias on the front of the detector attracts up to about 60% of the generated secondary electrons to the detector. Some SE initially moving away from the detector can be attracted by the bias. A positive field of up to +12 keV then accelerates the SE. The accelerated electrons lose their energy in the scintillator (S), and the emitted light from the scintillator goes down the light guide (LG) to a photomultiplier (PM) tube. The detection of SE from all directions from the specimen allows for the three-dimensional appearance of the specimen. Image interpretation is simple and intuitive since the SE image is closely analogous to the images we see using reflected light.

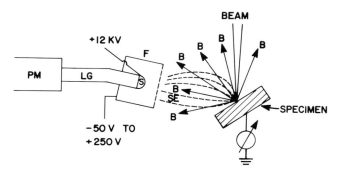

Figure 2.4 Schematic drawing of the Everhart–Thornley electron detector. Courtesy of Plenum Press.

Unfortunately some high-energy backscattered electrons (B) moving in the direction of the scintillator are also detected. These electrons form a background signal in the final picture. The intensity of the measured signal (SE plus B) at each point is displayed on the TV tube of the display unit (Figure 2.3). Once an entire area is scanned, the intensity variation, point to point, is used to give the final picture of the specimen surface as recorded on film (Figures 2.1 and 2.2).

Figure 2.5 shows schematically how a picture is formed and how magnification is controlled. The area on the specimen that is scanned has a length l. The area on the display screen is also scanned point to point with a one-to-one correspondence with

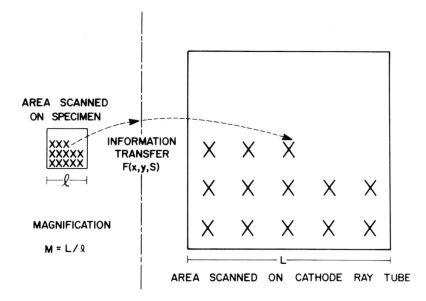

Figure 2.5 Schematic drawing to illustrate how magnification is achieved without lenses in the SEM. Courtesy of Plenum Press.

the specimen. If the length of the display screen is L, the magnification M of the picture is the ratio L/l ($M = L/l$). The operator has control over the area scanned on the specimen, l. By this means, a range of magnifications, M, usually 25–300,000×, can be obtained. The SEM was the first magnifying instrument to use this form of "lens-less" magnification system. This approach has since been used in many other scanning imaging instruments such as the scanning ion (Chapter 3), scanning Auger, scanning acoustic (Chapter 4), and scanning tunneling (Chapter 8) microscopes. Further information on the SEM instrument and its operation can be found in various texts on the subject, for example Goldstein et al. (1981).

The following sections of this chapter give examples of the use of the SEM to examine various types of materials. The first section describes applications to materials in which little or no specimen preparation is needed. The second section describes applications that require the use of metallography, i.e., the preparation of flat polished surfaces of specimens. The third section provides examples of the ever-increasing role of the SEM to study reactions *in situ*. In the fourth section the use of the SEM is described for crystallographic studies.

APPLICATIONS TO MATERIALS

The numbers of applications of the SEM to materials research is almost endless. This section will show specific applications to specimens that are examined in the as-received condition, without special preparation. All the specimens, unless noted, were covered with a thin film of carbon or gold to prevent charging and to allow electrons a direct path to ground.

Geologic Materials

Figure 2.6 shows a light optical microscope photograph of metal particle 63344.1, a unique particle consisting of metal globules (Goldstein et al., 1975), which was found in the Apollo 16 lunar soil. There are about 400 metal globules in a restricted size range from 0.15 to 0.60 mm in diameter welded together into a cohesive mass. The pale material that fills in between the globules is fine dust consisting of fragments of plagioclase crystals and very fine-grained anorthositic-rich rocks. The sample measures about 5.9 mm top to bottom and 5.6 mm across. It is by far the largest metal specimen found to date on the lunar surface.

Because of the limited depth of field of the LOM, the entire particle is not in focus when the photograph is taken. Figure 2.7 is an SEM enlargement of some of the globules near the top of the particle that are free of plagioclase crystals. The importance of the greater depth of field of the SEM is clear. Apparently some of the globules have become detached from one another, as can be observed in Figure 2.7. The material between the globules is mainly low-melting metal phosphide and metal sulfide that once held the particles together. This unique particle is an excellent

Figure 2.6 Light optical micrograph of metal particle 63344.1 extracted from the Apollo 16 coarse fines (Goldstein et al., 1975). Courtesy of *Earth and Planetary Science Letters.* The sample measures about 5.9 mm top to bottom and 5.6 mm across.

Figure 2.7 SEM enlargement of metal globules in the central area of the top surface of Apollo 16 particle 63344.1 (Goldstein et al., 1975). Courtesy of *Earth and Planetary Science Letters.* Field of view is $1.55 \times 1.55 \ mm^2$.

example of the major shock processes that occurred on the moon's surface some billions of years ago.

Polymers

Polymer (plastic) latexes, also known as emulsion polymers, are colloidal dispersions of polymer particles in a continuous medium. Uses of emulsion polymers are quite common, including synthetic rubber, adhesives, latex paints, printing inks, etc. Most of the latex particles are less than 0.1 mm (100 μm) in size and require the SEM to study the morphology of the particles. Figure 2.8(a) shows the morphology of 0.15 μm-diameter cationic acrylic latex spheres on the surface of 19 μm-diameter monodispersed polystyrene latex spheres. Figure 2.8(b) shows the crystal-like structure in a film composed of 0.44-μm-diameter monodisperse polystyrene latex particles. Variations in processing will change the morphologies and relative sizes of the individual particles. The preparation of "large" (2–100 μm diameter), monodisperse latex particles has been undertaken by a research group at Lehigh University using the microgravity environment of the space shuttles Columbia and Challenger. The gravitational effects of settling, which had proved destructive to the polymerization process, have thereby been alleviated, allowing the successful preparation of monodisperse particles in the size range of 5–30 μm diameter. The 10 μm latex has been accepted by the National Institute of Standards and Technology, formerly the National Bureau of Standards, as a standard reference material—the first commercial product to be made in space. These particles have been used as calibration standards, as a means of measuring pore sizes, etc.

Because polymer particles and fibers are prone to beam damage at high operating voltages and high currents, they require coating to prevent charging. Therefore observations of the individual particle surfaces at high magnification (greater than 10,000×) have been very difficult. Figure 2.9 shows an SEM photomicrograph of the surface of a 6 μm-diameter latex particle. The field-emission SEM was used because the specimen can be observed uncoated at an operating voltage of 1 kV, which minimizes electron beam charging. The sample surface can also be observed with an electron beam resolution better than 10 nm (0.01 μm). Figure 2.10 shows another example of the advantage of using a field-emission SEM at low voltages. Here a photoresist, etched on a silicon substrate, was examined without coating at 1.1 kV. In this case as well as the emulsion polymer example, coatings on the specimen often obscure the surface detail that is to be examined.

Another technique to prevent charging besides low kV field-emission gun operation is to use an environmental chamber that allows examination of specimens uncoated in their natural state–wet, moist, or dry. Figure 2.11 shows a secondary electron picture at an original magnification of 750× of silk fabric in an environment of a partial vacuum with 1.5 torr of N_2. No charging is observed, even though the specimen is uncoated and the instrument is operated with a conventional electron gun and at a 15 kV operating voltage.

Ceramics

The SEM has also been applied to problems in ceramic materials. Figure 2.12 shows SEM photographs of intragranular failure of an alumina (Al_2O_3) ceramic during bending. The extraordinary depth of field of the instrument again allows one to view the fracture details at this low magnification. The path of the crack is intragranular, following the cleavage planes in abnormally large-sized grains in the alumina. The normal-sized grains can be observed in Figure 2.12(a) surrounding the fracture region and are generally 10 μm in diameter. The unexpectedly large grain size developed during sintering must be controlled to minimize this form of failure.

Figure 2.13 shows an SEM micrograph of individual ZSM-5 zeolite crystals. ZSM-5 and other zeolites are important catalysts for the petroleum industry. Very tiny channels within the zeolite crystals allow for reactions with the chemicals and the zeolites themselves. It is very difficult to observe the voids or channels themselves with the SEM because of specimen charging, which requires either a coating that obscures the surface or low-voltage operation, which decreases the resolution of the SEM. Some progress has been made in observing voids or channels in zeolites using field-emission instruments, although the scanning transmission electron microscope is more useful for this application (See Chapters 5–7).

Figure 2.14 shows SEM images of pores or holes found at internal boundaries in lithium fluoride (LiF). The pores developed along the boundary that was formed when two single crystals of LiF were joined by heat treatment (Wang et al., 1986). The heat treatment process is called hot pressing, in which pressure is applied to the two single crystals when heated to 780°C in vacuum. Pores were present because the hot pressing was terminated before completion of the joining process. Figure 2.14(a) shows an SEM micrograph of the leading face of a pore. The trailing face of a pore is shown in Figure 2.14(b). The diffusion of atoms along the surface of the pore leads to a quasi-equilibrium pore shape based on a 26-face polyhedron. Stereo pairs of these pores lead to direct measurement of the planar faces shown in Figure 2.14(b).

Laser-induced damage is a material failure resulting from the interaction with intense laser radiation. Due to its complex nature, the understanding of this type of damage process is still incomplete. In a recent study, Wang et al. (1988) have investigated interior damage generated in transparent ceramic single crystals of LiF, KCl, and MgO using a focused, short pulse laser. These ceramics are extremely susceptible to thermal shock fracture, a failure due to a rapid change in temperature and resultant thermal stresses. The damage region in the LiF and KCl specimens consisted of irregular voids at the center and seared traces spreading away from the voids. The SEM micrograph in Figure 2.15(a) showed that these seared traces were markings characteristic of laser-induced liquid flow prior to solidification. The damage region in the MgO specimen exhibited no such seared traces. A void was located at the center of the damage region, surrounded by a disrupted and fragmented lattice as shown in the SEM image in Figure 2.15(b). In both figures the outer portions of damage regions showed penny-like cracks in river-type patterns typical of cleavage fracture surfaces. The laser-induced damage in these materials

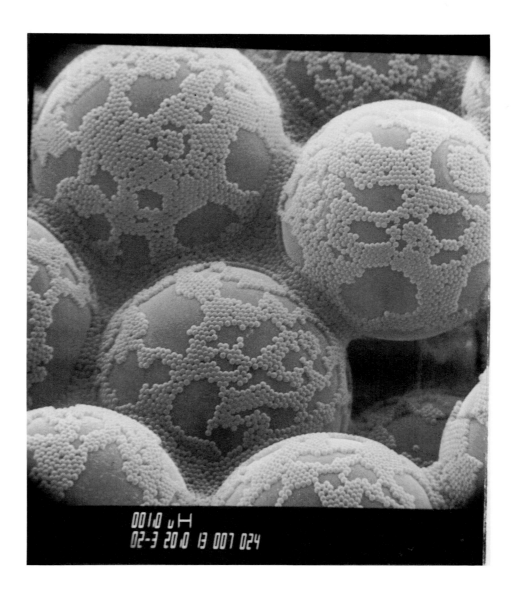

a

Figure 2.8 (a) SEM micrograph showing flocculation of a 154-nm-diameter cationic acrylic latex (positively charged) on the surface of 19-μm-diameter anionic monodisperse polystyrene latex particles (negatively charged). Courtesy of Olga Shaffer, Emulsion Polymers Institute, Lehigh University. (b) SEM micrograph of a crystal-like structure in a dried film of monodisperse polystyrene latex particles of 400 nm diameter. Courtesy of Daniel P. Durbin, *Interfacial Phenomena Occurring in Drying Surfactant Droplets and Polymer Latex Films* (Ph.D. Thesis. Lehigh University, 1980).

64

300nm

b

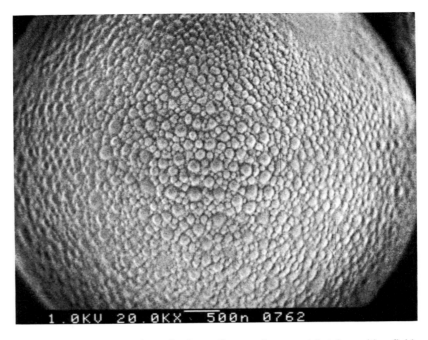

Figure 2.9 SEM image of the surface of a 6-μm-diameter latex particle taken with a field-emission instrument (1.0 kV, uncoated sample, 20,000× original magnification). The magnification marker is 500 nm (0.5 μm).

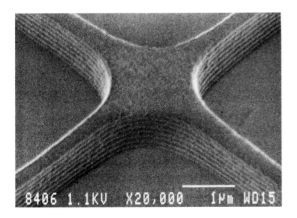

Figure 2.10 SEM micrograph of photoresist examined uncoated in a field-emission instrument (1.1 kV, 20,000 × original magnification). Courtesy of Nippon Iron Powder Co., Ltd.

Figure 2.11 SEM micrograph of a silk fabric, uncoated, in an environment of 1.5 torr (N_2). The original magnification was 750× and the photomicrograph was taken at 15 kV. Courtesy of ElectroScan, Danvers, MA.

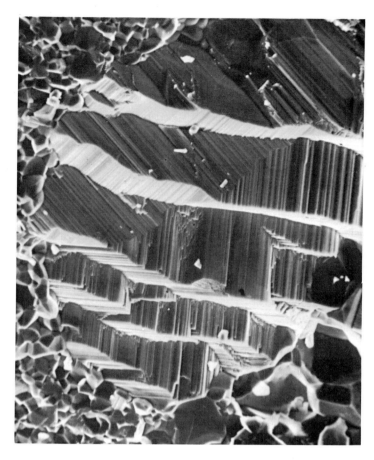

Figure 2.12 SEM micrographs of an intragranular failure of an alumina ceramic during bending. The widths of the photomicrographs are (a) 0.25 mm and (b) 0.1 mm. Courtesy of Martin Harmer, Lehigh University.

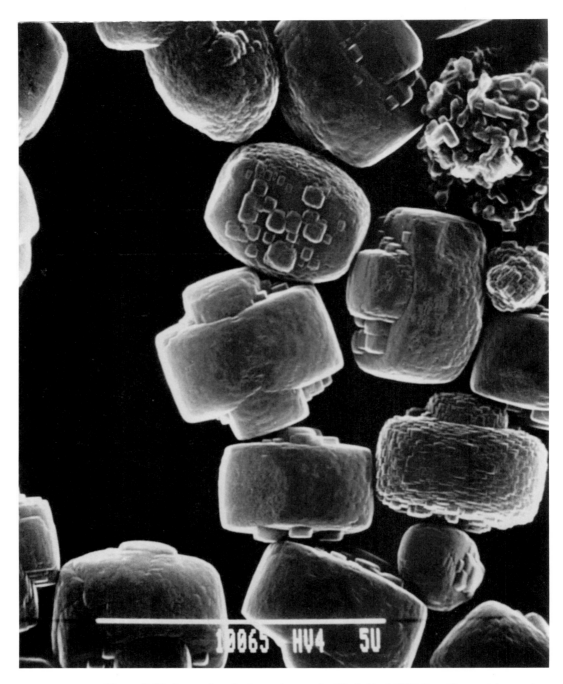

Figure 2.13 A scanning electron micrograph of individual ZSM-5 zeolite crystals, courtesy of Charles Lyman, Lehigh University. The SEM micrograph was taken with an analytical electron microscope operated at 100 kV. The secondary electrons were measured through the probe-forming lens, thus minimizing the contribution from backscattered electrons. The marker denotes 5 μm.

68

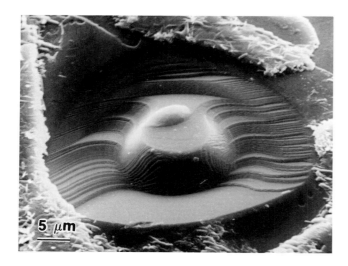

a

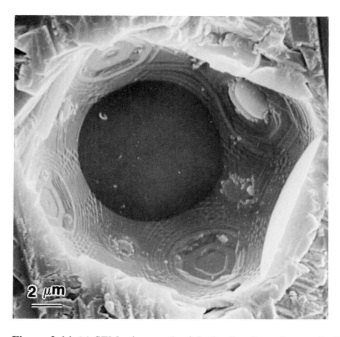

b

Figure 2.14 (a) SEM micrograph of the leading face of a pore in lithium fluoride, (Wang et al., 1986). Courtesy of the *Journal of the American Ceramic Society*. (b) SEM micrograph showing interior surface of a detached pore in lithium fluoride (Wang et al., 1986). Courtesy of the *Journal of the American Ceramic Society*.

a

was produced by a thermal spike process whereby the temperature of the spots where the laser was focused was much higher than the melting temperature of the material.

Metals

The SEM is most often used to examine metals because these materials are more robust and the surface is not usually subject to electron beam charging. Many of the applications of SEM are to fracture surfaces, in which the nature of the metal failure can be determined. Figure 2.16 shows an example of the fracture surface of steel, where microvoid coalescence has occurred. Microvoid coalescence occurs by a process involving fracture of inclusions or failure along the interface between the metal matrix and an inclusion. The associated stress concentration triggers localized

b

Figure 2.15 (a) SEM micrograph of a laser-induced damage region in a LiF crystal (Wang et al., 1988). Courtesy of Z. Wang, M. P. Harmer, and Y. T. Chou of Lehigh University. (b) SEM micrograph of a laser-induced damage region in a MgO crystal (Wang et al., 1988). Courtesy of Z. Wang, M. P. Harmer, and Y. T. Chou of Lehigh University.

plastic deformation, which generates a small hole, called a *microvoid,* within the material. When a large number of microvoids have formed along a particular macroscopic plane, they coalesce into round or elongated voids and cause failure as seen in Figure 2.16 (Hertzberg, 1987).

Another type of fracture surface is shown in Figure 2.17. A crack has propagated through a steel sample, which has a well-defined pearlitic structure. Pearlite is composed of two phases (α-iron and Fe_3C) that nucleate and grow parallel to one another in regions called *colonies*. The parallel markings shown in Figure 2.17 reflect fast fracture through several pearlite colonies in a low-carbon steel alloy. The large depth of field of the SEM is necessary to observe the rather rough fracture surfaces in both Figures 2.16 and 2.17.

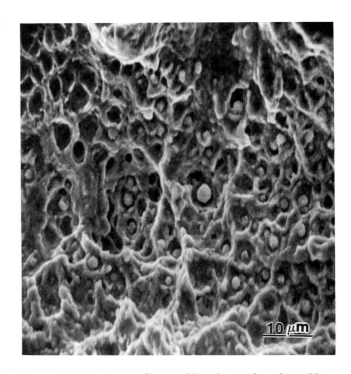

Figure 2.16 SEM photomicrograph of fracture surface markings in metals, microvoid coalescence in steel (Hertzberg, 1987). Courtesy of American Society for Testing and Materials.

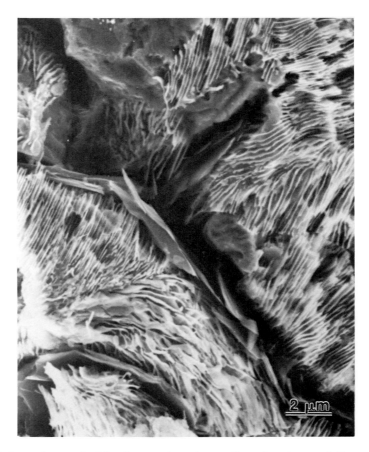

Figure 2.17 SEM photomicrograph of fast fracture through a pearlite colony in a steel alloy (Hertzberg, 1987). Courtesy of American Society for Testing and Materials.

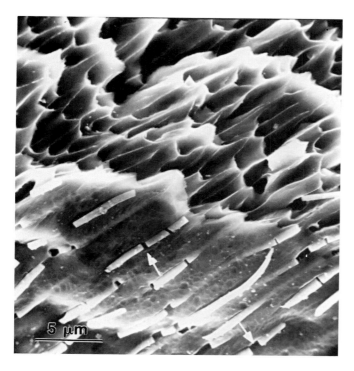

Figure 2.18 SEM photograph of the fracture surface (top) and microstructural appearance (bottom) of Al-Al$_3$Ni eutectic composite. Courtesy of Kenneth Vecchio, University of California, San Diego.

Composites

Composites are materials that are combinations of metals and ceramics, polymers and ceramics, etc. Composites are developed to take advantage of the properties of each material to maximize, for example, strength and ductility at high temperatures. The SEM is particularly useful for observing fracture surfaces of composites. These surfaces are commonly very rough due to the varying mechanical properties of each component. Figure 2.18 shows the fracture surface (top) and the internal structure (bottom) of an aluminum (Al)–nickel aluminide (Al$_3$Ni) eutectic composite. The fracture surface reveals elongated microvoids which have coalesced. The microvoids were nucleated by the rupture of Al$_3$Ni whiskers (see arrows). In this example the microstructure was obtained by masking part of the fracture surface with a lacquer and then polishing the exposed surface electrolytically. After removing the lacquer to reexpose the remaining fracture surface, the specimen was analyzed in the SEM so that one can observe the aluminum matrix with the embedded Al$_3$Ni whiskers as well as the fracture surface itself.

An example of a fracture surface in a polymer–ceramic composite, a hollow

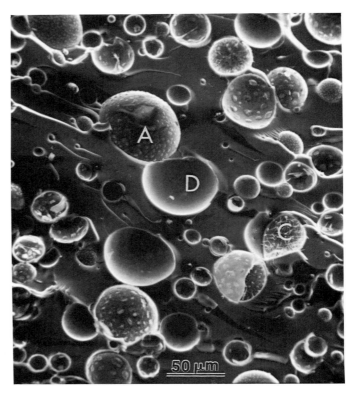

Figure 2.19 Fracture surface in hollow glass-sphere-filled epoxy resin. Glass particles located at A, detached at D, and cracked at C (Hertzberg, 1987). Courtesy of American Society for Testing and Materials.

glass-sphere-filled epoxy resin, is shown in Figure 2.19. In this case interfacial fracture occurred between the glass particle and the matrix epoxy resin. Some of the glass beads were retained on this half of the fracture surface, whereas other particles were either pulled out or shattered. A silicon x-ray map (see Chapter 10) was used to reveal the presence of both undamaged and fragmented hollow glass spheres on the fracture surface (Hertzberg, 1987). In both these examples, the large depth of field of the SEM was critical in order to observe the very rough fracture surfaces.

Semiconductors

One out of every three SEMs sold goes to the semiconductor industry because the instrument provides unique capabilities with which to examine both semiconductor materials and finished devices. The advent of high-resolution SEMs has made it

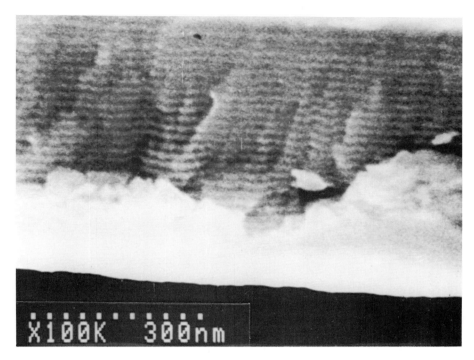

Figure 2.20 Layered Si-Ge structure formed by continuous vapor deposition process. The repeat distance is 150 Å. Imaged with a field-emission SEM at 5 keV. Original magnification 100,000×.

possible to observe directly structures that previously were only visible in the TEM. Figure 2.20 shows a layered silicon–germanium structure formed on a silicon substrate by a chemical vapor deposition (CVD) process. The sample was fractured to reveal the structure, which consists of alternating layers of Si and Ge, each 150 Å (15 nm) thick. The SE image can distinguish between the two materials because of their different secondary electron emission coefficients. Even though these differ only by a few percent, the on-line image-processing abilities of the SEM make this visible (Krause et al. 1989).

The semiconductor industry uses materials that range from metals through ceramics to polymers, often in close proximity to each other. Another important attribute of the modern SEM is thus its ability to image such diverse components simultaneously while preserving meaningful data about each. This is illustrated in Figure 2.21, which shows an uncoated highly dispersive x-ray mirror at an intermediate stage of fabrication using current device technologies. The upper structure is a 3000 Å (300 nm) grating, lithographically produced in a 5000 Å (500 nm) thick layer of poly-methyl methacrylate (PMMA). The x-ray multilayer mirror is composed of 30 alternating layer pairs of molybdenum (45 Å or 4.5 nm thick) and

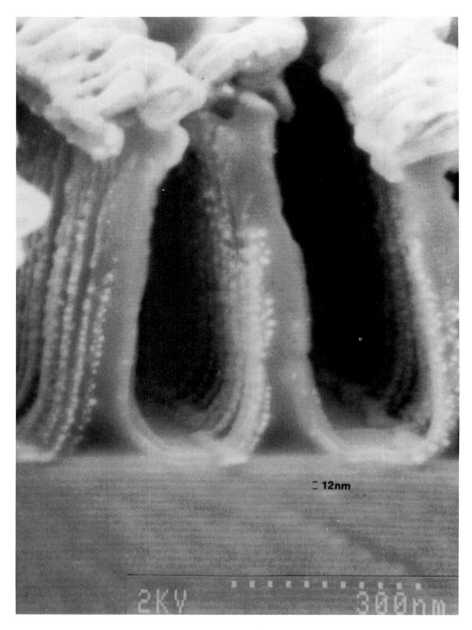

Figure 2.21 Uncoated x-ray mirror imaged at 2 keV with a field-emission SEM. The grating is PMMA; the substrate is alternate layers of molybdenum and silicon. Image courtesy of T. Reilly, Hitachi Ltd.

silicon (70 Å or 7 nm thick). By operating the instrument at a low beam energy (2 keV), it is possible to produce simultaneously a stable image of the uncoated resist while preserving enough spatial resolution to delineate clearly the 115 Å (11.5 nm) repeat of the layered structure of the mirror. Since it would not be feasible to prepare a cross section of such a complex structure for TEM observation, the SEM provides the only tool for monitoring the fabrication progress.

At a somewhat lower level of resolution, the SEM is also widely used to provide an accurate measurement of the spacing and width of structures within integrated circuits. While previously this job has been undertaken by specialized optical microscopes, now that the physical size of features in an integrated circuit has been pushed below 1 μm, this is no longer possible, and instead specially calibrated SEMs must be used to perform "critical dimension metrology" (Postek and Joy, 1987). In most cases these same structures were initially produced by electron beam lithography because the SEM is the only device capable of either fabricating a structure at this scale or examining its perfection once finished.

METALLOGRAPHY OF MATERIALS

In many analysis situations it is more important to examine the structure of the interior of a specimen than its outside surface. This is certainly the case in determining which phases or minerals are present in a specimen or if a reaction occurs between the surface of the specimen (coating, oxide layer, etc.) and the interior of the material. A typical approach to studying the interior of a specimen is to cross section the material at an appropriate place, mount the specimen in epoxy or another type of holder, and polish the surface with a series of finer and finer abrasives. This technique is known as *metallography* because of its origin with the preparation of metal specimens but is equally applied to ceramics, rocks and geological materials, semiconductors, composites, etc.

In most cases, after the specimen is polished flat and the surface scratches are submicrometer in size, the sample is etched with one or more chemicals to develop some surface topography between the phases. This technique was first developed for sample preparation for LOM and is discussed in some detail in Chapter 1. Since the number of secondary electrons produced when a focused electron probe impinges on the sample varies with surface topology or tilt, one can take advantage of the high magnification capability of the SEM and extend the magnification range of the LOM so that one can examine polished and etched samples at magnifications above 1000×. In many practical examples, the SEM image contrast can be enhanced by deeper or more severe etching. An example of this capability is shown in Figures 2.22 and 2.23. In Figure 2.22 a lamellar structure (L) is observed in the LOM at an original magnification of 500× in the high-Ni taenite (T) phase of the Toluca iron meteorite. The matrix structure is low-Ni kamacite (K) phase. The gray region surrounding the lamellar structure (L) is a high-Ni ordered phase FeNi called clear taenite 1 (CT1). This specimen was polished flat and etched with nital to reveal

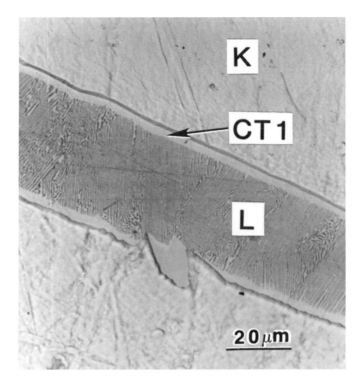

Figure 2.22 Light optical micrograph of the lamellar structure of the taenite phase in the Toluca iron meteorite. Courtesy of Janis Kowalik, Lehigh University.

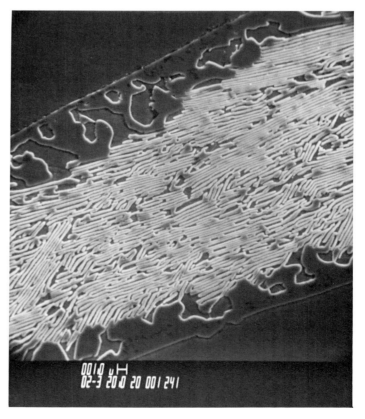

Figure 2.23 SEM micrograph of a similar lamellar structure as shown in Figure 2.22. Courtesy of Janis Kowalik, Lehigh University.

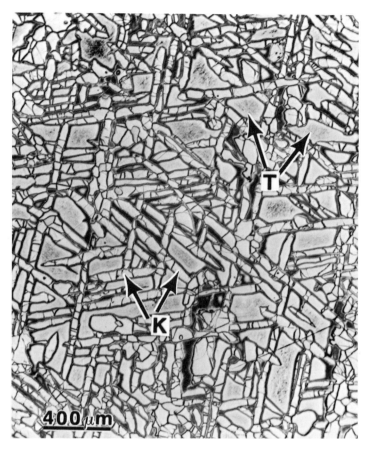

Figure 2.24 Light optical micrograph of the fine Widmanstätten structure of the Dayton iron meteorite.

the structure. Unfortunately the lamellar structure is hardly visible in this LOM picture. Figure 2.23 is an SEM micromicrograph of a similar region in the meteorite taken at 2000× original magnification in an SEM operated at 20 kV. The lamellar structure composed of aligned K and T phases is clearly observed.

The range of magnifications available using metallographic specimen preparation for SEM is truly remarkable. The LOM photograph (Figure 2.24) shows the fine Widmanstätten structure of the Dayton iron meteorite. The low-Ni phase, kamacite (K), forms the Widmanstätten pattern, and the high-Ni phase taenite (T) is the resultant matrix. The fine structure within the taenite regions and at the K–T interface is clearly revealed with SEM techniques. SEM micrographs, Figure 2.25(a,b,c), show the structure in the center of taenite in the Carlton meteorite at magnifications ranging from 700× to 50,000×. The latter picture was taken with a field-emission SEM. The iron–nickel matrix transforms by a diffusionless shear

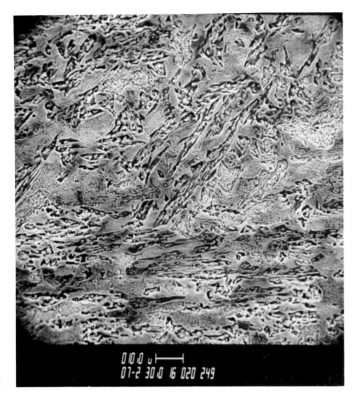

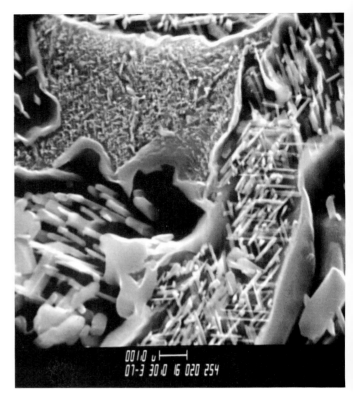

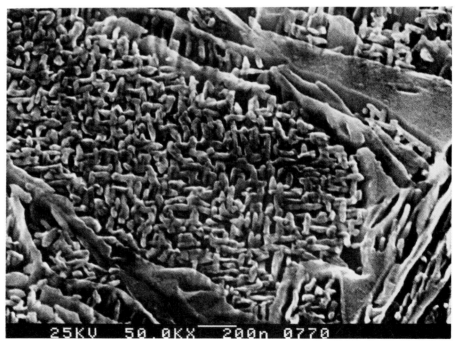

Figure 2.25 SEM micrographs of the fine structure within the central taenite regions of Carlton, a typical iron meteorite with a Widmanstätten structure similar to that shown in Figure 2.22. Original micrographs increase from 700× in (a) to 7000× in (b) to 50,000× in (c).

transformation, called a *martensitic transformation,* upon cooling from high temperatures. At lower temperature the diffusionless transformation product decomposes into finer and finer particles of taenite in a kamacite matrix. Such a process is confirmed by TEM analysis of the same regions of the meteorite.

At the K–T interface the number of reactions is increased, and the size of the phases that are formed is quite small. Figure 2.26 shows a TEM photomicrograph of the K–T interface. Within the taenite is a 0.5 μm region of CT1 and then a 1 μm region of cloudy zone (CZ) with bright regions of CT1 surrounded by a low-Ni phase (Reuter et al., 1988). A similar cloudy-zone region was unresolvable by LOM, see Plate 1.8 in Chapter 1. The same structure has been revealed recently by field-emission SEM using a metallographically prepared specimen (Figure 2.27). The increased resolution of the field-emission SEM now allows the investigator to overlap the magnification range of the conventional TEM. This is a major advantage, as specimen preparation for TEM involves making transparent foils from bulk materials, which is often a tedious and difficult problem.

The metallographic technique can also be used to determine the average atomic number of phases or particles in a specimen using the backscattered high-energy electrons (B) in the SEM. If the specimen is polished flat but not etched, the surface topography of the specimen will be minimized. Therefore the number of SE generated will be approximately constant across the sample surface. As was mentioned at the beginning of this chapter, if the atomic number of the specimen varies from point to point on the specimen, the number of backscattered electrons will vary. The backscattered signal can be measured by the Everhardt–Thornley detector or more efficiently by several types of detectors placed just above the specimen and below the final lens. In this technique, it is important of course not to etch the specimen and induce topography.

Figure 2.28 shows a backscattered electron micrograph of an unetched 27 at. % Au–5 at. % Cu–68 at. % Sn alloy specimen annealed for 1000 hours at 170°C (Roeder, 1988). Three distinct phases are present in chemical equilibrium: (1) the dark phase, K, (2) the white phase, $AuSn_2$, and (3) the gray phase, $AuSn_4$. The three phases, which are in equilibrium, give the tie triangle at the 170°C isothermal section of the Au-Cu-Sn alloy. Figure 2.29 shows the more complex as-cast structure of a 15 at. % Au–10 at. % Cu–75 at. % Sn alloy. In this backscattered electron photomicrograph six individual structures can be identified. X-ray microanalysis can be performed on the various phases present in these specimens without further metallographic preparation. Thus the combination of SEM and x-ray microanalysis leads to a full characterization of the elemental chemistry and surface microstructure of the specimen.

The metallography of semiconductor materials is usually simpler than that of metals because they are specifically chosen and grown to be both single-crystal and single-phase materials. However, subtle variations of the local chemistry due to the presence of trace amounts of impurities such as carbon or oxygen can have very adverse effects on the electronic properties of the material. Such effects can be monitored with the imaging modes of the SEM, which rely on the special properties of a semiconductor. The most powerful of these is the electron-beam-induced current (EBIC) signal. This relies on the fact that although, by definition, a semicon-

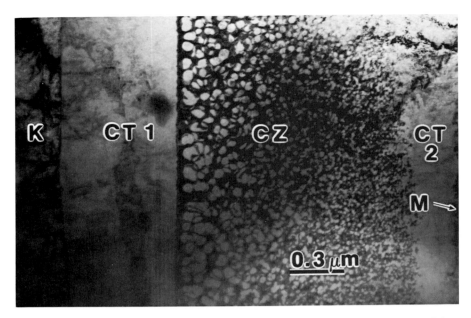

Figure 2.26 TEM photomicrograph of the kamacite-taenite interface in the Tazewell iron meteorite (Reuter et al., 1988). Four distinct regions are observed in the taenite close to the kamacite–taenite interface. Courtesy of *Geochimica et Cosmochimica Acta*.

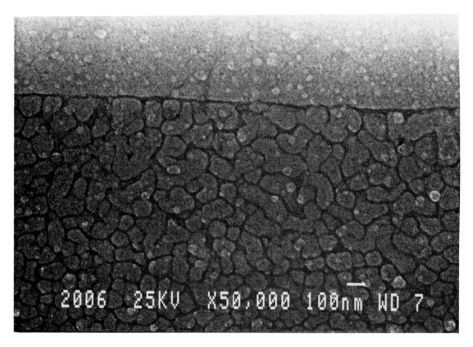

Figure 2.27 SEM photograph of the clear taenite 1–cloudy zone interface taken with a field-emission-electron microscope. The original magnification is 50,000× and the scale bar is 100 nm.

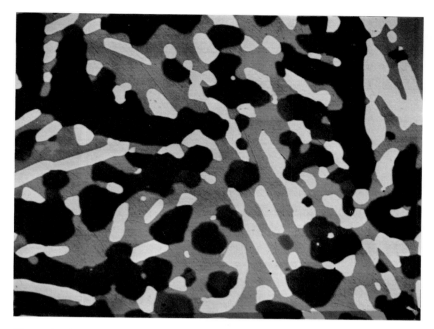

Figure 2.28 Backscattered electron micrograph of a 27 at. % Au–5 at. % Cu–68 at. % Sn alloy annealed for 1000 hours at 170°C. Three distinct phases are observed in the unetched specimen. Courtesy of J. Roeder, Lehigh University. Field of view is 330 × 250 μm²

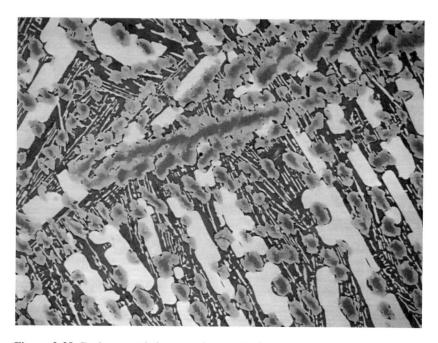

Figure 2.29 Backscattered electron micrograph of an as-cast 15 at. % Au–10 at. % Cu–75 at. % Sn alloy. Six distinct structures are observed in the unetched specimen. Courtesy of J. Roeder, Lehigh University. Field of view is 275 × 210 μm².

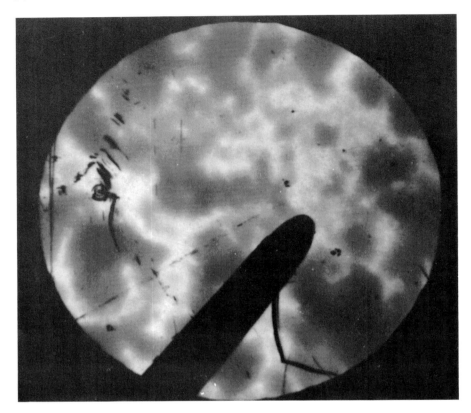

Figure 2.30 EBIC image of GaAs crystal. The dark regions show local accumulations of oxygen and silicon at the 1 part-per-million level. Field of view is 1 mm; beam energy 30 kV.

ductor contains only a few free electrons that can transport charge across it, extra free electrons are produced when the material is irradiated by the electron beam of the SEM. If an electric field is applied to the semiconductor, then these excess charge carriers can be collected, and the resultant current used to form an image (see, for example, Chapter 3 in Newbury et al., 1986). The magnitude of this signal is a sensitive function of the electronic state of the material, which, in turn, depends on the local chemistry. Figure 2.30 shows an EBIC image from a crystal of gallium arsenide (GaAs). The feature crossing the lower half of the micrograph area is the contact used to collect the current. The prominent bright and dark areas in the field of view arise from local variations in the concentration of oxygen and carbon in the crystal. It is noteworthy that the average concentration of either of these impurities, which enter the molten phase while the crystal is being grown, is only about 1 part per million by weight. This particular technique therefore provides an extremely sensitive indicator of the purity and cleanliness of the semiconductor crystal and is an invaluable indicator as to the likely success of devices subsequently fabricated on the material.

Figure 2.31 Tensile stage designed for use inside the specimen chamber of an SEM. A maximum load of 300 g, and a maximum extension of 100% are possible. Unit designed by D. J. Dingley, University of Bristol.

IN SITU EXPERIMENTS

Because the specimen chamber of an SEM is often quite large, typically having the ability to accommodate specimens up to 10 or 15 cm in diameter, it is readily possible to build into it stages that can heat, cool, bend, or strain a sample as it is being observed. Such *in situ* experiments are of great value in the study of the behavior of materials because it is no longer necessary to try to deduce how the sample reached its final condition as a result of the conditions to which it was subjected. Instead the process of change can be studied and recorded as it happens. Figure 2.31 shows a tensile stage designed for use in an SEM chamber. The stage takes a sample about 1 cm long and a few mm thick and can be driven from outside the vacuum to apply a maximum load of 300 g and give an extension of up to 100%. A solid-state strain gauge provides a continuous readout of the applied load. While this is in progress, any of the imaging and analysis modes of the SEM described above can be applied to the specimen, so that its response to the applied stress can be fully characterized. Figure 2.32 shows three images from a sample of a Pb–1.5 wt % Sn superplastic alloy at increasing levels of strain, illustrating the formation of the internal hole in the sample that ultimately causes the sample to fracture (Newbury, 1971). Since most modern SEM's can form images at TV rates, the entire experiment can be recorded on video tape as it happens. Events of interest can then

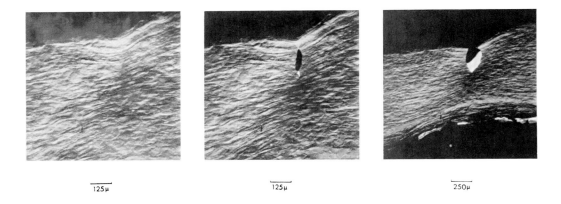

125μ   125μ   250μ

Figure 2.32 SE images of Pb–1.5 wt % Sn superplastic alloy deformed *in situ* at increasing levels of strain. Courtesy D. E. Newbury, NIST.

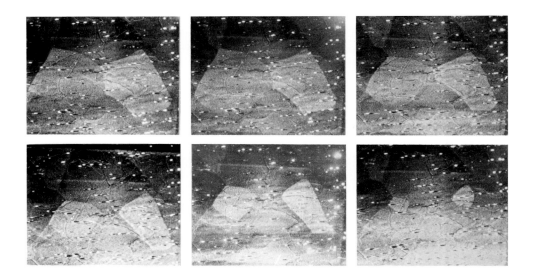

Figure 2.33 An *in situ* study of the annealing of an Sn–0.3 wt % Pb sample held at 125°C in the SEM. The images were taken at time intervals of 1 min. The process of grain growth and recrystallization is evident. Field of view 250 μm. Courtesy D. E. Newbury, NIST.

86

be studied both forwards and, with the benefit of hindsight, backwards to elucidate the details of the failure. A single *in situ* experiment thus provides both more, and more meaningful, data than a comparable set of static experiments.

In many instances the temperature of the specimen is an important variable, and so heating or cooling stages are of value. Figure 2.33 shows details from an *in situ* study of annealing in a Sn–0.3 wt % Pb sample held at 125°C in the microscope. Successive frames were recorded at 1 min intervals. Initially the specimen shows small, equiaxed grains of random orientation. But within 5 min at 125°C some recrystallization has already occurred replacing the fine grains with several large grains. Within another 5 minutes these grains too are seen to have recrystallized. Similarly liquid-nitrogen-cooled stages are now playing an important part in the study of high-T_c superconductors, permitting these materials to be observed as they are cycled through their transition temperature in the microscope. In either case a wide range of temperatures can be accessed, the ultimate limits being set by excessive thermal emission of electrons at the high-temperature end, and condensation of gases on the surface at the low-temperature extreme.

CRYSTALLOGRAPHIC STUDIES

The transmission electron microscope played a major role in advancing our knowledge of materials science, not simply because it could show the finest details of the microstructure, but because the techniques of electron diffraction made it possible to study the crystallography of the specimen and the crystallographic defects, such as dislocations and stacking faults, which determine its basic properties. Examples of electron diffraction patterns are illustrated in Chapter 6 by Williams and Vecchio. The SEM can also provide this type of information through the technique of electron channeling patterns (ECP). As discussed earlier in this chapter, some fraction of the electrons emitted by the specimen under the impact of the beam are high-energy or backscattered electrons. For an amorphous material the yield of backscattered electrons depends only on the atomic number of the specimen, but for a crystalline material the yield is also found to depend on the angle made by the incident beam with the lattice. This is because, as shown in Figure 2.34, if the electron enters the crystal at an arbitrary angle, there is a high probability that it will soon come close to a nucleus and be scattered back out of the lattice, but, if the beam is traveling along one of the symmetry directions of the lattice, then the electron "channels" deep into the crystal before coming close to a nucleus and hence has a lower backscattering probability (Joy et al., 1982). The angles θ at which this occurs are those that satisfy Bragg's law, i.e., $\sin \theta = \lambda/2d_{hkl}$, where λ is the wavelength of the electron and d_{hkl} is the spacing of the lattice planes of Miller indices (*hkl*). For 20 kV electrons λ is of the order of 0.1 Å (0.01 nm); so for a typical lattice spacing of 3 Å (0.3 nm), the Bragg angle is 1–2°. If the angle of incidence of the beam to the surface is changed by this degree of angle, then the backscattering yield will vary.

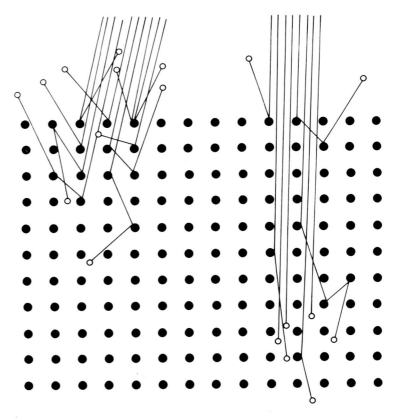

Figure 2.34 Origin of the electron channeling contrast effect in the backscattered (B) signal from a crystal.

Moreover, if the angle of incidence is changed in two dimensions rather than in just one, then an image formed from the B electrons will show a map of these variations, which constitutes the ECP. These variations will hence display the crystallographic symmetry of the lattice about the beam direction. The necessary variation in beam direction is achieved by observing the specimen at low magnification, since, as the beam scans from one side of the sample to the other, its angle of incidence changes by several degrees.

Figure 2.35 shows an ECP recorded from a single crystal of indium phosphide InP. The pattern is composed of *bands,* the angular width of which is equal to twice the appropriate Bragg angle, which intersect at *poles.* In this example the pole is seen to have a fourfold axis of symmetry with mirror planes at 45° to each other. The beam is therefore traveling along the [011] axis of the crystal in this example. The indices of the bands are of the type {220} and {400}, and in each case the bands are running parallel to the trace of the respective index lattice planes on the surface of the crystal. If the sample is moved laterally, then the pattern will not change since

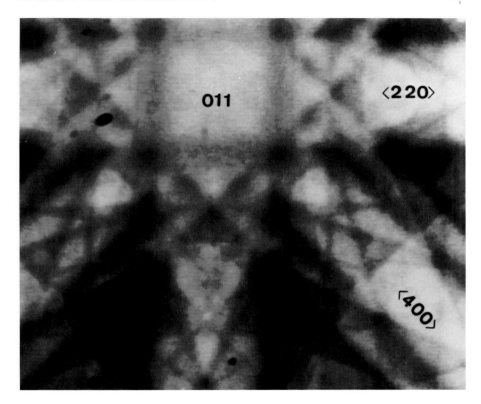

Figure 2.35 Channeling pattern from a single crystal of InP showing the symmetry of the {011} index crystallographic pole. Recorded at 20 kV.

the symmetry of a crystal does not alter under translation. However, if the crystal is rotated or tilted, then the pattern will move as if rigidly fixed to the lattice. By recording a sequence of patterns as the crystal is tilted, a montage channeling map can be generated that shows all possible unique symmetry conditions for the crystal. It has been shown (Joy et al., 1982) that the poles in the ECP correspond to zone axes in the stereographic projection of the crystal. Thus, the orientation can be determined by comparing the ECP from a crystal of unknown orientation with the known montage.

While this technique is valuable, it is applicable only to large single crystals, since the whole area scanned by the beam must be of the same orientation. To overcome this limitation a selected area channeling pattern (SACP) technique has been developed (van Essen et al., 1970). This allows the ECP to be obtained from micrometer-sized regions of the sample. The beam no longer scans over the surface of the specimen but rocks about a fixed point on the surface. Since the ECP comes only from the change in the angle of incidence, this procedure still produces a

Figure 2.36 (a) SE image of artificial diamond crystal with well-developed facets about 300 μm in size. (b) Selected area channeling pattern from the crystal in (a) showing the orientation of the facet to be {111}. Courtesy GATAN, Inc.

90

 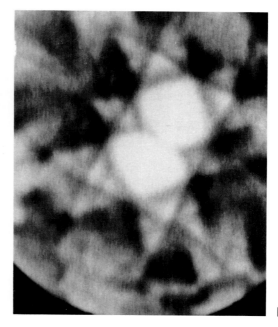

a   b

Figure 2.37 (a) Backscattered image of amorphous silicon film 0.5 μm thick thermally regrown on an insulator (SiO₂) layer showing the grain structure revealed by channeling contrast. (b) Selected area channeling pattern from region shown in (a) indicating the regrown film has an average {110} orientation but with some random misalignment between successive grains.

pattern but now from an area defined by the accuracy with which the beam can be held stationary on the surface. Figure 2.36 shows a simple application of this technique. The sample consists of small crystals of artificially grown diamonds. Working in the standard SE imaging mode described above, one of the crystals with a well-developed facet structure [Figure 2.36(a)] is selected. The SEM is then switched to SACP mode to give the pattern shown in Figure 2.36(b), which in this case can be seen by its clear threefold symmetry to be a {111} pole. To demonstrate this result by a conventional TEM or x-ray diffraction technique would require considerable sample preparation and processing. This technique is of special value in studies of fracture, as it permits a direct determination of the orientations associated with each mode of fracture (Newbury et al., 1974; George et al., 1989).

Figure 2.37(a) shows the backscattered electron image of a 0.5-μm-thick layer of amorphous silicon thermally regrown on a 2-μm-thick layer of oxide that is, in turn, on a substrate of silicon. The image shows the "finger-like" grain structure produced as a consequence of the regrowth procedure in which a hot wire is moved across the surface. As the molten zone follows the heater, the solid–liquid interface is pulled out parallel to the movement of the wire. The changes in contrast between the different grains arise because of the ECP effect, since depending on its exact orientation each grain will backscatter slightly more or slightly less signal. The image shows that the material is not a single crystal but gives no quantitative evidence as to how large the angular misorientations between grains might be. However, an SACP examination [Figure 2.37(b)] of this area produces a pattern that can be identified as a {110} pole. On closer examination it can be seen that the pole is actually the superimposition of two separate {110} poles with an angular offset of about 1° between them. This occurs because the selected area is straddling the boundary between two grains, each nominally of {110} orientation but with a random misalignment between them. Studies of this kind demonstrate that this type of rapid regrowth produces a strongly textured crystal with a well-defined surface normal orientation but with a random misorientation in the plane of the film. To obtain this same type of information using a TEM would require a lengthy preparation procedure to backthin to the regrown film. Furthermore, the result would be less statistically valid because the available field of view in a TEM micrograph is much smaller than that attainable in the SEM; so far fewer grains can be sampled. X-ray techniques would also be less satisfactory because reflections would be obtained from both the film and the substrate, whereas the ECP contrast comes only from the top 1000 Å (100 nm) of the surface and thus contains no complicating information from the substrate.

SUMMARY

The scanning electron microscope combines much of the resolution and analytical power of the transmission electron microscope with much of the ease of operation and interpretation of the optical microscope. Images can be formed from a very wide range of materials, from metals to ceramics and from semiconductors to polymers. These materials can be examined with low-energy secondary electrons, which display the surface topography of the sample, with high-energy backscattered electrons, which produce images of chemical inhomogeneities within the sample, or with other emissions such as light, heat, and sound. The high depth of field of the SEM image makes it especially suitable for the study of fracture surfaces and complex microstructures such as those found in composite materials. Since all of these images are formed in real time, and because there is generally good access to the specimen area in the SEM, in situ experiments such as straining, heating, or cooling are readily possible.

Through the use of the electron channeling pattern mode of operation, the SEM can also determine whether a sample is crystalline and, if it is, its crystalline type, its orientation, and the degree of perfection of its crystal lattice. This information, which can be obtained from areas of micrometer size, is comparable with that provided by the selected-area-diffraction mode of the TEM. Finally, chemical microanalysis, again on a micrometer scale, is possible through the use of energy- or wavelength-dispersive x-ray systems (see Chapter 11 by Newbury, et al.) The SEM therefore provides a comprehensive set of techniques with which to characterize materials of all types.

REFERENCES

George, E. O., Porter, W., and Joy, D. C. (1989). Mater. Res. Soc., **133,** 311.

Goldstein, J. I., Axon, H. J., and Agrell, S. O. (1975). *Earth and Planetary Science Letters* **28,** 217.

———, Newbury, D. E., Echlin, P., Joy, D. C., Fiori, C. E., and Lifshin, E. (1981). *Scanning Electron Microscopy and X-Ray Microanalysis.* Plenum, N.Y.

Hertzberg, R. W. (1987). *Fractography of Modern Engineering Materials: Composites and Metals,* ASTM STP 948, J. E. Masters and J. J. Au (eds.), American Society for Testing and Materials, Philadelphia, p. 5.

Joy, D. C. Newbury, D. E., and Davidson, D. L. (1982). *J. Appl. Phys.* **53,** R81.

Krause, S. J., Maracas, G. N., Varhue, W. J. and Joy, D. C. (1989). *Proc. 47th Ann. Meeting Electron Mic. Soc. Amer.,* G. W. Bailey (ed.), San Francisco Press p. 82.

Newbury. D. E. (1971). D.Phil. Thesis. University of Oxford.

———, Christ, B. W., and Joy, D. C. (1974). *Metall. Trans.* **5,** 1508.

———, Joy, D. C., Echlin, P., Fiori, C. E., and Goldstein, J. I. (1986). *Advanced Scanning Electron Microscopy and X-Ray Microanalysis.* Plenum, N.Y.

Postek, M. T., and Joy, D. C. (1987). *J. Res. Nat. Bur. Stand.* **92,** 205.

Reuter, K. B., Williams, D. B., and Goldstein, J. I. (1988). *Geochimica et Cosmochimica Acta* **52,** 617.

Roeder J. (1988). Ph.D. Thesis, Lehigh University.

van Essen, C. G., Schulson, E. M., and Donaghay, R. H. (1970). *Nature* **225,** 847.

Wang, Z. Y., Harmer, M. P., and Chou, Y. T. (1986). *J. Amer. Ceramic Soc.* **69,** 735.

———, Harmer, M. P., and Chou, Y. T. (1988). *Materials Letters* **7,** 224.

3

Analytical Imaging with a Scanning Ion Microprobe

R. LEVI-SETTI, J. M. CHABALA, Y. WANG, P. HALLÉGOT, AND C. GIROD-HALLÉGOT

A number of modern microanalytical techniques, many of which are illustrated in this book, describe the chemical composition of materials by means of images. One technique, ion microscopy, has been advanced in recent years as a tool for the sensitive determination of the isotopic and molecular content of any sample surface. Micrographs depicting the two-dimensional distribution (also called *maps*) of selected constituents over small areas of a sample are one of the essential means by which the retrieved information is recorded. Such images embody a wealth of interwoven information in a most compact form and remain one of the most efficient vehicles for the intelligent interpretation of material microstructure and composition. Compositional maps can be thought of as two-dimensional views of an abstract multidimensional space encompassing the spatial dimensions, the mass or atomic number dimension, and, as a further variable, the concentration for each constituent. Often, an additional unquantifiable attribute enriches these already multifaceted images: their aesthetic appeal.

We hope to convey these notions through the ion microscopy images presented in this chapter.

FUNDAMENTALS OF ION MICROSCOPY

Ion microscopy exploits the interaction of fast ions with matter to gather information about the structure and chemical composition of the bombarded material. The ion bombardment can be performed either with an ion gun that illuminates a broad surface area or with focused ion beams that can be steered to selected locations or rastered over a microscopic field of view. Secondary electronic and atomic processes, initiated by these impinging primary probe ions, provide the signals that carry the structural and chemical information. In fact, secondary electrons, neutral atoms, and ions of either charge are copiously emitted from the topmost atomic layers of the sample material as a result of the bombardment. The secondary electrons and ions can then be detected in order to obtain images of the target's surface topography or to yield material contrast (Levi-Setti, 1983); these images are very similar to those produced by a scanning electron microscope (SEM), as discussed in Chapter 2 by Joy and Goldstein. However, in the ion microscope additional images can be constructed that visualize the chemical makeup of a sample. This feat is accomplished by recourse to an analysis technique called secondary ion mass spectrometry (SIMS), in which the emitted secondary ions are filtered by a mass spectrometer before detection (for a review, see Benninghoven, 1975).

During the observation of a solid with the ion microscope, the surface of the material is continually eroded, at a controllable rate, and new layers of the sample are sequentially exposed. This process, which is effectively tomography on a microscale, permits the three-dimensional reconstruction of the chemical and structural constitution of a volume of the target object.

Two types of instrument are available, depending upon whether broad or focused ion beams are used to illuminate the sample. The stigmatic, analytical emission ion microscope (EIM) (Castaing and Slodzian, 1962), in analogy to the electron emission microscope as well as to the optical reflection microscope, is based on the broad beam method. Here, the mass-resolved image is obtained by simultaneously collecting with an objective lens the secondary ions emitted by each point on the sample surface, filtering these ions through a mass spectrometer system, and focusing them, with a projector lens, onto an image plane while retaining the original spatial relationships. The image resolution of the EIM is, at best, comparable to that of the optical microscope (1–0.5 μm).

The scanning ion microprobe (SIM) (Liebl, 1967), also called the scanning ion microscope or microanalyzer, parallels in design the SEM. In this type of instrument, a sharply focused probe is used to scan a region of the sample surface, usually in a raster pattern. The resulting secondary emission signal originating from each point of the scanned surface (analyzed or not by the mass spectrometry system) is separately recorded, either as an intensity-graded spot appropriately positioned on photographic film or as an element in a two-dimensional array of computer memory. The data stored in the computer can subsequently be processed and reconstructed as an image. Using liquid metal ion sources, a recent innovation implemented in scanning ion microprobes, superior image resolution is obtained that approaches the size of a minimum probe diameter, approximately 20 nm (Levi-Setti et al., 1985).

Secondary ion mass spectrometry is a very sensitive microanalytical technique. In principle, all chemical elements and many compounds can be separately detected (McHugh, 1975). In contrast to x-ray and electron-beam-based methods of surface analysis, which are sensitive to atomic charge, SIMS gives mass and isotopic information. In practice, not all elements can be detected by SIMS with equal efficiency. Those elements that in the emission process (sputtering) are inherently ejected in ionized states (as opposed to uncharged states) are more sensitively detected. Among these favorable elements are the alkali metals (lithium, sodium, potassium, etc.) and the halogens (fluorine, chlorine, etc.). Assorted other elements and compounds, including calcium, aluminum, magnesium, and oxygen, and a host of organic molecules, can also be easily detected. Not coincidentally, these elements and compounds are chemically active and therefore are commonly present in a large number of materials. Special methods have been developed to ionize, and hence to detect, sputtered neutral atoms, thus extending the applicability of the ion microprobe mass spectrometry method to the analysis of all chemical species (Reuter, 1985). In this article we limit ourselves to a discussion of the simpler, unmodified SIMS technique.

A HIGH-RESOLUTION SCANNING ION MICROPROBE

The images presented here were obtained using our high-resolution scanning ion microprobe (Levi-Setti et al., 1984), conceived and developed at the University of Chicago in collaboration with Hughes Research Laboratories. A simplified exploded view of this instrument is shown in Figure 3.1. Conceptually, this instrument consists of two integrated subsystems. The first of these, the primary ion optical column, is the vertical assembly shown on the right-hand side of the figure. A strong electric field (~ 1 V/Å) extracts, by a field evaporation process, heavy ions from a liquid metal ion source. This source is composed of a tungsten needle coated with a layer of liquid gallium (Clampitt et al., 1975), which protrudes through a tungsten ribbon that acts as a heater and reservoir for the gallium. The primary ions emerging from this high-brightness source (about 10^6 A cm^{-2} sr^{-1}) are drawn through a beam-defining aperture that serves to collimate the ions, reducing the optical aberrations of the system. The ion current extracted from the source is in the microampere range; only a small fraction, in the picoampere (10^{-12} A) range, ultimately passes through this aperture to bombard the sample.

The ion beam is accelerated to 20–60 keV energy and is focused by two electrostatic lenses. The probe is maneuvered in a controlled manner by two sets of octupole deflectors. The length of the primary optical column is 25 cm. Two channel electron multiplier detectors (CEMs), positioned near the sample, collect secondary electrons or non-mass-resolved secondary ions. These signals yield topographic and material contrast information, as will be described in the following. The diameter of the probe in this system can be reduced to 20 nm by controlling the width of the beam-defining aperture. For a number of practical and theoretical reasons, it is not worthwhile to reduce the probe diameter below this value (Chabala et al., 1988b).

The secondary ion optical column lies horizontally in Figure 3.1. Secondary ions emerging from the sample are first accelerated by a small (about 100 V) potential,

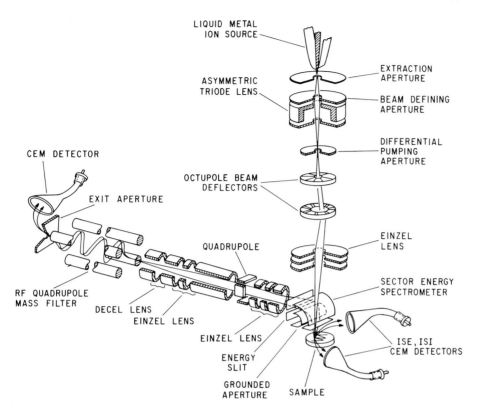

LIQUID METAL
ION SOURCE

EXTRACTION
APERTURE

ASYMMETRIC
TRIODE LENS

BEAM DEFINING
APERTURE

CEM DETECTOR

DIFFERENTIAL
PUMPING
APERTURE

EXIT APERTURE

OCTUPOLE BEAM
DEFLECTORS

EINZEL
LENS

QUADRUPOLE

SECTOR ENERGY
SPECTROMETER

RF QUADRUPOLE
MASS FILTER

DECEL LENS
EINZEL LENS

EINZEL LENS

ENERGY
SLIT

ISE, ISI
CEM DETECTORS

GROUNDED
APERTURE   SAMPLE

Figure 3.1 Exploded view schematics of the University of Chicago–Hughes Research Laboratories scanning ion microprobe (Levi-Setti et al., 1988b, courtesy of Elsevier Science Publishers).

then analyzed by a compact 90° sector energy spectrometer. The energy-selected ions subsequently are transported through a series of lenses and deflectors to enter an RF quadrupole mass filter. This filter allows the transmission of ions with only one mass-to-charge ratio. Finally, the remaining ions are again accelerated and detected by a third CEM operated in a pulse counting mode (one detected ion yields one count). The resulting digital signal is routed to photo-oscilloscopes or to arrays of digital image memory in an image processor. Positively or negatively charged secondary ions can each be detected by reversing the polarity of the optical elements in the secondary ion system. Once several mass-resolved images (each composed of 512 × 512 or 1024 × 1024 picture elements) have been sequentially accumulated, they can be color coded and superimposed using the image processor; this procedure is valuable for the correlation of compositional features.

IMAGE FORMATION MECHANISMS OF ION MICROSCOPY

The formation of images in all microscopies is based on the existence of basic processes that create image contrast. In emission microscopy with any probe, contrast arises because of the different secondary emission properties of discrete object areas, the result of their nonuniform physical or chemical structure. The most frequently detected contrast-carrying signals of heavy ion microscopy are the emitted ion-induced secondary electrons (ISE) and secondary ions (ISI). Many physical contrast mechanisms contribute to the formation of SIM images. They can be grouped into two broad categories: emission contrast and analytical contrast (Levi-Setti, 1983, 1988).

Emission contrast contributes to both secondary electron and non-mass-resolved secondary ion images. In turn, various aspects of emission contrast are at the origin of (1) topographic contrast, in which the surface relief of an object modulates the collected signal intensity by either geometric effects (such as shadowing), or due to the dependence of the secondary emission intensity on the orientation of the surface relative to that of the incident probe; (2) crystallographic or material emission contrast, in which the emission intensities are modulated by the phenomenon of primary ion channeling, according to the orientation of the material's crystal planes relative to the direction of the incident beam; and (3) chemical emission contrast, related to the dependence of the secondary emission yields on the atomic species composition and chemical state of a surface.

Analytical contrast is encountered in images obtained with mass-resolved secondary ion signals, such as those emerging from a secondary ion mass spectrometer, and is due to the presence of compositional variations on the surface of an object. One can distinguish two subsets: (1) elemental or isotopic contrast, in which the contrast signal is provided by monatomic species, and (2) chemical analytical contrast, in which emitted molecules or clusters of atoms determine the contrast.

In addition to the existence of contrast, signal intensity is obviously an important factor that governs image formation. When the signal recorded for each picture element is small, as may occur in SIMS imaging, it becomes the overriding determinant of image resolution. In scanning microscopy, the image resolution can be made to approach a limiting value equal to the probe size only when sufficient signal statistics is available (Chabala et al., 1988b; Levi-Setti et al., 1988b).

SCANNING ION MICROSCOPY IMAGES OF SURFACE TOPOGRAPHY AND MATERIAL EMISSION CONTRAST

The performance of a scanning ion microscope is best evaluated, and optimized, by imaging finely structured objects using the topographic contrast mode. The abundant secondary electron and non-mass-resolved secondary ion signals employed for this purpose permit real-time primary ion optical column alignment, probe focusing, and correction of astigmatism. Furthermore, such images provide immediate evidence of the instrumental resolution that can be attained.

Objects with sufficient detail to challenge the resolution limits of our SIM can often be found in the realm of biology, as illustrated in the micrographs collected in Figure 3.2. Figure 3.2(a) is a closeup (20×20 μm^2) ISE image of an eye of the fruit fly (*Drosophila melanogaster*). The grainy appearance of the spherical lens surfaces is due to the corneal nipple array, which serves as an antireflection coating and is found on most insect eye lenses. The protruding spikes (*setae*) seem to cast shadows across the lenses, as if illuminated by a spotlight. In reality, these shadows occur when secondary electrons emerging from the eye surface are intercepted by the *setae* and are thus prevented from reaching the detector, which is positioned at a shallow angle with respect to the surface.

The lattice-shaped object in Figure 3.2(b) is an ISI image (10×10 μm^2) of a butterfly wing scale. Extremely fine details can be seen in this micrograph, which has an estimated image resolution of 25 nm.

Remarkable structures and tessellated patterns are displayed by the pair of images of fossil diatom skeletons, composed of SiO_2, shown in Figure 3.2(c),(d). Aside from the technical function served by acquiring these pictures, one cannot help but be captivated by the intrinsic beauty of these structures of nature.

In the metallurgical domain, one can also find exciting visual patterns, as depicted in the images of Figure 3.3. The two secondary ion topographic micrographs in Figures 3.3(a),(b) are views of the nitric acid–etched surface of recrystallized pure copper. The striated texture provides an insight into the lattice orientation of twin structures within the metal.

The same polycrystalline material, when etched more lightly, lends itself to the study of crystallographic contrast, another form of emission contrast. Figures 3.3(c),(d) are micrographs of the same area of the sample, produced with the sample surface normal tilted relative to the probe at slightly different angles [11° in Figure 3.3(c); 22° in Figure 3.3(d)]. When the incident ions enter a crystal along particular axes, they may penetrate more deeply and with less interaction than if they had entered from a random direction. This channeling phenomenon occurs only for angles departing by at most 5°–10° from the preferred crystal axes, under the conditions of this experiment. When channeling occurs, secondary emissions are dramatically reduced; consequently, regions in both secondary electron and ion images will appear dark. If, however, the crystal is rotated by an angle larger than the critical range, the emission intensities will rapidly increase. The dark–light inversions seen for individual, randomly oriented, crystallites in the copper sample, and the contrast reversals at grain boundaries observed when comparing Figure 3.3(c) and (d), can be readily interpreted in terms of this mechanism. It is remarkable that, in ion microscopy, crystallographic contrast is so pronounced. In fact, the ratio of the emission intensities between the nonchanneling and channeling conditions may be as large as four in secondary electron, and up to about ten in secondary ion, images (Levi-Setti, 1983).

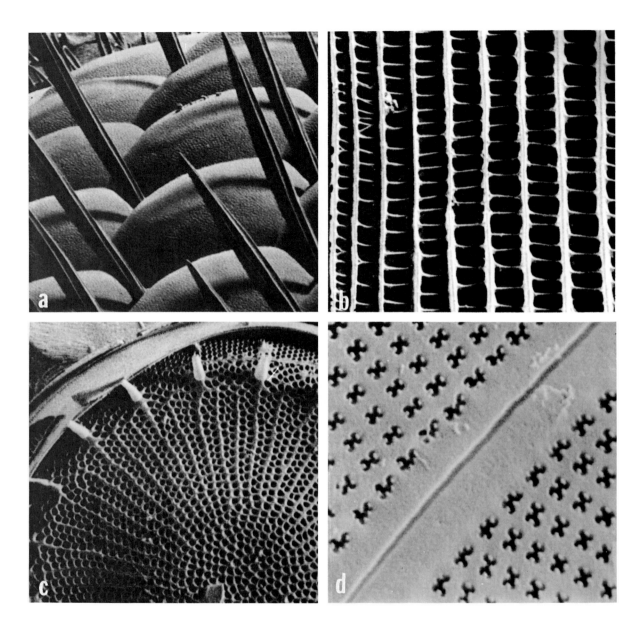

Figure 3.2 Ion microscope images of surface topography (Levi-Setti et al., 1988b, courtesy of Elsevier Science Publishers). (a) Fruit fly eye (*Drosophila melanogaster*); 20 μm full scale. (b) Butterfly wing scale; 10 μm full scale. (c) Fossil diatom skeleton; 20 μm full scale. (d) Fossil diatom (*Pleurosigma angulatum*) skeleton; 10 μm full scale.

Figure 3.3 Ion-induced secondary ion images of polished copper, etched by nitric acid, revealing crystallographic structure (Levi-Setti et al., 1985), 40 keV indium probe. (a) and (b) Surface topography of crystal twin structures; (a) 40 μm full scale, (b) 20 μm full scale. (c) and (d) Crystallographic contrast. The contrast reversals are due to ion channeling effects. (c) Sample normal tilted 11° with respect to the probe; (d) 22°; 40 μm full scale.

ION MICROSCOPY IMAGES: TOPICS
IN THE BIOLOGICAL SCIENCES

We begin the survey of analytical images obtained by high-spatial-resolution secondary ion mass spectrometry with a selection of micrographs obtained during current studies of biological materials and physiological processes (Chabala and Levi-Setti, 1987a; Levi-Setti, 1988). Most elements of biological interest are emitted, under bombardment by the gallium probe, with a sufficiently large fraction of atoms or molecular fragments in an ionized state so as to be easily detected by SIMS. Of the positive secondary ions, Na, K, Ca, and Mg are among the most frequently observed, and of the negative ones, C, O, F, Cl, C_2, CN, PO, PO_2, PO_3. Of the vast array of biological materials, a broad classification can be made in terms of hard (mineralized) and soft tissues.

Hard Tissue

Several systematic studies of bone physiology have been undertaken with our high-resolution ion microprobe. In collaboration with D. A. Bushinsky at the University of Chicago, we investigated the metabolism of Na, K, and Ca in neonatal (newborn) mouse skull bone (*calvariae*), which is a convenient and representative model of mammalian bone and its function (Bushinsky et al., 1986). This incompletely mineralized bone is interwoven with soft tissue; therefore, during preparation for analysis, it must be freeze-dried to preserve the distributions of diffusible elements in their natural state. The application of a thin coating of gold is sufficient to prevent electrical charging of the sample, which might otherwise adversely affect SIMS imaging.

During the course of our investigations, we have accumulated a large number of intriguing maps that reveal the unsuspectedly complex structure of this growing bone. We compare in Figure 3.4 the appearance and differing composition of live and dead bone. The elemental images in Figures 3.4(*a*),(*b*) show the K and Ca content, respectively, of one area of the live bone. Surprisingly, the surface appears much richer in K than in Ca. By contrast, the two images of K and Ca obtained from one area of dead bone [Figures 3.4(*c*),(*d*)] exhibit comparable mineral content. Apparently, an active control mechanism maintains an ionic disequilibrium in live bone that ceases to function upon death (Bushinsky et al., 1989).

In additional to their analytical significance, these images reveal in extreme detail the morphology of this hard tissue. The ridges and valleys in the bone surfaces alter the secondary ion emission intensities in such a way as to impart to the images a subtle three-dimensional appearance.

An element of discovery always accompanies the observation of new samples with our SIM, because until now no other instrument has combined simultaneously, at a comparable level, its elemental sensitivity with spatial resolution. For example, much new information has been extracted from studies of dental tissue. We are analyzing the mineralization of healthy and pathological teeth in collaboration with

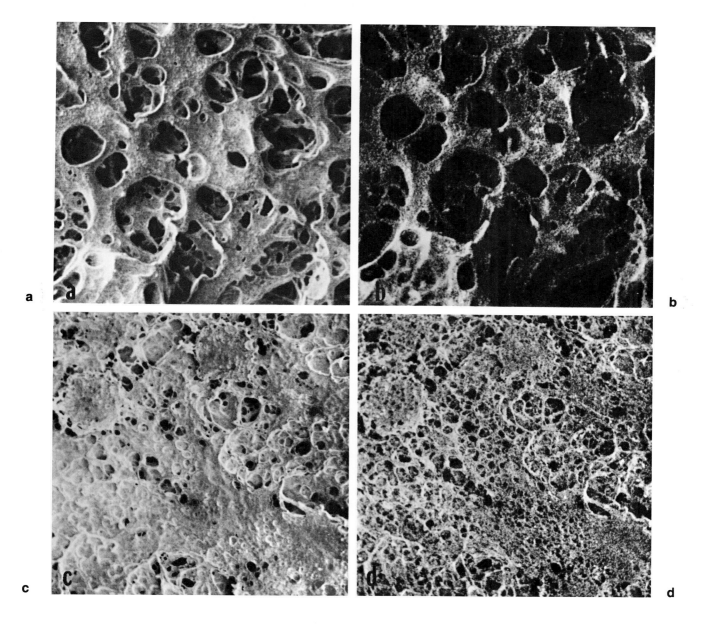

Figure 3.4 Ion microscope mass-resolved images of neonatal mouse calvariae. (a) and (b) K and Ca distributions, respectively, of one area of live bone (Bushinsky et al., 1989, reprinted by permission of Am. Physiol. Soc.); 20 μm full scale. (c) and (d) K and Ca distributions, respectively, of one area of a dead bone. Notice the larger concentrations of Ca relative to that in the live bone; 40 μm full scale.

A. Lodding of Chalmers University of Technology and T. Lundgren and J. G. Norèn of the University of Göteborg, Sweden (Chabala et al., 1988a). Diverse mineral and organic components (e.g., K, Na, Ca, O, F, CN, PO_2) are mapped throughout polished sections of various regions of these teeth. For reference during the following discussion, enamel is the exposed tooth component, familiar to all brushers, that envelops the interior dentine.

Figures 3.5(a),(b) show the distribution of CN (carbon–nitrogen molecule) and PO_2 (phosphate) in a cross section through the dentine of a healthy human tooth. Prominent features in these images are the oblique teardrop-shaped sections of dentinal tubules. The CN signal originates from organic material in the tooth. Clearly, there are at least three levels of organic content in this area. The core of each tubule is the richest, and also contains a large concentration of PO_2 (b). This core is surrounded by a heavily calcified sheath that appears dark (a), being devoid of CN. The dentinal matrix, principally composed of the mineral hydroxyapatite, contains relatively large amounts of CN and PO_2.

Figures 3.5(c),(d) show, at two different magnifications, CN distribution images of the enamel in the same tooth. The bright rings and elongated contours outline stacks of rodlike enamel prisms, which in Figure 3.5(c) are oriented both perpendicular to and in the plane of the sample section. These CN rings are telltale indicators of the presence of an organic layer surrounding the prisms, information that was, to our knowledge, lacking before this investigation (Lodding et al., 1988).

Soft Tissue

There exists an extensive body of work covering analysis of soft biological tissues with heavy ion microscopy (for a review, see Levi-Setti, 1988). In an effort to convey the flavor of this research, we present a sampling of images gathered from a number of disparate specialties.

We start with an examination of a botanical system, onion root cells. The quartet of images in Figure 3.6 are topographic and elemental micrographs of root tip cells from *Allium cepa,* a variety of onion. For microanalysis, the cells were embedded in plastic resin, then longitudinally sliced with a microtome. The pictures are configured so that down corresponds to the downward direction the root encountered during growth.

The first micrograph [Figure 3.6(a)], a secondary electron topographic image, is similar to classic optical microscope photographs of these cells. Cell boundaries are visible as slightly raised, angular walls separating large areas of featureless cytoplasm. Ovoid, somewhat corrugated nuclei are within three cells. The prickly looking structures visible in both the leftmost and uppermost cells, below their nuclei, are amyloplasts. These amyloplasts are starchy geosensing organelles that guide the root's downward growth; notice that the amyloplasts have "fallen" to the bottom of each cell.

By comparing the topography micrograph with the sodium (Na) elemental distribution image [Figure 3.6(b)], one notices that the amyloplasts contain signifi-

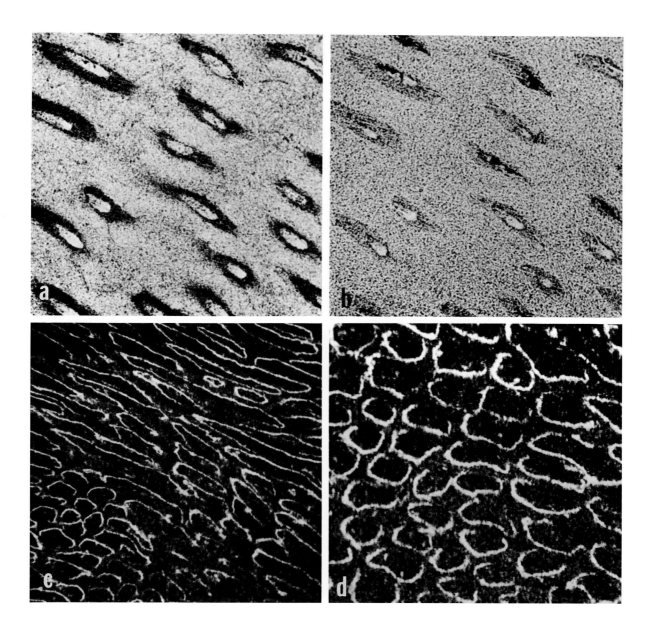

Figure 3.5 Mass-resolved images of polished sections through a healthy adult human tooth. (a) and (b) CN and PO_2 molecular distributions, respectively, across one area of dentine. The teardrop-shaped objects are transverse cross sections through dentinal tubules; 40 μm full scale (Chabala et al., 1988b). (c) and (d) CN map of the enamel. The enamel prisms are outlined by CN-emissive material. (c) 80 μm full scale; (d) 40 μm full scale (Lodding et al., 1988, reprinted by permission of John Wiley & Sons, Ltd.).

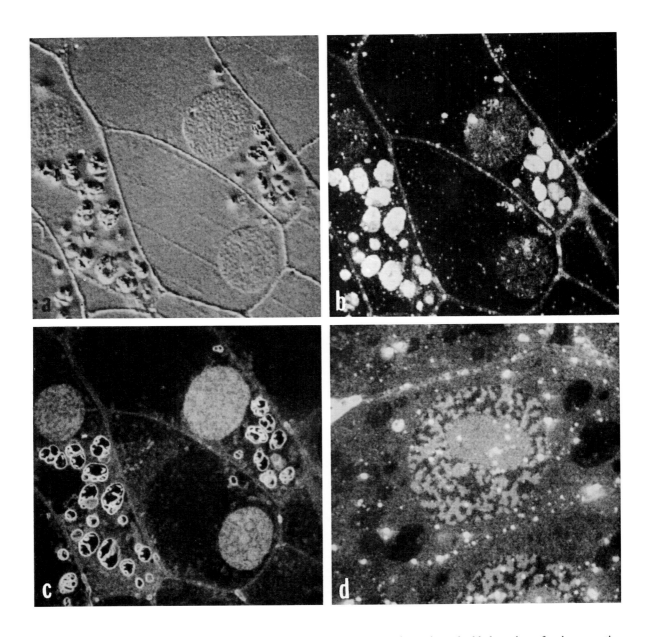

Figure 3.6 Topographic and SIMS images of a resin-embedded section of onion root tips (*Allium cepa*) (Levi-Setti et al., 1988b, courtesy of Elsevier Science Publishers). (a) Secondary electron image of surface topography; 40 μm full scale. (b) Na distribution, showing bound concentrations in cell membranes, nuclei, and amyloplasts. (c) CN map of same area, showing distribution of proteins. (d) CN map of a cell in telophase, a final stage of mitosis; 20 μm full scale.

106

cantly more Na than the other components of the cells. The nuclei and cell walls contain lesser amounts of Na, the cytoplasm still less. The embedding process used to prepare this sample has rinsed the cells of all diffusible elements, preserving the distribution only of macromolecules and elements tightly bound to the rigid cellular framework. The corresponding CN image [Figure 3.6(c)] gives a fuller representation of the organelles. The CN mass-resolved signal is a marker for protein-rich material. The internal, curdled morphology of the nuclei is here revealed. The amyloplasts, which have been opened by the ion bombardment, are seen to have a CN-rich shell and a compartmentalized interior that contains, as seen above, abundant Na (and also K).

In the higher-magnification CN image [Figure 3.6(d)], a cell has been captured during telophase, a final stage of mitosis. Condensed and shortened chromosomes, appearing as a white irregular mesh in the center of the image, have begun to uncoil, forming the uniform egg-shaped body within the mesh (which is possibly a nucleolus). The accompanying daughter cell, of course at the same stage of development, lies at the bottom of the micrograph, its nucleus is bisected by the lower edge of the picture. Nuclear membranes and the cell wall dividing these cells are almost completely formed. In a short time, the cells would appear as the others in this figure, with continuous, mottled nuclei.

In order to understand the spatial intercorrelation of the distributions of different elements and compounds better, we chose to superimpose pairs of SIMS images by image-processing techniques. The reader is referred to Chapters 10–12 for detailed discussions of the methods of image processing. Each separate micrograph was color coded and combined with another image of the same area; then they were photographed together.

In collaboration with the laboratory of P. Galle at the University of Paris (Creteil), we have investigated selective absorption of toxic metals by several organs of the rat. A wide spectrum of pathological reactions is known to occur in humans and animals following accidental or controlled administration of beryllium-bearing compounds. It is important to establish at the subcellular level the localization sites of this metal. In the example shown in Plate 3.1(a), we show a section of a proximal tubule of the kidney of a rat that had been injected with a dilute solution of beryllium sulfate. Green shows the distribution of CN and blue the distribution of beryllium. The latter preferentially concentrates as spheroidal inclusions within the nucleus of several cells (Levi-Setti et al., 1988a).

Small concentrations of Be (blue) are also seen in a section of rat bone marrow [Plate 3.1(b)], where the cytology of the material is strikingly depicted by the CN distribution, here color coded as yellow.

Indium, an element frequently used in nuclear medicine, has also been traced to the kidney of the rat, as shown in Plate 3.1(c). The CN distribution (yellow) clearly outlines a rounded glomerulus, at the center, and surrounding tubules. Indium, labeled in blue, in contrast to the situation for beryllium, is seen to concentrate as widely scattered dots, which, at higher magnification, appear to fall exclusively in the cytoplasm.

The effect of silver poisoning on the rat kidney is convincingly demonstrated in the composite CN (red) and Ag (blue) image in Plate 3.1(e). Silver, which is found in several medications as well as in photographic materials, has been deposited, as a consequence of the filtering action of the glomerulus, onto the basal membrane surrounding the capillaries. This silver plating gives rise to the convoluted blue filaments seen here.

We must emphasize that the scanning ion microprobe, despite the destructive character of its beam, can still image fragile biological samples. Among these, the retina of the eye presents fine and complex structures, such as illustrated in Plate 3.1(d) and (f). The samples shown here are segments of rat retina, cryogenically prepared and resin embedded (using a technique that minimizes the redistribution of diffusible elements) by our collaborator, M. Burns of the University of California, Davis. The distributions of K and CN are compared, giving insight into the ion emission from various tissue areas (Burns et al., 1988).

ION MICROSCOPY IMAGES: TOPICS IN THE MATERIALS SCIENCES

We now turn to subjects in the nonorganic, materials sciences. Again, we have selected images not only because they demonstrate the utility and wide-ranging applicability of microanalysis with the scanning ion microprobe, but also for their visual appeal.

Stoney meteorites (chondrites) are generally believed to have been formed by condensation from primordial material in the interstellar nebula. Because of their resulting diverse microstructure and chemical composition, these objects provide a good test of the analytical capabilities of the ion microprobe. The micrographs in Plates 3.2(a,b) are SIMS composite images of a chondrule, the basic subunit of stoney meteorites. Chondrules consist primarily of crystalline silicates such as olivine and pyroxene, cemented by a glassy matrix rich in Si and Al, among other elements. The similar green and red regions in these images represent the distributions of Si and Al, respectively, and imply that these areas are portions of the glass component. The complementary blue areas in both micrographs represent the Mg distribution, indicative, in conjunction with similar maps for Fe, of olivine, while the lacy blue veins in the glassy matrix are attributed to pyroxene (because of their calcium content, not shown). The dendritic shape of the pyroxene suggests that the chondrule, during formation, was rapidly quenched from molten material.

Secondary ion mass spectrometry imaging is frequently used in metallurgical studies. All alloys are composed of crystallized material; often it is useful to determine what, if anything lies sandwiched as precipitates between these crystals. As an example, the micrographs in Plates 3.2(c) and (d) are elemental distribution images of different phases in an Al-based casting alloy. Copper [green, Plate 3.2(c)], which has been added to the alloy, forms hexagonal crystals and branched precipitates, probably intermetallic $CuAl_2$. Some Cu is also in solution with the bulk Al (Al fills the black areas), visible as a gradient that decreases toward the interior of the Al phases. Silicon (blue) also forms separate branched eutectic precipitates. The silicon distribution is duplicated in Plate 3.2(d), now displayed as yellow. Cesium (blue), which coexists with the Si and $CuAl_2$, is possibly a residue of the aluminum refining process.

The yellow-blue micrograph (shown in Plate 3.2(e) is a composite SIMS image of a section through an Al-Si-Mg casting alloy. This alloy contains large tree-like dendrites of aluminum and an interdendritic eutectic lattice containing silicon precipitates. The large circular blue areas are transverse sections through branches of the Al dendrites; the interposed yellow and blue network is the eutectic. Magnesium and other elements (Cr, O, Fe, Na) that are present in trace amounts form additional intermetallic precipitates not shown here.

Aluminum-lithium alloys are high strength-to-weight alloys that are attractive candidates for the next generation of airframe components. In contrast to the difficulties encountered in the detection of lithium by electron-beam probes, SIMS is extremely sensitive to Li. Different aspects of an ion microprobe study of these alloys are illustrated in Figure 3.7. These results originate from a study undertaken with D. B. Williams of Lehigh University and D. E. Newbury of the National Institute of Standards and Technology (Williams et al., 1987, 1989). The surface of a recrystallized Al-Li alloy (containing 12.7 at. % Li) is described by the secondary electron topography image [Figure 3.7(a)]. The crystal grains are differentiated by strong crystallographic contrast and are delineated by raised ridges (composed of Li oxide) that have, seemingly, been extruded from some grain boundaries. When the sample was analyzed again after 7 months storage, the lithium was seen to have diffused from the grain boundaries to cover adjacent surfaces [Figure 3.7(b)]. Crystallographic contrast, usually present in similar images of freshly cleaned samples, has been suppressed by an oxide layer that has blanketed the sample during the long storage in air. The two micrographs in Figures 3.7(c) and (d), acquired by mass selecting the secondary Al-Li molecular signal, show two precipitate phases that have grown during the aging of the alloy at elevated temperature. The brighter, somewhat diffuse areas correspond to an actual AlLi alloy phase called δ, and the dark spots arranged in rows and asterisk patterns represent the δ' (Al_3Li) phase. Prior to this analysis, precipitates such as these could only be visualized with the transmission electron microscope (see, for example, Chapters 5–7); with the high-resolution SIM, however, chemical information can also be acquired.

Like the three alloys just discussed, silicon nitride ceramic is an advanced material with a number of desirable properties. For example, some types of silicon nitride ceramic can retain mechanical strength up to about 1200°C ($\approx$2190°F), enabling the construction of efficient automobile engines. The white and blue terrazzo Plate 3.2(f) is a composite mass-resolved image of the surface of silicon nitride that was sintered (fused) with Y_2O_3, an oxide of the rare-earth element yttrium. The sample was cracked in half, gold coated, and the fractured surface positioned for SIM observation. The fracture occurs preferentially along grain boundaries, accentuating the crystal structure, seen here as white submicrometer rods and patches. An yttrium compound, indicated by the yttrium secondary ion signal (blue), fills the interstices between the silicon nitride grains, binding them together (Chabala et al., 1987b). As can be appreciated, this "glue" must be strong enough to prevent the fragmentation of the ceramic under the extreme conditions to which it will be subjected.

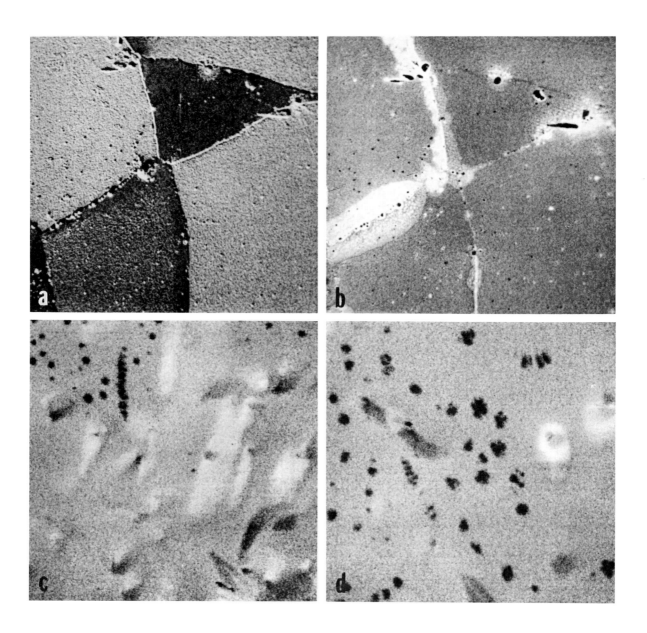

Figure 3.7 Micrographs of polished surfaces of Al-Li alloys. (a) Topographic image of a 12.7 at. % Li alloy, aged 100 hours at 190°C; 80 μm full scale. (b) Li map of the same area, after 7 months dry storage at room temperature and pressure (Williams et al., 1987, by permission of Blackwell Scientific Publications Ltd.). (c) and (d) Al-Li molecular distribution across two areas of 10 at. % Li alloy, aged 4 hours at 290°C. The diffuse white regions are the δ (Al-Li) phase; the black spots δ' (Al$_3$Li); 10 μm full scale (Williams et al., 1989, by permission of Elsevier Science Publishers).

110

The family of micrographs in Plate 3.3 is representative of images obtained while studying integrated circuits, another type of engineered material. Different views of two circuits are presented. The first circuit in this collection shown in Plate 3.3(*a*) is a portion of a 2.5 μm feature size NMOS test circuit manufactured by Digital Equipment Corporation. The exposed surface of the chip is either first- or second-level aluminum-silicon alloy (1 at. % Si), shown in red, or silicon dioxide, green. Intricate circuit details are visible; in particular, Al and Si mixing occurs around the short vertical aluminum spurs, giving a yellow hue. It is possible that aluminum has been implanted into the silicon in these regions.

The second circuit [Plate 3.3(*b*)–(*f*)], constructed by electron beam lithography at International Business Machines, consists of a rectilinear patterned bilayer of aluminum on titanium, both deposited over a uniform silicon wafer. In Plate 3.3(*b*) a portion of this device is shown, where the oxygen originating from the oxidized aluminum metalization is displayed in yellow, and the silicon substrate in blue. At higher magnification [Plate 3.3(*c*)], the aluminum wires (blue) are edged by the protruding underlying titanium layer (white). Silicon, not separately displayed in this picture, fills the black regions. At still higher magnification [Plate 3.3(*d*)], the central portion of Plate 3.3(*c*) is shown with greater detail. These images are valuable for the diagnostic evaluation of circuit quality.

The SIMS micrograph in Plate 3.3(*e*) requires a moment's discussion. Remember that ion microscopy is a destructive process that gradually erodes the surface being examined. This Al image was acquired after previous scans; consequently, it has inherited features not present in the pristine circuit. Using a pseudocolor image-processing scheme, the Al wires appear red, as in Plate 3.3(*d*). Additional aluminum, sputtered from the wires by the probe bombardment, has been redeposited on the adjacent silicon areas. This migrant material is detected and displayed as the blue halo surrounding the wires.

The last micrograph of this circuit, Plate 3.3(*f*), exhibits resolution better than 30 nm. We believe that no other instrument is capable of equally describing the compositional morphology of these devices.

OBSERVATIONS OF DYNAMIC PROCESSES

So far, we have only presented microanalytical images of static objects, that is, materials that do not change during observation (except for the slight inevitable erosion resulting from the primary ion beam bombardment). It is also possible to monitor dynamic processes. Among these processes, controlled surface chemical reactions can be studied by the sequential acquisition of elemental distribution maps. A picturesque example is presented in Plate 3.4, where the progressive growth of gallium oxide over a clean drop of liquid gallium is described by a series of time-resolved "snapshots" (Wang et al., 1989).

First, in an ultrahigh vacuum, we use the probe to sputter a square window through the thin shell of oxide that naturally forms, in air, on liquid gallium. Then, oxygen is admitted into the microprobe chamber and, shortly afterwards, the image in Plate 3.4(*a*) is obtained. This Ga elemental map has been assigned colors, by image processing, so that the newly emerging oxide areas (green–yellow–red, in order of increasing intensity) are differentiated from the cleansed Ga area (blue). The oxide growths form branched patterns, anchored to the border surrounding the clean window (where nucleation first occurs). These patterns are characteristic of fractal growth and can be modeled, theoretically, by a diffusion-limited-aggregation process. Oxygen molecules, uniformly adsorbed onto the liquid surface, randomly migrate until they react and merge to the edge of the advancing oxide fern.

In this sequence of images, the oxide coverage increases (departing from a simple fractal pattern), creeping across the clean expanse of Ga until it eventually fills the entire square window. At the same time, the color of the oxide changes toward red, indicating a thickening of the crust. In the end, the surface returns to its original, static, oxidized state.

A CLOSING REMARK

As the micrographs in this chapter attest, scanning ion microscopy with a heavy ion probe is eminently capable of yielding high-resolution, undistorted images of the surface chemical composition of many materials. The wide variety of graphic patterns encountered in this endeavor are a continual source of wonder. These patterns, often enhanced by color, bestow to otherwise technical micrographs an artistic element than can be appreciated by both scientists and nonscientists alike.

ACKNOWLEDGMENTS

We are grateful to the many collaborators we have mentioned throughout this chapter who, by their comments and assistance, have made this research, and these images, possible. We also thank S. Chandra of Cornell University for preparing the onion specimen, and E. Olsen of the Field Museum of Natural History, Chicago, for loaning us the meteorite sample. The IBM test circuit was provided by D. Kern of the IBM Watson Research Center. This work was supported by the National Science Foundation under grants BBS-8610518 and DMR-8612254 and by the Materials Research Laboratory at the University of Chicago.

REFERENCES

Benninghoven, A. (1975). Surf. Sci. **53**, 596.

Burns, M. S. Chabala, J., Hallégot, P., and Levi-Setti, R. (1988). In *Microbeam Analysis— 1988*, D. E. Newbury (ed.), 445, San Francisco Press, San Francisco.

Bushinsky, D. A., Chabala, J. M., and Levi-Setti, R. (1989). *Am. J. Physiol.* **256,** E152.

————, Levi-Setti, R., and Coe, F. L. (1986). *Am. J. Physiol.* **250,** F1090.

Castaing, R., and Slodzian, G. J. (1962). *Microsc. Paris* **1,** 395.

Chabala, J., Edward, S., Levi-Setti, R., Lodding, A., Lundgren, T., Norèn, J. G., and Odelius, H. (1988a). *Swedish Dental J.* **12,** 201.

————, and Levi-Setti, R. (1987a). In *Microbeam Analysis—1987,* D. C. Joy, (ed.), 203, San Francisco Press, San Francisco.

————, Levi-Setti, R., Bradley, S. A., and Karasek, K. R. (1987b). *Appl. Surf. Sci.* **29,** 300.

————, Levi-Setti, R., and Wang, Y. L. (1988b). *Appl. Surf. Sci.* **32,** 10.

Clampitt, R., Aitken, K. L., and Jefferies, D. K. (1975). *J. Vac. Sci. Technol.* **12,** 1208.

Levi-Setti, R. (1983). *Scanning Electron Microsc.* O. Johari, (ed.) **1,** 1, AMF O'Hare, IL.

————, Berry, J. P., Chabala, J. M., and Galle, P. (1988a). *Biology of the Cell* **63,** 77.

————, Chabala, J. M., and Wang, Y. L. (1988b). *Ultramicroscopy* **24,** 97.

————, Chabala, J. M., Wang, Y. L., and Hallégot, P. (1988c). In *Microbeam Analysis— 1988,* D. E. Newbury (ed.), 93, San Francisco Press, San Francisco.

————, Crow, G., and Wang, Y. L. (1985). *Scanning Electron Microsc.* O. Johari, (ed.) **2,** 535, AMF O'Hare, IL.

————, Wang, Y. L., and Crow, G. (1984). *J. Phys. Paris* **45,** C9.

————, Wang, Y. L., and Crow, G. (1986). *Appl. Surf. Sci.* **26,** 249.

———— (1988). *Ann. Rev. Biophys. Biophys. Chem.* **17,** 325.

Liebl, H. (1967). *J. Appl. Phys.* **38,** 5277.

Lodding, A., Norèn, J. G., Petersson, L. G. (1988). In *Secondary Ion Mass Spectrometry SIMS VI,* A. Benninghoven, A. M. Huber, and H. W. Werner (eds.), 865, Wiley, N.Y.

McHugh, J. A. (1975). In *Methods of Surface and Interface Analysis,* A. W. Czanderna (ed.), 481, Elsevier, N.Y.

Reuter, W. (1985). In *Secondary Ion Mass Spectrometry SIMS V,* A. Benninghoven, R. J. Colton, D. S. Simons, and H. W. Werner (eds.), 94, Springer-Verlag Berlin.

Wang, Y. L., Levi-Setti, R., and Raval, A. (1989). *Scanning Electron Microsc.* O. Johari, (ed.) **3,** 731, AMF O'Hare, IL.

Williams, D. B., Levi-Setti, R., Chabala, J. M., and Newbury, D. E. (1987). *J. Microsc.* **148,** 241.

————, Levi-Setti, R., Chabala, J. M., and Newbury, D. E. (1989). *Appl. Surf. Sci.* **37,** 78.

4

Acoustic Microscopy

G. A. D. BRIGGS AND M. HOPPE

The principle of the scanning acoustic microscope is becoming increasingly well known (Lemons and Quate, 1979; Briggs, 1985). For any wave motion, the wavelength is equal to the wave speed divided by the frequency. Thus by increasing the frequency sufficiently the wavelength can be made as small as you like. The speed of sound in water is approximately 1500 ms^{-1}, or in more convenient units for microscopy 1.5 μm ns^{-1}. This means that, with a frequency of 2 GHz, the wavelength is 0.75 μm, or roughly the wavelength of red light. If a lens with a large numerical aperture and free from serious aberrations can be used, a microscope can have a resolution slightly better than the wavelength. Thus there is a possibility of an acoustic microscope with a resolution approaching that of a light optical microscope (LOM); the reader is referred to Chapter 1 for a discussion of the LOM.

It turns out not to be possible to make an acoustic microscope objective that can form an image of an extended object at a given moment. However, it is possible to make a high-resolution lens that has excellent focusing properties on its axis. Such a lens is illustrated in Figure 4.1. At the top is an epitaxially grown zinc oxide transducer, sandwiched between two gold electrodes. The transducer is excited by a radio-frequency pulse, which it converts into plane acoustic waves. The lens itself is made of sapphire, accurately oriented so that the lens axis is the crystallographic c axis. In the opposite face of the sapphire a spherical cavity is ground, subtending an angle of typically 60° at the center of curvature. A drop of liquid, usually water, couples the acoustic wave in the lens to the solid. The wave is reflected from the specimen and propagates back through the lens to the transducer. At the transducer it is converted back to an electrical signal, which occurs a fraction of a microsecond or so after the original excitation. By this time the connection has been switched over so that the signal passes to the input of sensitive receiving and detection

114

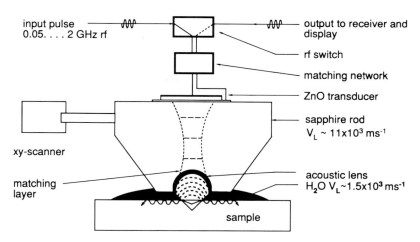

input pulse ——————— output to receiver and
0.05. . . . 2 GHz rf   display

—— rf switch

—— matching network

—— ZnO transducer

—— sapphire rod
$V_L \sim 11 \times 10^3$ ms^{-1}

xy-scanner

matching ——   —— acoustic lens
layer H_2O $V_L \sim 1.5 \times 10^3$ ms^{-1}

                                    sample

Figure 4.1 Schematic view of acoustic microscope objective, showing transducer, lens, water droplet, and specimen. The wave fronts are also shown schematically: Inside the lens they are planar and relatively widely spaced, in the water droplet they are spherical and closer together.

circuitry. In this way the strength of the acoustic reflection from that spot on the specimen can be measured (Atalar and Hoppe, 1986).

To build up an image, the spot is scanned over the specimen in a raster pattern, just as in any other scanned image system; for example, see Chapter 2. This is achieved by moving the lens mechanically in a plane parallel to the specimen surface. Typically scan rates of 25 or 50 Hz are used, so that a 512 line image can be built up in 10 or 20 seconds; faster frame rates are possible if fewer lines are recorded. These rates can be achieved for scans of up to a millimeter or so; larger areas generally take longer. If the image is stored digitally, then sophisticated image-processing and analysis techniques can be used (see Chapters 10–12). Often the contrast from features of interest can be enhanced by defocusing the specimen along the axis of the lens. The distance between the specimen surface and the focal plane of the lens is designated z, with negative z denoting displacement toward the lens. Because of the sensitivity of the contrast to variations in z (especially in the surface imaging mode described later), it is usually desirable to have specimens whose surfaces are flat to a tolerance of about a tenth of the acoustic wavelength being used. For high-resolution surface imaging at 2 GHz, this means better than 0.1 μm; such surfaces can be prepared by lapping with soluble diamond paste using a soft metal wheel and holding the specimen in a precision polishing jig. Specimens for acoustic microscopy should never be etched if the purpose is to image elastic properties, and not simply topography.

The resolution that can be achieved is determined by the frequency of the acoustic waves, and this is limited by attenuation in the coupling fluid. Extensive studies have been made of the best fluid to use, and it turns out that almost the only fluid that is better than water is hot water! Liquid helium below about 200 mK offers the prospect of almost unlimited resolution, but at the expense of most of the interesting contrast mechanisms illustrated in this chapter, and also at considerable experimental inconvenience. Liquid gallium and mercury offer advantages in certain applications, but special precautions must be exercised. Methanol alone is useful as an alternative to hot water for certain regular applications. The attenuation in these fluids increases as the square of the acoustic frequency. This provides a very severe limit to increasing the frequency indefinitely. For example, consider a lens designed to have an attenuation of 40 dB at its upper frequency, and that this gave a 20 dB signal-to-noise ratio. Then doubling the frequency in order to try to improve the resolution by a factor of two would result in a signal 100 dB below the noise; this would not be detectable. The only solution is to make the lens smaller, but this is limited by the need to be able to separate in time the signal from the specimen and an unwanted echo from the lens surface. In practice frequencies up to 2 GHz can be used routinely, and microscopes are commercially available covering the range from this frequency down to approximately 60 MHz.

The reasons for using an acoustic microscope cannot lie in the resolution alone, although to make an ultrasonic imaging system with submicrometer resolution is a major achievement. The benefits lie in being able to image, with such good resolution, the interaction between the acoustic waves of the microscope and the elastic properties of the specimen. This interaction can take many forms; indeed, the complexity of the interaction gives rise to the wealth of information available, but for simplicity three distinct kinds of interaction can be identified. The first is interaction with subsurface features, exploiting the ability of acoustic waves to penetrate opaque materials. The second is variation of reflection from the surface depending on its acoustic impedance. Third, surface acoustic waves can be excited in some specimens, and these can give rise to great sensitivity to surface features. Each of these contrast mechanisms will be illustrated.

SUBSURFACE IMAGING

Ultrasonic waves are exploited in medical ultrasound and in ultrasonic nondestructive testing for their ability to penetrate opaque media for interior imaging. Subsurface imaging in the acoustic microscope is an extension of this to higher frequencies. This enables higher resolution to be obtained and features nearer to the surface to be examined. A classic example of this is the ability to image subsurface inclusions in steel, illustrated in Figure 4.2. The first image, Figure 4.2(a), was obtained with the microscope focused on the surface. The specimen had not been polished, and parallel scratches can be seen. The second image, Figure 4.2(b), was obtained with the lens moved 0.47 mm closer to the specimen ($z = -470$ μm); because of refraction this corresponds to a paraxial focus about 120 μm below the surface. The scratches are still visible (albeit out of focus), but in addition inclusions inside the steel appear as bright features.

Simple disbonds generally give strong contrast, because the change in acoustic

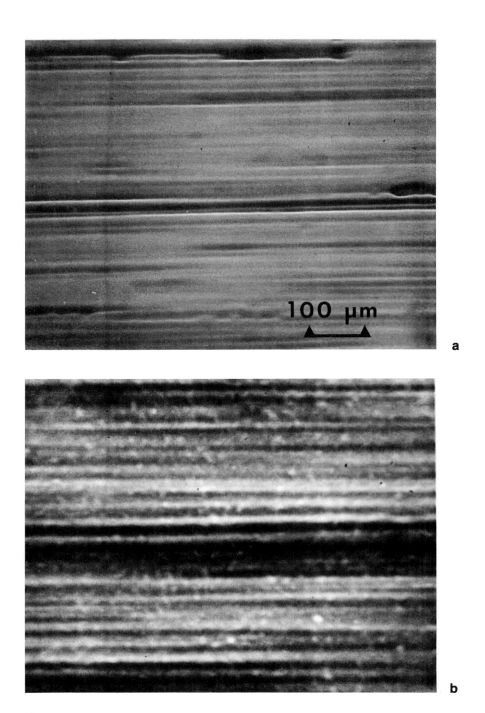

100 μm

a

b

Figure 4.2 Subsurface inclusions in steel, 100 MHz. (a) $z = 0$, (b) $z = -470$ μm.

117

properties at an interface between a material and air gives rise to a strong reflection. The ability to detect subsurface disbonding is illustrated in Figure 4.3, which shows a series of acoustic images of a layer of paint on a carbon-fiber-reinforced polymer substrate. In Figure 4.3(a) the microscope is focused on the top surface. A certain amount of topographic detail can be seen, but nothing definitive about the paint–substrate interface. In Figure 4.3(b) the microscope is focused on the interface between the paint film and the polymer substrate. The bright areas in this image indicate strong reflection where the paint is disbonded. Model experiments can be conducted with transparent protective coatings to confirm this interpretation, but with opaque paints only the acoustic microscope can give the information. Finally, in Figure 4.3(c) the microscope is focused further down still, revealing contrast from the fibers in the composite itself.

The acoustic microscope is a confocal imaging system (Wilson and Sheppard, 1984). Both the illumination and the detection optics are focused (indeed, the same lens and transducer are used for both jobs). This gives enhanced depth discrimination compared with the simple depth of focus of a conventional microscope. If a scatterer is out of focus, then it is only weakly illuminated in the first place, and so it contributes very little to the reflected signal. This gives very sharp depth discrimination in favor of the plane being imaged: Scatterers in other planes do not appear out of focus; they simply do not appear at all. This has been exploited for imaging of fiber distributions in composites, and also for detecting subsurface damage (Khuri-Yakub et al., 1984).

Another important application of subsurface imaging is the bond between a die and its heat sink in semiconductor packaging (Smith et al., 1985). An example of this is seen in Figure 4.4. Figure 4.4(a) is an optical image, showing the die surface and the surrounding heat sink. Figure 4.4(b) is an acoustic image, focused on the die surface; this shows similar detail to the optical image. The resolution is limited by the relatively low frequency used. In Figure 4.4(c) the lens has been moved 0.96 mm closer to the die surface. Because of refraction at the water–die interface, this corresponds to focusing between 200 and 250 μm below the surface. Delaminations between the die and the heat sink show up as bright areas, again because of strong reflection.

Die attach bonds can, of course, also be examined by radiography. In many cases radiography is more convenient, but acoustic microscopy may be preferable where the device is extremely radiation sensitive or where a non-radio-opaque bonding agent, such as epoxy, is used. Also, while radiography can detect disbonds due to failure of solder to wet the bond area uniformly (because of the relatively thick void that results), it cannot detect subsequent cracking parallel to the interface of a solder bond that was initially uniform (because there would be a difference in the net attenuation of the x rays). This is illustrated in Figure 4.5, which shows images of three devices from a series that showed various degrees of failure due to cracking of the die attach bond (Wakefield and Videlo, 1989). In each case an acoustic image is shown, taken from the outside of the package through the heat sink, together with the corresponding radiograph. It is difficult to reproduce the radiographs with satisfactory dynamic range, but no significant difference between the quality of the bonds could be deduced from them. The acoustic images, on the other hand, show

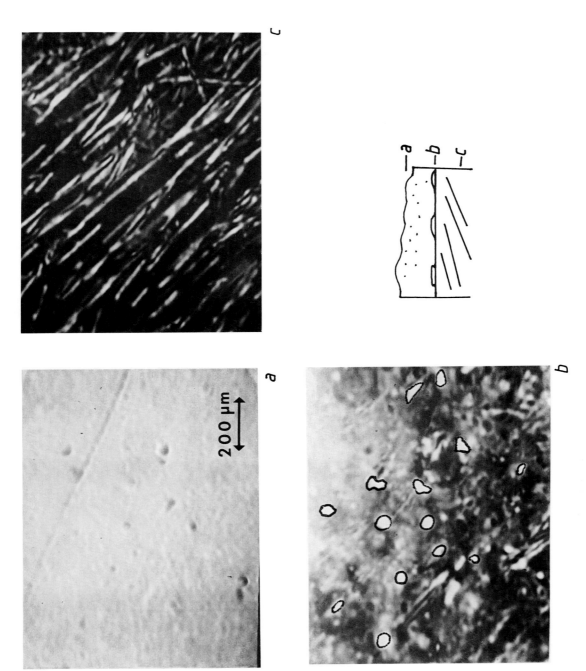

Figure 4.3 Paint on a carbon-fiber-reinforced polymer substrate, 100 MHz, focused (a) on the paint surface, (b) on the interface between the paint and the substrate, and (c) within the substrate.

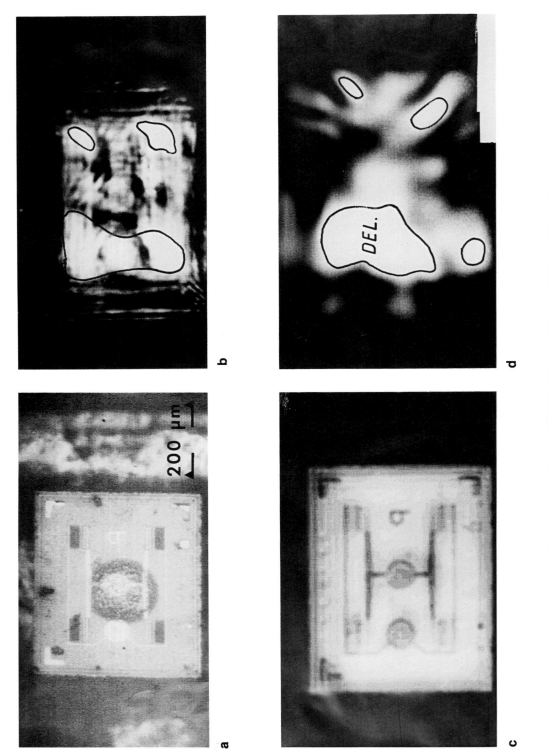

Figure 4.4 Semiconductor die: (a) optical image; acoustic images, 100 MHz, (b) $z = 0$, (c) $z = -960$ μm, (d) showing delaminations.

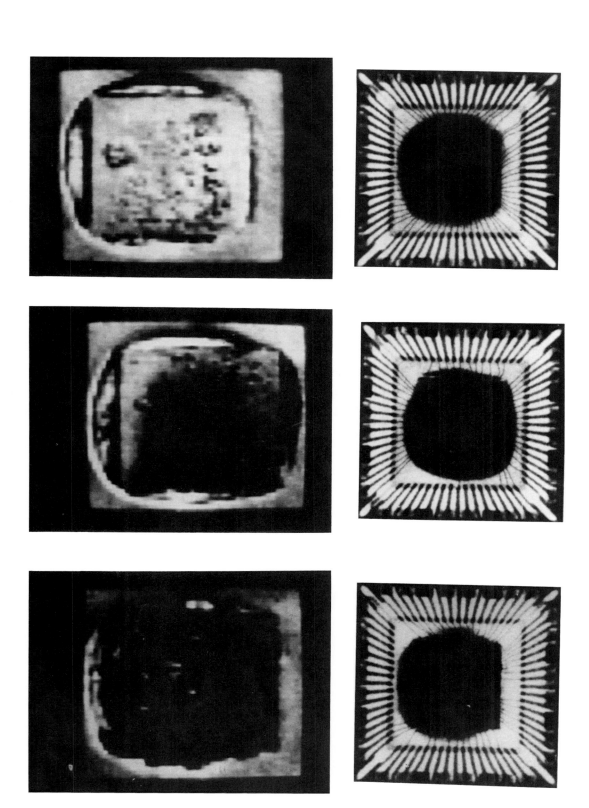

Figure 4.5 A series of radiographs (lower) and acoustic images (60 MHz, upper) of semiconductor chip bonds with varying amounts of cracking of the bonds (courtesy of Mr. I. D. E. Videlo, British Telecom Research Laboratories).

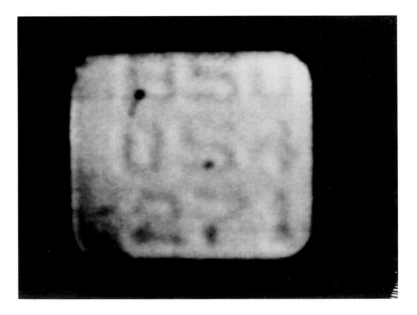

a

c

Figure 4.6 The lid of a ceramic semiconductor device package intended to be hermetically sealed, 60 MHz. The sketch indicates the surface on which the microscope was focused for each picture: (a) the top of the lid, (b) the bottom of the lid (at higher gain), (c) slightly above the bottom of the lid, (d) the top of the solder seal (courtesy of Dr. B. Wakefield, British Telecom Research Laboratories).

b

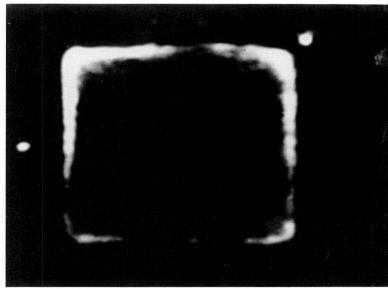

d

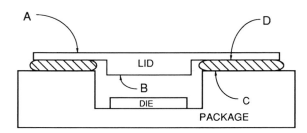

123

quite strong differences between good bonds (which appear dark) and cracked bonds (which show reflections).

The acoustic microscope has also proved useful industrially in the examination of lid seals in hermetically sealed components. Figure 4.6 shows acoustic images of the metal lid of the ceramic package of a semiconductor device. In Figure 4.6(a) the microscope is focused on the top of the lid. In Figure 4.6(b) it is focused at a depth corresponding (allowing for refraction) to the bottom of the lid. The lid has a rim, and so the area of the bottom is smaller than the top. The bright reflections outside the lid area are from the edges of the ceramic package [the gain is much higher in Figure 4.6(b) than in Figure 4.6(a) because the signals of interest have suffered transmission losses each way at the interface between the water and the top of the lid]. In Figure 4.6(c) the microscope was focused at a slightly shallower depth, and there are some reflections from outside the cavity area where the rim of the lid should be soldered onto the package. These reflections are even stronger in Figure 4.6(d), where the microscope was focused on the bottom of the rim of the lid. It turned out that in this case an incorrect soldering temperature had resulted in a noneutectic solder alloy that did not properly wet the surfaces to be joined. The correct temperature depends on the thickness of the gold coating on the surfaces to be joined, and a slight adjustment in temperature cured the problem.

It would be very nice to be able to distinguish between strong bonds and weak bonds. However, the strength of a bond is essentially a plastic property, depending on the local deformation behavior at failure. The stresses associated with the acoustic wave in the acoustic microscope are very small and are well within the elastic limit. Thus it is unlikely that the acoustic microscope can directly image the strength of a bond as such. The difference between a good bond and a poor bond can be detected only insofar as it corresponds to differences in the elastic contact, and this may be rather specific for individual bond systems. The difference between bonded and unbonded regions, however, generally shows up with strong contrast.

Interference effects can often be exploited in acoustic microscopy, especially because it is a coherent system, and for subsurface features this is illustrated in Figure 4.7. This figures contains optical and acoustic images of a metallic paint. The paint contains metallic particles in a transparent matrix. In this case the property of interest is the orientation of the metallic particles. The optical image shows their position satisfactorily, but not their orientation. In the acoustic image, there is interference between the waves reflected from the surface of the paint film and from the interface between the matrix and the metallic particles. Adjacent fringes correspond to $\lambda/2$ height variation, so the spacing of these fringes gives a measure of the tilt of that interface with respect to the paint surface, rather like contour lines on a map. Such fringes are amenable to automatic image analysis. Interference effects will be seen to be particularly important in the section on surface wave contrast.

In most interior imaging applications it is necessary to use the lower frequencies of the microscope in order to obtain the penetration that is required. When examining surface or near-surface features, the highest frequencies can be used, and since the acoustic velocity in water is lower than in almost all solids, the highest resolution of the microscope is then available.

Figure 4.7 Metallic paint: (a) optical, (b) acoustic, 1 GHz.

125

IMPEDANCE CONTRAST

The acoustic impedance of a medium (Z) is equal to the acoustic wave velocity (c) multiplied by the density (ρ):

$$Z = \rho c. \tag{4.1}$$

With ρ and c in SI units, the unit of acoustic impedance is the *rayl*. When a wave is incident on the interface from one medium to another (with impedances Z_1, Z_2), the ratio of the reflected amplitude to the incident amplitude is the reflection coefficient:

$$R = \frac{Z_2 - Z_1}{Z_2 + Z_1}, \tag{4.2}$$

and the proportion of the power reflected is R^2. Thus if the microscope is focused on the surface of the specimen, then variations in the impedance will show up as variations in the contrast. The impedance of water is approximately 1.5 Mrayl. Most solids have an impedance greater than this, so that generally in impedance imaging a stronger reflection corresponds to a region of the specimen with higher impedance.

Impedance contrast can be particularly strong in two-phase polymers. Figures 4.8(a,b) show optical and Figures 4.8(c,d) show acoustic images of two such materials. It is quite difficult to distinguish the phases in the optical images, but the contrast between the phases is relatively strong in the acoustic images. The second material [Figures 4.8(b,d)] had been subject to deformation in the manufacturing process, and this is manifest in the elongation of individual particles. The ability of the acoustic microscope to image the dual-phase structure in such materials is particularly important. Although not shown here, it is also possible to check for disbonds between constituents because they give strong reflections.

Figure 4.9 shows images of two different magnetic recording tapes. In an optical image, Figure 4.9(a), it is not easy to discern how the magnetic particles are distributed. But the acoustic images Figures 4.9(b,c) reveal marked differences in the homogeneity of distribution of the magnetic particles. This gives considerably more information about any variations in the manufacturing process than, for example, simply checking the final performance by measuring the occurrence of dropouts in recorded test signals.

A final example of impedance imaging is the examination of contact holes in semiconductor processing for unwanted residues of photoresist. Photoresist has a much lower impedance than silicon or oxide (or indeed most other semiconductors), and so any photoresist present greatly reduces the reflectivity, causing a hole to appear dark relative to its neighbors. Examples are shown in Figure 4.10. A relatively large area is scanned in Figure 4.10(a), and a suspect hole (indicated with an arrow) is seen at higher magnifications in Figures 4.10(b,c). Another case is shown in Figure 4.10(d), this time with deflection-modulated video enhancement. These contact holes are only 0.8 μm in diameter, so that an acoustic frequency in the vicinity of 2 GHz is necessary (the images in Figure 4.9 were taken at 1.9 GHz).

In Equation (4.2) it can be seen that the coefficient of reflection, R, is most sensitive to changes in the impedance of the specimen, Z_2, when it is not too different from the impedance of the coupling fluid, Z_1. This is why impedance contrast is so suitable for imaging the differences between two phases of polymer, or between a polymer and a semiconductor or an oxide. But many materials, including most metals, ceramics and semiconductors, have an impedance very much greater than that of water. In that case $Z_2 \gg Z_1$ in Equation (4.2), and R has a value of almost 1, and varies only very slightly, even for quite large changes in Z_2. For these materials, therefore, only weak contrast is obtained when the microscope is focused on the surface. Fortunately, strong contrast can be obtained by exploiting surface-wave phenomena.

SURFACE-WAVE CONTRAST

Unlike fluids, in which only longitudinal waves can propagate, solids can support both longitudinal and transverse waves, the transverse wave velocity being typically about half the longitudinal velocity. At a surface, a special wave can propagate that has a mixture of longitudinal and transverse wave components, each of which decays exponentially away from the surface. The wave is thus bound to the surface, and it has a velocity slightly slower than the transverse wave velocity in the solid. Such waves are known as surface acoustic waves (sometimes abbreviated SAW), or as Rayleigh waves after Lord Rayleigh, who discovered them theoretically in 1885.

The excitation of waves in a specimen in an acoustic microscope is subject to Snell's law which states that the ratio of the sines of the angles of incidence and refraction equals the ratio of the velocity of the radiation in each medium, i.e.:

$$\frac{\sin \theta_1}{\sin \theta_2} = \frac{c_1}{c_2} \tag{4.3}$$

Since $\sin \theta$ cannot exceed 1, this gives critical angles of incidence for longitudinal and transverse waves in the specimen. The Rayleigh wave can only propagate parallel to the surface of the specimen, i.e., with $\sin \theta_2 = 1$. Thus Rayleigh waves will be excited in the specimen by waves in the fluid incident at the Rayleigh angle:

$$\sin \theta_R = \frac{c_0}{c_R} , \tag{4.4}$$

where c_0 is the velocity in the coupling fluid and c_R is the Rayleigh velocity. If the coupling fluid is water and the Rayleigh velocity is 3 μm ns^{-1} (a typical value for metals), then the Rayleigh angle is 30°. For many semiconductors and ceramics the Rayleigh velocity is higher, so that the Rayleigh angle is correspondingly smaller. Most high-resolution acoustic objective lenses have an aperture that subtends a semiangle at the focus of about 50°, so that they include rays at the Rayleigh angle. Therefore Rayleigh waves can be excited in the specimen by waves from the lens. When Rayleigh waves propagate along the surface of a solid that is in contact with a

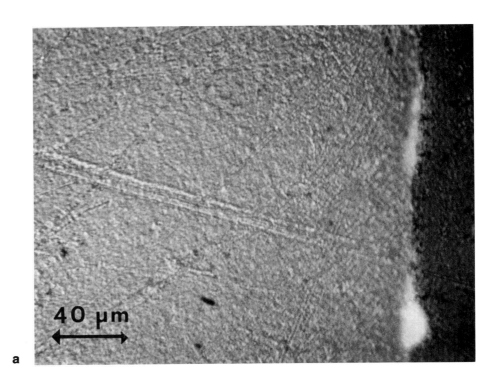

a

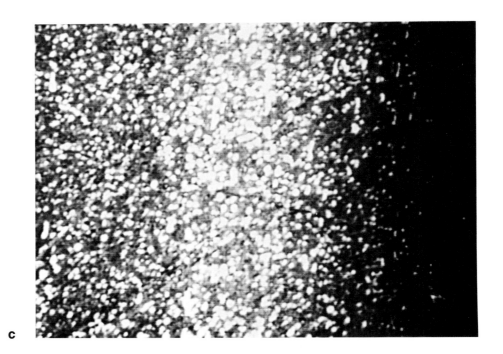

c

128

b

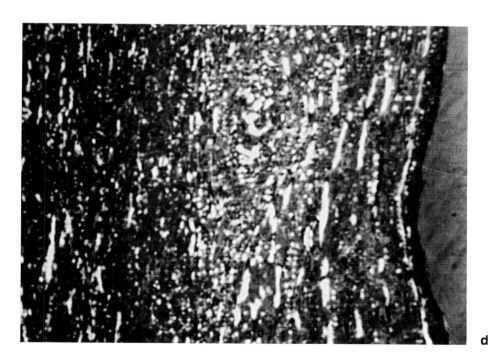

d

Figure 4.8 Dual-phase polymers: (a,b) optical, (c,d) acoustic, 1 GHz, $z = 0$. The specimen in (b) and (d) had been subject to deformation during the manufacturing process.

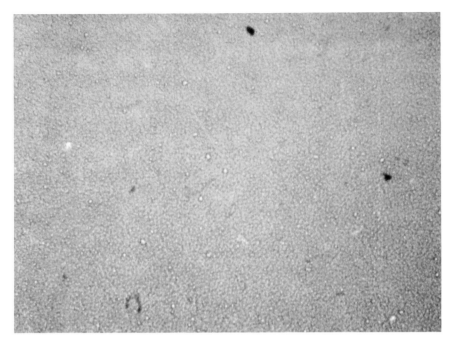

a

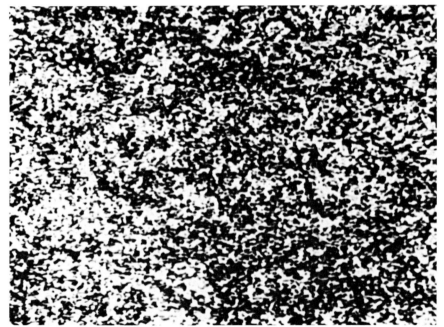

b

130

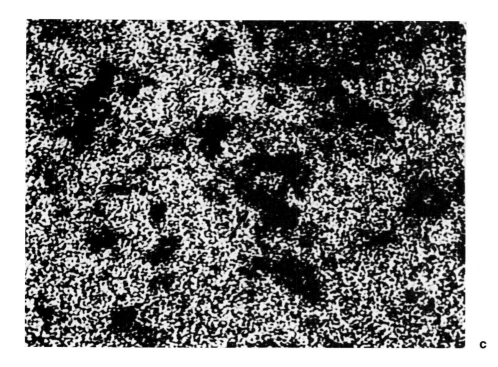

c

Figure 4.9 Magnetic recording tapes: (a) optical; (b,c) acoustic, 2 GHz, $z = -1$ μm.

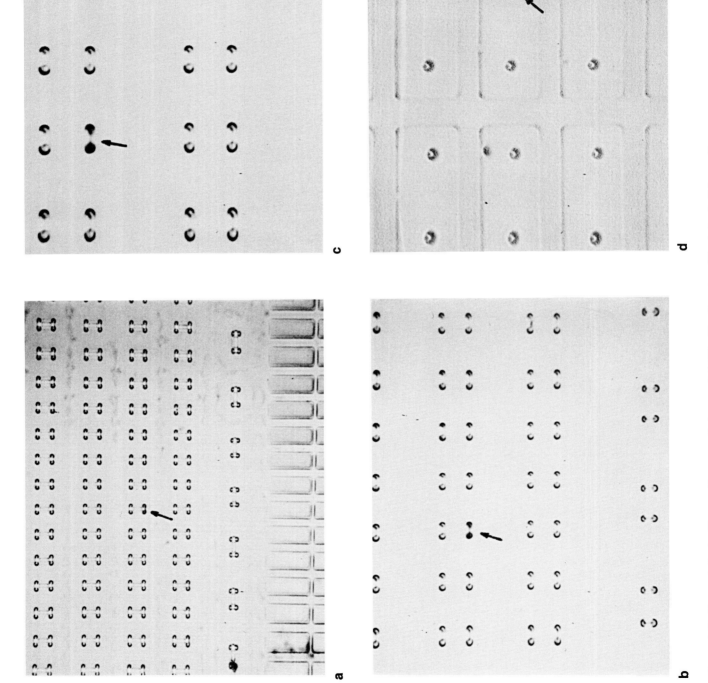

Figure 4.10 Contact holes in silicon wafers, acoustic images, 1.9 GHz. The diameter of the holes is 0.8 μm, the arrows indicate holes with residual resist.

fluid, they excite waves in the fluid, again at the Rayleigh angle. Because this causes the energy to leak out of the Rayleigh wave into the fluid, they are sometimes described as leaky Rayleigh waves (Bertoni and Tamir, 1974).

When the lens is moved towards the surface of a specimen in which leaky Rayleigh waves can be excited, oscillations are found in the video signal, V, which depend on the distance, z, by which the lens is moved toward the specimen relative to focus. It is found (Bertoni, 1984) that the periodicity (Δz) of these oscillations in $V(z)$ is:

$$\Delta z = \frac{\lambda_0}{2(1 - \cos \theta_R)} \tag{4.5}$$

where λ_0 is the wavelength of the waves in the coupling fluid. These oscillations are caused by interference between rays reflected geometrically from the surface of the specimen and rays that are incident at the Rayleigh angle, excite Rayleigh waves in the specimen, and then radiate back into the coupling fluid and travel back to the lens along a path symmetrical to the rays responsible for the excitation. By careful choice of the defocus, z, the interference conditions can be made very sensitive to factors that affect the propagation of the Rayleigh wave, such as small changes in its velocity or attenuation, or scattering by inhomogeneities. For high-modulus materials, this mechanism can give very useful contrast from surface features.

One of the first problems in which this kind of contrast mechanism was identified was the imaging of polycrystalline materials. It was found that the grain contrast varied, and indeed could reverse, depending on the defocus z. The contrast cannot be explained in terms of simple variations of impedance alone. In a single-phase polycrystalline material the material is the same in each grain. What changes is not the material but the crystalline orientation, and therefore the orientation of the anisotropic elastic properties. This would present a small difference in the reflection coefficient for normally incident waves, but it could not account for either the strength of the contrast or the reversals. What is more important is the excitation of Rayleigh waves, and also for certain orientations pseudo surface waves (Farnell, 1970). These propagate with different velocities and attenuations that depend both on their azimuthal angle on the surface, and on the crystallographic orientation of the surface itself. Thus different grains, presenting different crystallographic surfaces, support surface waves that interfere in different ways with the geometrical reflections from the surface, giving the strong contrast that is found at suitable values of defocus (Somekh et al., 1984).

Optical and acoustic micrographs of a polycrystalline specimen are presented in Figure 4.11. This is a specimen of lead lanthanum zirconate titanate. This is a transparent ferroelectric ceramic that is used as an electro-optic modulator. It happens to be difficult to etch (like many ceramics), so that it is not easy to reveal the grain structure by conventional metallographic techniques. No grain structure is visible in the optical image Figure 4.11(a), but in the acoustic image Figure 4.11(b) it can be seen without difficulty (Yin et al., 1982). There is a gold electrode on the specimen that appears bright in the optical image because of its higher optical reflectivity. It may appear either bright or dark in the acoustic image, depending on the defocus; at the defocus in Figure 4.11(b) it appears dark. If an acoustic frequen-

Figure 4.11 Grain structure of PLZT (lead lanthanum zirconate titanate) ceramic: (a) optical, (b) acoustic, 2 GHz, $z = -1.2$ μm; the specimen had been polished but not etched.

cy is chosen so that the Rayleigh wavelength is somewhat greater than the thickness of the gold film, then the grain structure can be seen through it, again by judicious choice of the defocusing conditions. This enables the grain structure to be seen through the optically opaque coating of gold (Hoppe and Bereiter-Hahn, 1985), but there is an important difference between this kind of surface-wave imaging and the interior imaging described above. In this case adjustment of the distance between the lens and the specimen must not be thought of as focusing below the surface, for example on the interface between the gold and the ceramic. Rather, varying z changes the interference conditions between the surface waves (albeit perturbed by the gold layer) and the geometrically reflected waves. The subsurface information comes from the fact that the profile of a Rayleigh wave decays into the material in a finite way, so that a depth approximately equal to a Rayleigh wavelength is sampled. This applies to all applications of acoustic microscopy in which surface-wave contrast dominates.

The sensitivity of the acoustic microscope to anything that affects the propagation of surface waves enables it to detect surface cracks with greatly enhanced contrast (Ilett et al., 1984; Somekh et al., 1985). A model example is shown in Figure 4.12, which presents optical and acoustic images of the microcracks around an indent in glass. In this case the cracks can be seen quite well in the optical image Figure 4.12(a), indeed slightly better than in the acoustic image at 1 GHz taken at focus in Figure 4.12(b). But in the acoustic image with 5 μm defocus towards the lens in Figure 4.12(c), the cracks show up with exceptional clarity. They can be recognized by the characteristic fringe pattern either side of them. This is caused by interference between the Rayleigh waves reflected by the crack and the waves geometrically reflected from the surface. As the lens is scanned in a direction normal to the crack, the phase of the former changes by 2π when the lens moves an amount $\lambda_0/2$, while the phase of the latter remains constant. This gives rise to fringes with spacing $\lambda_0/2$, which incidentally is extremely strong evidence confirming that Rayleigh waves are responsible for the contrast. It is also apparent that the acoustic contrast in Figure 4.12(c) stops at a well-defined distance along each crack, even with the shorter crack just below the main one to the right. The contrast mechanism seems to operate however thin the crack is, even when it is much less than an acoustic wavelength wide (in which case it would be below the resolution limit for waves that were simply geometrically reflected from the surface). Comparisons have been made between scanning acoustic microscope images and scanning electron microscope images of the same crack, and it seems that however thin the crack is, provided atomic contact has been lost between the faces, strong contrast is found in the acoustic microscope. This means that acoustic microscopy can be used to measure unambiguously the length of cracks from microhardness indents, for example for fracture toughness determination. Of course in glass the crack can be seen quite well optically, but for some ceramics the acoustic fringe pattern gives a much better defined measure of the length of microcracks than is possible by any other technique (Fatkin et al., 1988).

Cracks from hardness indents in brittle materials are introduced deliberately, and it is known where they will appear, but the high detection sensitivity arising from the scattering of Rayleigh waves by surface cracks can also be exploited to inspect a

a

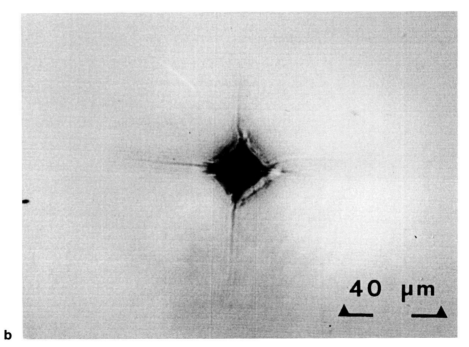

40 μm

b

136

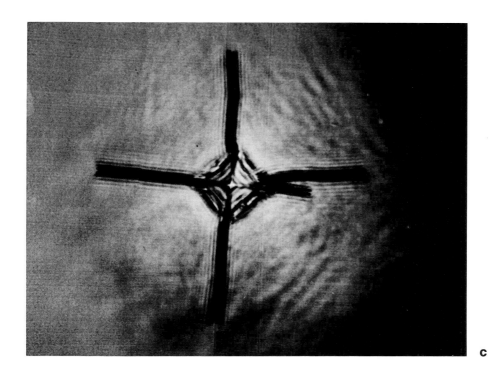

c

Figure 4.12 Microcracks in glass around a hardness indentation: (a) optical image; acoustic images, 1 GHz, (b) $z = 0$, (c) $z = -5$ μm.

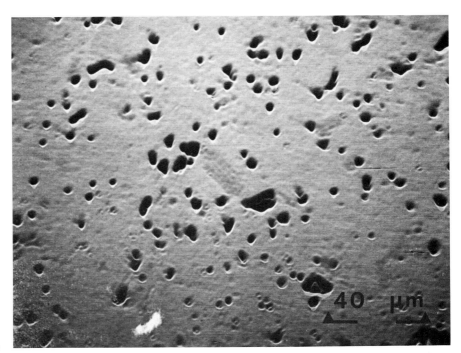

a

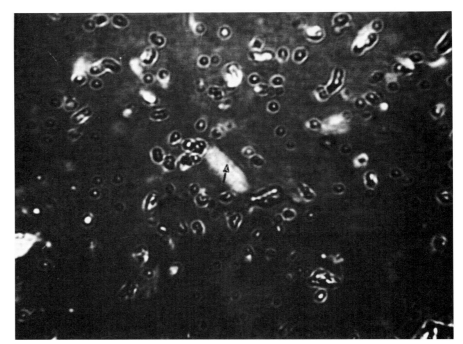

b

138

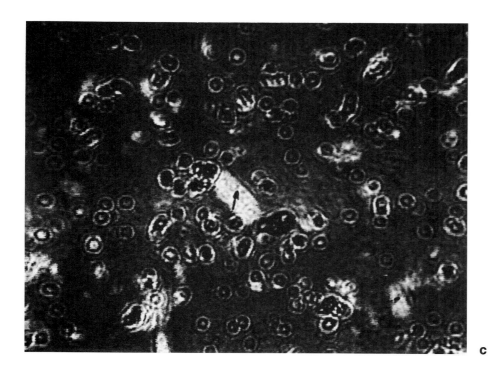

c

Figure 4.13 Fatigue cracks in an engineering ceramic, 1.6 GHz, (a) $z = 0$, (b) $z = -3$ μm, (c) $z = 3$ μm image processed with a convolution fiber.

surface for fine defects. Figure 4.13 shows acoustic images of a ceramic surface that has been subject to impact testing. In the image at focus in Figure 4.13(*a*) various voids are apparent; such voids may be intrinsic to the material or they may be artifacts produced during the polishing process. There are also one or two grey features of uncertain nature. Their identity becomes clearer at modest defocus in Figure 4.13(*b*), enhanced by image processing with a convolution filter in Figure 4.13(*c*). The bright feature indicated with an arrow is a crack that runs obliquely to the surface. Although not fully reproduced in the picture here, fringes that were seen suggested that the contrast may be due either to perturbation of surface waves or to reflection of bulk waves, or perhaps some interaction of the two. But it should be clear that at such a modest defocus ($z = 3$ μm) there is no question of having focused below the surface onto the fatigue crack: In all these examples of surface-wave imaging the effect of defocus is to change interference conditions. The capability of the acoustic microscope to detect surface cracks in this way should prove useful both in the examination of ceramics for surface flaws and in the study of short fatigue cracks in metals and alloys (Briggs et al., 1986).

SUMMARY

The range of applications of acoustic microscopy in research and evaluation of materials and components is growing so rapidly that the examples given here are only a small illustrative selection, though all of them represent applications that are in regular use in science and industry. The kind of contrast to be expected depends both on what one is hoping to see, and on the acoustic properties of the material. Polymers in general have Rayleigh wave velocities slower than the speed of sound in water, so that Rayleigh waves cannot be excited in them in the acoustic microscope (although in some circumstances surface skimming compressional waves can give contrast similar to surface-wave contrast). But most polymers also have a relatively low density, so that the reflection coefficient is small and a high proportion of the incident energy is transmitted into the bulk. Therefore interior imaging is very easy in such materials. Also, because the acoustic impedance is relatively close to that of water, the amount of energy reflected can depend sensitively on changes in the elastic properties, and therefore good impedance contrast is obtained with the microscope focused on the surface of the specimen from different polymer phases, and *a fortiori* from oxide or other high-impedance particles within a polymer. On the other hand, many metals, semiconductors, and ceramics have a Rayleigh velocity that is roughly double the speed of sound in water or higher, and therefore Rayleigh waves can be excited in their surface with a high numerical aperture acoustic lens. When such materials are imaged with a small amount of defocus toward the specimen, the contrast can be dominated by interference between geometrical reflections and reflections involving Rayleigh wave excitation, and therefore is very sensitive to factors that affect Rayleigh wave propagation. Such factors can include surface layers, whether implanted, deposited, or bonded; anisotropic effects such as grain structure; attenuation mechanisms; simple changes in material composition; or surface discontinuities such as interfaces or cracks.

Sophisticated analytical techniques are available for measuring elastic properties by acoustic microscopy. For coatings of polymers and other low-impedance materials, time-resolved measurements can be made to determine the thickness, acoustic velocity, density, and frequency dependence of the attenuation (Weaver et al., 1989; Sinton et al., 1989). For complete measurements of the angular-dependent reflectance function, Fourier inversion techniques can be used (Liang et al., 1985; Fright et al., 1988). In the case of higher velocity materials (and this is a large category), ray methods are available for analyzing the $V(z)$ curve to find the Rayleigh wave velocity and attenuation. These were pioneered with a cylindrical, or line-focus-beam, lens to give high accuracy and azimuthal resolution on anisotropic surfaces (Kushibiki and Chubachi, 1985), but they can also be used with a spherical imaging lens (Rowe, 1988). Thus elastic measurements can be made with the spatial resolution of the acoustic microscope. There is great scope for development of comparable techniques to measure lateral features found in acoustic microscope images.

REFERENCES

Atalar, A., and Hoppe, H. (1986). *Rev. Sci. Instrum.* **57,** 2568.

Bertoni, H. L. (1984). *IEEE Trans.* **SU-31,** 105.

———, and Tamir, T. (1979). *Appl. Phys.* **2,** 157.

Briggs, G. A. D. (1985). *An Introduction to Scanning Acoustic Microscopy.* Royal Microscopic Society Handbook 12, Oxford University Press, Oxford.

———, de los Rios, E. R., and Miller, K. J. (1986). In *The Behavior of Short Fatigue Cracks.* EGF Pub. 1. K. J. Miller and E. R. de los Rios (eds.), Mechanical Engineering Publications, London, pp. 529–36.

Farnell, G. W. (1970). In *Physical Acoustics* W. P. Mason and R. N. Thurston (eds.), **6,** 109.

Fatkin, D. G. P., Scruby, C. B., and Briggs, G. A. D. (1989). *J. Mater. Sci.* **24,** 23.

Fright, W. R., Bates, R. H. T., Rowe, J. M. R., Spencer, D. S., Somekh, M. G., and Briggs, G. A. D. (1989). *J. Microsc.* **153,** 103.

Hoppe, M., and Bereiter-Hahn, J. (1985). *IEEE Trans.* **SU-32,** 289.

Ilett, C., Somekh, M. G., and Briggs, G. A. D. (1984). *Proc. R. Soc. Lond.* **A393,** 171.

Khuri-Yakub, B. T., Reinholdtsen, R., and Jun, K. S. (1984). *IEEE Ultrasonic Symposium,* 580.

Kushibiki, J., and Chubachi, N. (1985). *IEEE Trans.* **SU-32,** 189.

Lemons, R. A., and Quate, C. F. (1979). In *Physical Acoustics.* W. P. Mason and R. N. Thurston (eds.), **14,** 1.

Liang, K. K., Kino, G. S., and Khuri-Yakub, B. T. (1985). *IEEE Trans.* **SU-32,** 213.

Rowe, J. M. (1988). D.Phil. Thesis, Oxford University.

Sinton, A. M. Briggs, G. A. D., and Tsukahara, Y. (1989). In *Acoustical Imaging,* H. Shimizu, N. Chubachi, and J. Kushibiki (eds.), Vol. 17. Plenum Press, N.Y., pp. 87–94.

Smith, I. R., Harvey, R. A., and Fathers, D. J. (1985). *IEEE Trans.* **SU-32,** 274.

Somekh, M. G., Bertoni, H. L., Briggs, G. A. D., and Burton, N. J. (1985). *Proc. R. Soc. Lond.* **A401,** 20.

———, Briggs, G. A. D., and Ilett, C. (1984). *Phil. Mag.* **49,** 179.

Wakefield, B., and Videlo, I. D. E. (1989). *Br. Telecom. Tech. J.* **7** (4), 27.

Weaver, J. M. R., Daft, C. M. W., Peck, S. D., and Briggs, G. A. D. *IEEE Trans. UFFC* **36,** 554.

Wilson, T., and Sheppard, C. J. R. (1984). *Theory and Practice of Scanning Optical Microscopy.* Academic Press, London.

Yin, Q. R., Ilett, C., and Briggs, G. A. D. (1982). *J. Mater. Sci.* **17,** 2449.

5

Transmission Electron Microscopy

G. THOMAS

"Metallography" has been a major tool in metallurgy for centuries (Figure 5.1), and there is no doubt that the development of transmission electron microscopy (see Marton, 1968 for a history) originating from Ruska and his colleagues in 1932 (Figure 5.2) and its synergism with materials has led to dramatic developments in our understanding of structure–property relationships over a wide range of materials (Table 5.1). Now metallographers can achieve the dream of atomic resolution with the new generation of electron microscopes exemplified by the unique "ARM" (atomic-resolution microscope) installed in Berkeley in 1983 (see Chapter 7), by which interpretable point-to-point resolution of 1.6 Å is possible over 40° of tilt of the sample.

The primary advantage of transmission electron microscopy (TEM), which is truly a diagnostic tool, is of course its very high resolution, and the ability to detect and record the various scattering events that occur when materials are bombarded with electrons (elastic, inelastic, coherent, and incoherent), so that in one instrument it is possible to combine imaging, diffraction, and spectroscopy, *in situ,* for static and dynamic experiments, as shown schematically in Figure 5.3. The information so obtained provides great insights into materials behavior and helps enormously to understand and improve materials performance. It is necessary, however, to examine enough samples in the microscope in order to arrive at statistically significant conclusions, because the volume of material per micrograph is very small (≈ 0.4 μm^3). The most important problem is still specimen preparation—herein lies the microscopists' art!

In the following, I have selected representative examples of some of the major imaging methods in TEM that we have applied to various problems in materials. However, I have made no attempt to be comprehensive, but Table 5.1 summarizes

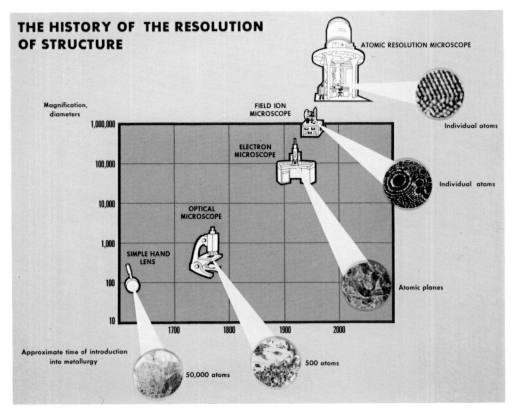

Figure 5.1 The history of the resolution of structure via microscopy.

Figure 5.2 The 1932 operational microscope built by von Ruska and colleagues, Berlin.

Table 5.1. Developments of TEM in materials science

Year	Specimens	Applications/Developments	Instrumentation	Resolution[a]
1932–	First operational transmission electron microscope, Berlin (Ruska, Knoll)			
1938	First commercial TEM, Berlin (Ruska, von Borries)			
1940–50	Replicas 1. Oxide 2. Carbon 3. Plastic	Surfaces Slip steps Extracted particles Fractography	50 kV instruments Single condenser Little or no theory	~100Å
1949	Heidenreich published first paper on TEM foils; basic theory outlined			
1950–60	Many developments in instrumentation, specimen preparation methods, and image contrast and diffraction theory for interpretation of data			
	Thin foil techniques 1. From bulk 2. Deposition	Defects Phase transitions	100 kV instruments Contrast theory developed	~5–20 Å
1960–70	Metals Nonmetals Semiconductors Ceramics Minerals	Dynamic, *in situ* studies: Information explosion on substructure of solids Ion thinning Radiation damage Microdiffraction	First HVEM built in Toulouse (1.2 MeV) First 3 MeV HVEM built in Toulouse Accessories for *in situ* studies Controlled experiments	3 Å
1970–80	Structure imaging to ~2 Å interpretable resolution; lattice imaging widely used			
	As above Catalysts	Theories for high- resolution interpretation developed	TEM/STEM analytical, convergent beam Spectroscopy EDXS, EELS Commercial HVEMs 0.5–1.5 MeV General acceptance	2 Å
1980–90	Virtually all materials	Atomic resolution in close- packed solids Surface imaging Small particles Fast computation facilities for interpretation Simulation Quasicrystals	Medium voltage HREM/AEM (100– 400 keV commercially available) Improved analytical capabilities Parallel detection in EELS Widespread applications in UHV microscopes Tunneling microscopes	1.5 Å
1986	Nobel Prize to Ernst von Ruska and Binnig and Rohrer (for scanning tunneling microscope (STM))			

[a]Interpretable point-to-point.

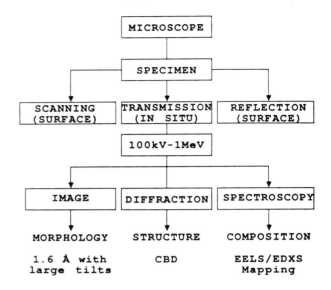

CHARACTERIZATION BY ELECTRON MICROSCOPY

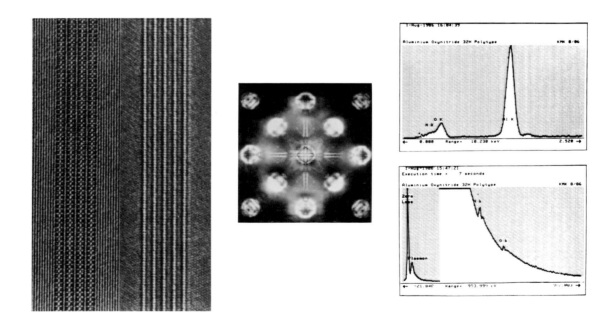

Figure 5.3 A diagram showing the types of electron microscopy and the information obtainable from them.

146

the major developments. The text is very brief and only alludes to the enormous versatility of the TEM as a tool for materials characterization. The images speak for themselves.

IMAGING AND INTERPRETATION

As is well known, the formation of images in the electron microscope is done by a system of lenses that allows fine probes of electrons to pass through a specimen. These are brought to diffraction focus in the objective back focal plane (see Chapter 6) and transformed to an image by magnifying lenses. Two general methods are used *viz.*, (1) amplitude contrast imaging, which is not limited by instrument resolution, and is emphasized in this chapter and (2) phase-contrast–high-resolution imaging, to be discussed in detail in Chapter 7. The basic principles of imaging in bright and dark field, use of multiple images, special effects at high voltage, diffraction, structure imaging, spectroscopy, etc., that is, the whole "bag of tools" for expert microscopy, are well described in the literature (e.g., Hirsch et al., 1977; Edington, 1974–77; Thomas and Goringe, 1979; Chescoe and Goodhew, 1984; Williams, 1984) and will not be discussed here. Routinely, choice of orientation and imaging conditions is done via the diffraction pattern, and interpretation involves constant interchanges between imaging and diffraction.

EXAMPLES

The earliest studies of metals were concerned with lattice defects, especially dislocations, their origin, motion, multiplication, etc. (Figure 5.4), verifying theories that were already two decades old. But new defects were also discovered, for example, the stacking fault tetrahedron (Figure 5.5). The development of contrast analysis, notably the invisibility criterion $\mathbf{g} \cdot \mathbf{R} = 0$, where $\mathbf{g}$ is the reciprocal lattice vector for a strong diffraction maximum and $\mathbf{R}$ the displacement vector of the defect, is illustrated for interfacial dislocations in Figure 5.6 and compositional faults in TaC (Figure 5.7). The resolution (i.e., image width) of such defects depends on the material and the effective extinction distance ξ_g, (see Figure 5.8). This enhanced-resolution concept is utilized by weak-beam dark-field (WBDF) (Figure 5.9), or high-order bright-field (HOBF) imaging (Figure 5.10) to achieve kinematical rather than dynamical contrast, and thus to minimize the width of the image of the defect.

In situ dynamic studies showed directly many of the predicted properties of lattice defects, and climb experiments (Figure 5.11) lead to useful diffusion data for aluminum alloys. Other defects common to a wide range of materials include accommodation twinning to minimize shape and strain deformation following phase transformations (e.g., Figure 5.12), ordered domains (in which antiphase domain

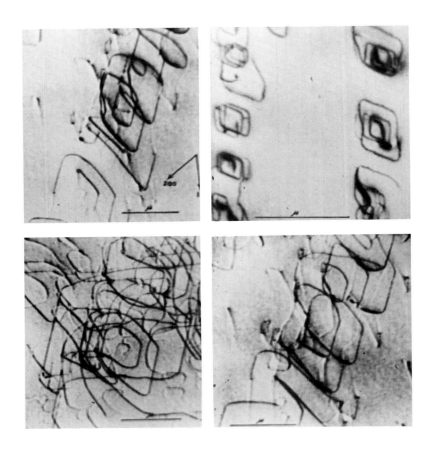

Figure 5.4 Dislocation sources and multiplication in Al–5 at. % Mg solid solution.

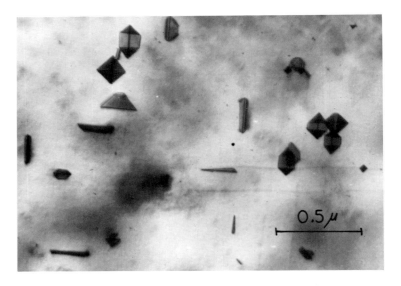

Figure 5.5 Tetrahedron of stacking faults in quenched gold. Courtesy J. A. Silcox.

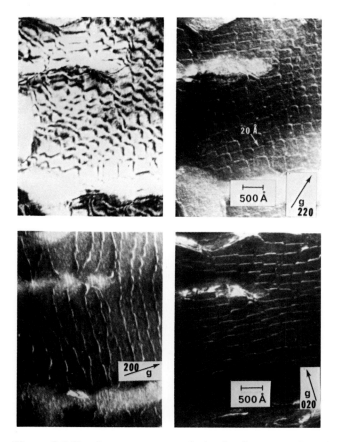

Figure 5.6 Two-beam contrast analysis showing pure edge dislocation networks $\mathbf{b} = \frac{1}{2}$ <110> at the interface between two fcc phases in Cu-Mn-Al. For certain $\mathbf{g}$ vectors the dislocations are invisible when $\mathbf{g} \cdot \mathbf{b} = 0$.

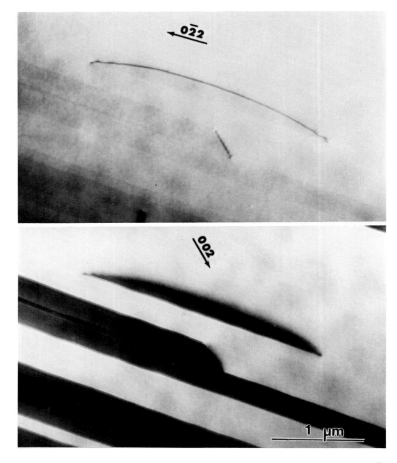

Figure 5.7 Shear faults in $TaC_{0.8}$; the lack of complete invisibility of faults with $\mathbf{g} = 0\bar{2}2$ shows there is residual structure factor contrast (composition change).

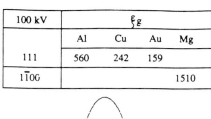

100 kV	ξ_g			
	Al	Cu	Au	Mg
111	560	242	159	
$1\bar{1}00$				1510

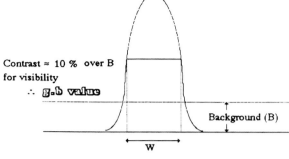

Contrast ≈ 10 % over B
for visibility

∴ **g.b value**

Background (B)

w

$$w = \frac{\xi_g^{eff}}{3} \quad (\text{many beam } \varepsilon_g)$$

make ξ_g smaller image is narrower

e.g. increase s >> WBDF
or use   ng > 2 HOBF

Figure 5.8 The variation in the extinction distance, ξ_g for different materials at 100 kV showing how the image resolution for a defect such as a dislocation can be improved by weak beam dark field or high order bright field imaging.

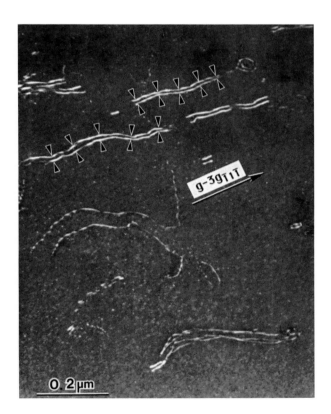

0.2 µm

g-3g$\bar{1}1\bar{1}$

Figure 5.9 The **g-3g** weak-beam image showing paired $\frac{1}{2}$ <110> slip dislocations in Al–2.4 at. % Li deformed 1% after 4 h aging at 150°C. Compare the image resolution with that in Figure 5.4 which shows dislocations imaged with conventional bright field techniques.

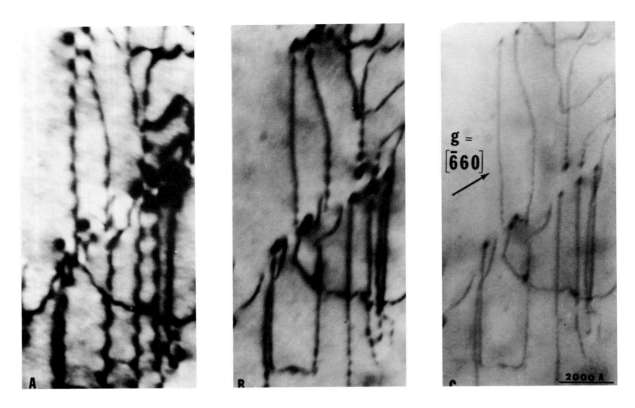

Figure 5.10 The **g**, 2**g**, 3**g** bright-field images in GaAs; the dislocations are not dissociated.

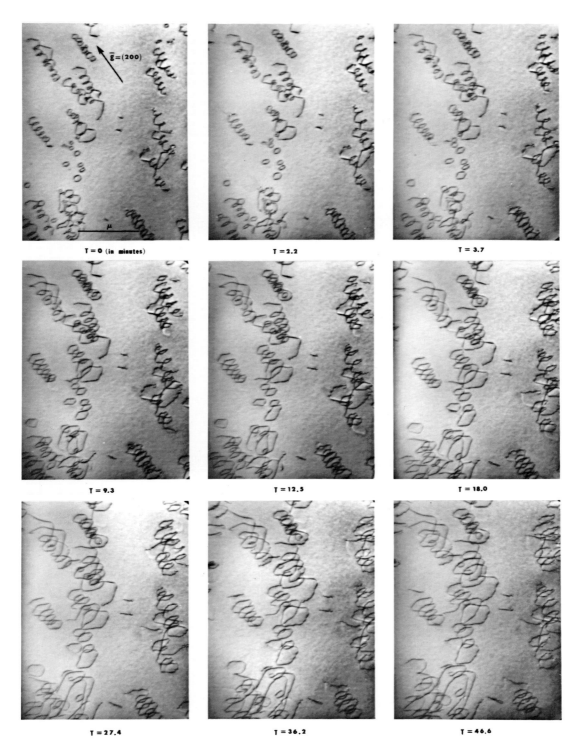

Figure 5.11 Direct observation of the growth of prismatic loops and helicoidal dislocations in an Al–5 wt. % Mg alloy during an in-microscope isothermal annealing experiment. Material was quenched from 550°C and aged at room temperature for about 5 years. The isothermal annealing temperature was 130°C. The diffraction conditions were kept nearly constant by employing a Valdrè double-tilt hot stage.

Figure 5.12 Dark-field images of twins formed during transformations in four materials.

154

boundaries can be analyzed) as in Figure 5.13, enantiomorphic domains in Li ferrite (Figure 5.14), interstitial ordering (Figure 5.15), and short-range order revealed as domains by lattice imaging (Figure 5.16). Magnetic domains and boundaries can be imaged by Lorentz microscopy but at low spatial resolution (Figure 5.17).

Much attention has also been given to studies of alloys. For example, Figure 5.18 shows how imaging of the diffuse streak due to thin-plate Guinier–Preston zones in Al-Cu alloys provides a dark-field, strain-free image with much better resolution than the bright-field image. The advantages of the TEM are clearly in defining local morphological details such as precipitate-free zones, common in nucleation and growth reactions (Figure 5.19) but not present in spinodals (Figure 5.20). Figure 5.21 shows the loss of coherency upon coarsening of a spinodal. In nucleation and growth transformations heterogeneous nucleation at defects is common, which leads to nonuniform structures and, of course, properties. Aluminum alloys are typical examples (Figures 5.22–5.27), but the Al-Li base alloys appear initially to undergo spinodal ordering (Figure 5.23), out of which the δ' (Al_3Li) ordered phase coarsens (Figures 5.24 and 5.25). Notice that the precipitation morphologies depend on alloy composition (e.g., the faceting of δ' in Figure 5.25 compared with no faceting in Figure 5.24). Al-Cu alloys exhibit heterogeneous nucleation of θ' (Al_2Cu), Figure 5.26 and Al-Li-Cu alloys similarly show heterogeneous nucleation of T_1 (Al_2CuLi) on dislocations (Figure 5.27).

Another form of heterogeneous precipitation is shown in Figure 5.28(a) for boron and phosphorus diffusion in silicon. In the [112] orientation the stresses due to diffusion must be accommodated both by precipitation on (111) and dislocation generation of Burgers vectors ($a/2$) [110] in the same region. By dark-field imaging in a systematic many-beam condition, the specimen can be selectively imaged near the top and bottom surfaces, showing in this [111] orientation that dislocations and precipitates form at different depths after diffusion [Figures 5.28(b),(c)].

As emphasized in the introduction, diffraction is a key aspect in imaging. In complex low-symmetry materials the use of Kikuchi maps (Figure 5.29) and convergent beam pattern maps (Figure 5.30) greatly facilitate control of imaging by allowing the operator to determine the sample orientation in a very rapid manner, as well as providing a wide range of structural information, as described in detail in Chapter 6.

In the past decade or two, electron microscopy of minerals and ceramics has become popular, since the Apollo missions and geological analyses of moon dust (Figure 5.31). Generic problems in ceramics relate to grain boundaries (Figure 5.32) such as the retention of glassy phases after processing with additives. These phases can be detected by the TEM imaging methods illustrated in Figure 5.33. Studies of glass crystallization are being done to try to avoid such phases (Figure 5.34), which often account for the generally poor mechanical properties of ceramics.

More recently, ceramic materials have become even more prominent, since the discovery of superconductivity at relatively high critical temperatures, T_c, in complex oxides, such as Y-Ba-Cu-O (Figure 5.35) and Bi-Sr-Ca-Cu-O, in which polytypoids of different T_c values form adjacent to grain boundaries (Figure 5.36). These may be modified by Pb additions, as seen in Figure 5.37. Such polytypoid structures are also commonly seen in sialons (ceramics composed of silicon, aluminum,

Phase Shift for Antiphase Vectors in the DO₃ Phase

phase factor

$$\alpha = 2\pi \bar{g} \cdot \bar{p}$$

h k l	type of reflection	phase shift (α)		extinction distance (Å)
		1/4<111>	1/2<100>	
111	S	±π/2	±π	2,467
200	S	±π	0	2,498
220	F	0	0	272
311	S	±π/2	±π	2,752
222	S	±π	0	2,832
400	F	0	0	404

F : Fundamental reflection
S : Superlattice reflection

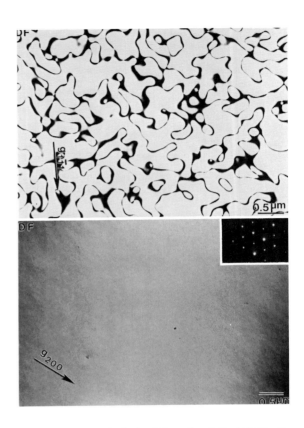

Figure 5.13 Sketch and table showing contrast analysis in DO_3-ordered $Fe_3(Al,Si)$; the antiphase boundary vectors must be $\frac{1}{2}$ <100>.

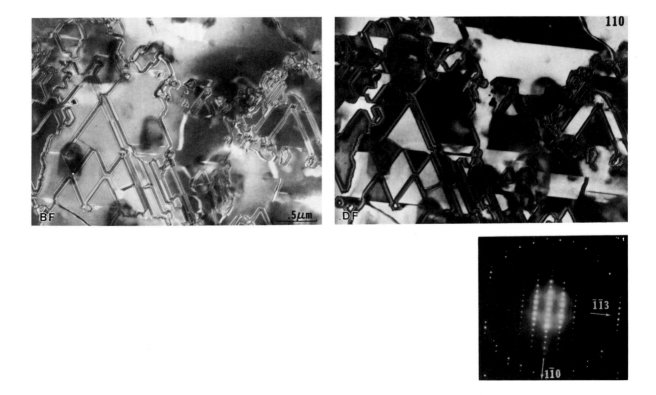

Figure 5.14 Bright-field and dark-field images of Li ferrite in multibeam nonsystematic diffraction conditions; the breakdown of Friedel's Law (see Chapter 6) allows enantiomorphs to be identified in the dark-field image.

Figure 5.15 *In situ* transformation of Ta to a Ta-O phase by interstitial ordering; the strain produces an almost four-fold relaxation. Foil normal is in the [001] direction.

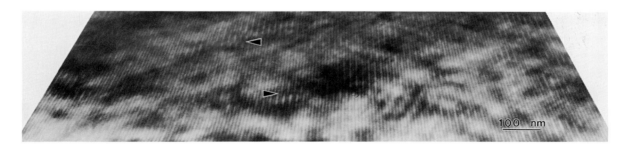

Figure 5.16 Lattice image using 000+**g**(superlattice) +2**g**(fundamental) conditions showing the presence of short-range order, that is, domains of order (double-spaced fringes) in a disordered matrix (oblique printing).

157

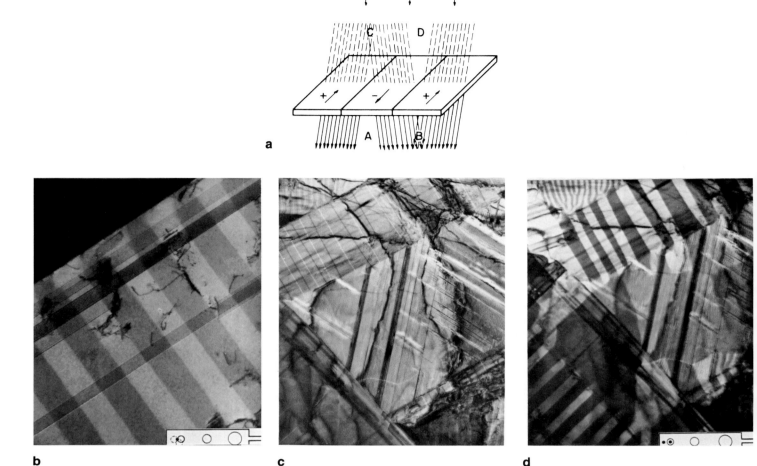

a

b   c   d

Figure 5.17 Background: Domains and domain walls are revealed by the deflection they produce in the electron beam. In cobalt the walls are about 150 Å thick and lie parallel to [0001], the direction of easy magnetization. The width and shapes of the observed domain images depend on the physical nature of the wall and angle which the [0001] direction makes with the plane of observation. Description of the out-of-focus method: The reversal of magnetization direction causes the electron beam to be deflected in opposite directions in adjoining domains. If one focuses on the image below the plane of the foil (overfocused image) the deficiency of electrons at A and the excess at B produces black and white lines whose width is proportional to the degree of defocusing. The contrast is reversed (C and D) if a virtual image is observed by focusing above the plane of foil (underfocused image). (a) Domains imaged by the modified displaced-aperture method: The transmitted beam is not "visibly" split; however, the aperture position is manipulated about the spot while observing the image until the domains appear. This method employs a specimen holder with high resolution plus double tilting for maximum contrast. Note also the faults and dislocations. (b) Domain walls imaged by the out-of-focus method: The walls are black and white lines. (c) Domains imaged by the displaced-aperture method: The objective aperture in the back focal plane is positioned so that it stops out one of the divided beams. Alternate domains appear black and white. A long focal length holder is necessary to split the spots.

158

Figure 5.18 Bright field, dark field, and selected area diffraction pattern from an Al-Cu alloy; dark-field imaging of streak from Guinier–Preston zones in the [001] direction. Notice the lack of resolution of the individual G. P. zones due to overlapping strains in the bright-field image.

Figure 5.19 Precipitate-free zone around three grain boundaries in an Al-Mg-Zn alloy. This microstructure is typical of early commercial aircraft alloys, which were intergranularly brittle. The precipitate-free zone arises because the coarse boundary precipitates deplete the surrounding matrix of solutes and/or vacancies. This depletion prevents precipitation of the fine precipitates.

Figure 5.20 Cu-Ni-Fe alloy that has undergone a spinodal decomposition reaction showing no precipitate-free zone at the grain boundary. See Figure 5.21 for additional details.

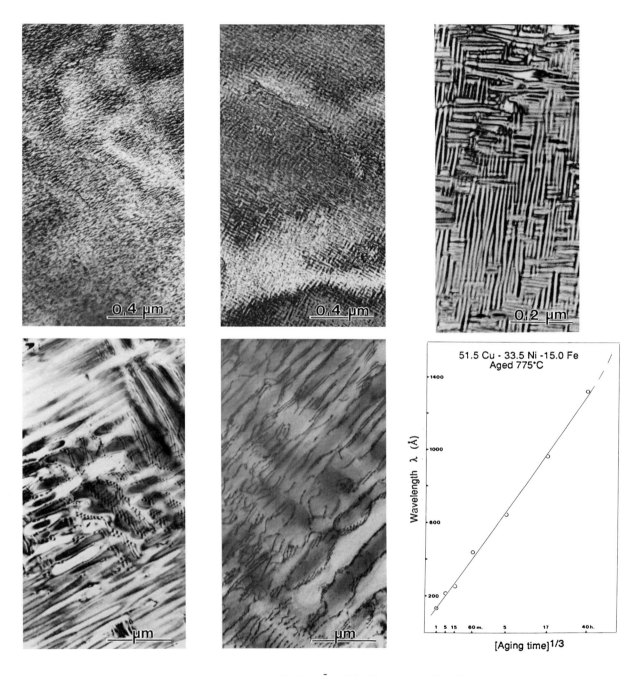

Figure 5.21 Decomposition initially follows a coarsening law $\bar{\lambda} = kt^{\frac{1}{3}}$ with copper-rich and copper-poor regions growing at approximately constant volume fraction and compositional amplitude. At wavelengths exceeding ~ 1000 Å interface dislocations are created, and the coarsening rate is accelerated.

161

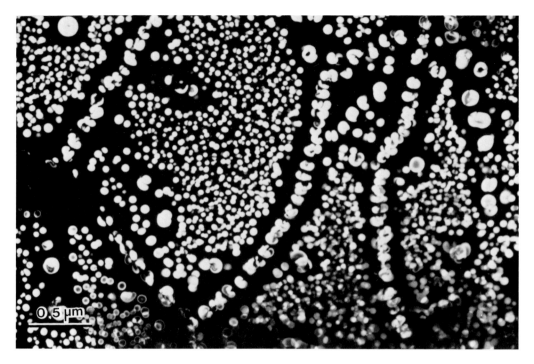

Figure 5.22 Al–2.4 at. % Li alloy, aged 5 min at 300°C, which shows coarse δ′ (Al₃Li) phase at subboundaries and fine δ′ in between. The dark-field image was formed from a diffraction [100] superlattice reflection from the ordered δ′ (LI_2) phase.

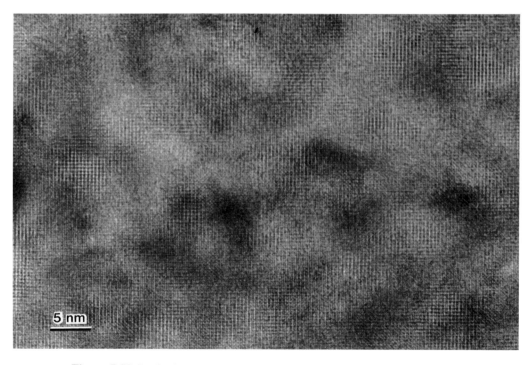

Figure 5.23 Lattice image showing the early stages of spinodal ordering in Al–2.5 at. %Li–1 at. %Cu–1.5 at. %Mg (cf. Figure 5.16). Clear ordered precipitates are not yet defined.

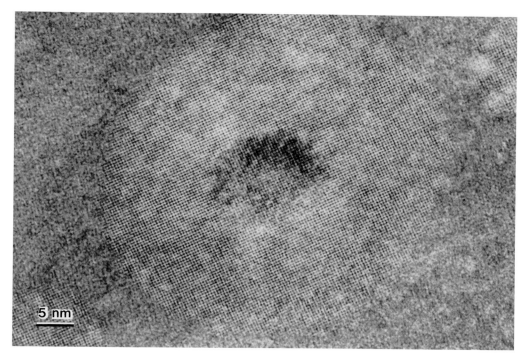

Figure 5.24 Lattice image showing a coarsened spinodally-formed δ′ particle in Al–2.5 at. %Li–1 at. %Cu–0.5 at. %Mg, which has been aged for a longer time than the sample in Figure 5.23.

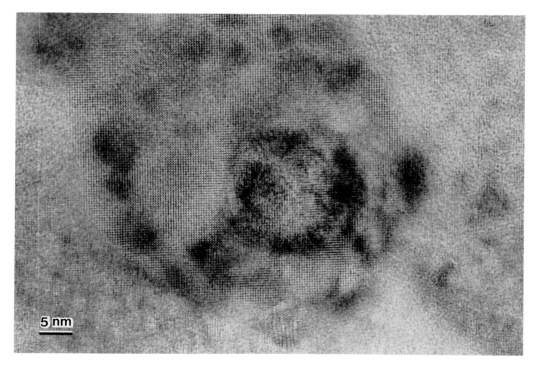

Figure 5.25 Lattice image similar to that in Figure 5.24 but in Al–2.4 at. % Li–2 at. % Cu; note the faceting of the δ′ particle. The dark region in the center of Figures 5.24 and 5.25 is contrast from the β′ Al_3Zr phase. (Zr is present in all Al-Li-Cu alloys to a level of ~0.1%).

Figure 5.26 Heterogeneous nucleation of θ′ in aged Al–4 at. %Cu. The θ′ plates have a strain **R** normal to the {100} habit. The habits form only for **R·b** ≠ 0, where **b** is the Burgers vector of the nucleating dislocation, that is, only two habits (maximum).

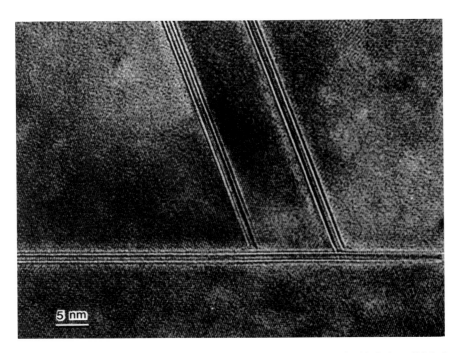

Figure 5.27 Heterogeneous nucleation of the variants of the $T1$ phase in Al–2.4 at. % Li–2 at. % Cu aged 8 h 175°C. As in Figure 5.26, only two out of four possible {111} variants can form on dislocations.

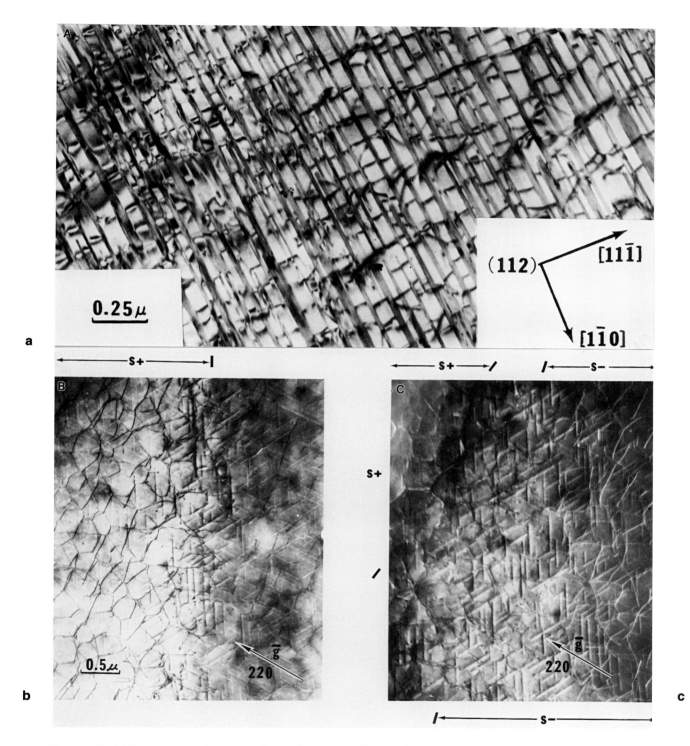

Figure 5.28 (a) Heterogeneous formation of precipitates on (111) and dislocations with **b** = $\frac{1}{2}$ [110] to relieve strain in Si due to P diffusion; notice spacings of the precipitates (**b** ≈ $\frac{1}{3}$ [111]) and dislocations. (b),(c) Image asymmetry in dark field at $s+$ and $s-$ allows contrast enhancement preferentially at the top (dislocations) and bottom (precipitates), giving a "pseudo" three-dimensional effect.

Figure 5.29 Kikuchi map of hexagonal close-packed lattice with c/a = 1.588.

166

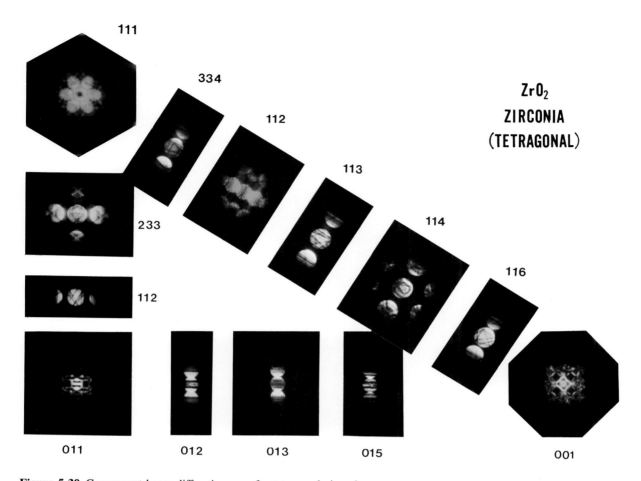

Figure 5.30 Convergent beam diffraction map for tetragonal zirconia.

167

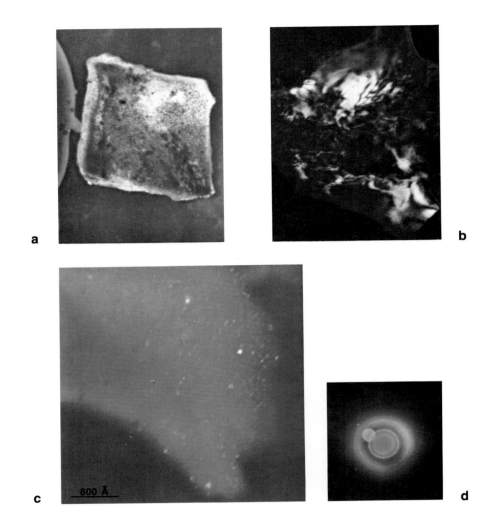

Figure 5.31 Investigation of moon dust from Apollo 11. (a),(b) Flakes of mineral showing radiation damage and heavy ion tracks. (c),(d) Identification of very small particles of α-Fe in amorphous matrix.

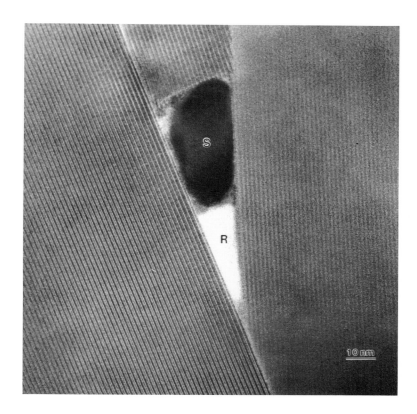

Figure 5.32 Glassy intergranular phase R and inclusion S at a triple grain junction of Ba$_2$ (Cu,Zn) ferrite.

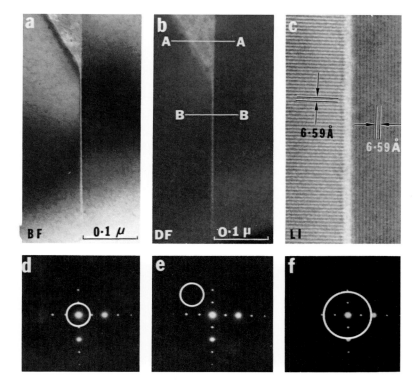

Figure 5.33 Imaging conditions to reveal intergranular glassy phases in silicon nitride sintered with Al$_2$O$_3$ and MgO. (a) conventional bright-field image, (b) image formed from the diffuse-scattered electrons only, and (c) lattice image formed from the reflections shown in the diffraction pattern.

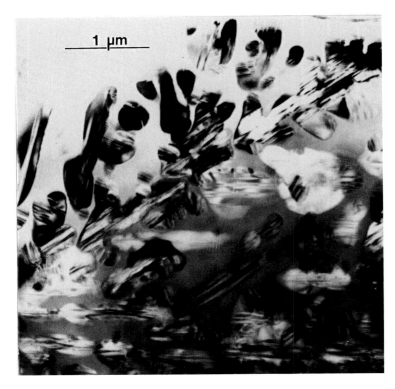

Figure 5.34 Dendritic crystallization in an Y-Si-Al-O-N glass annealed at 1100°C.

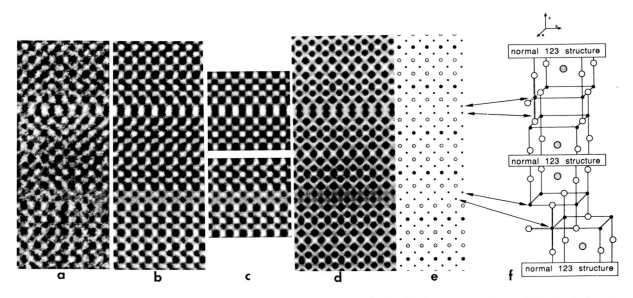

Figure 5.35 Atomic structural image of $YBa_2Cu_3O_{7-x}$ superconductor (123 phase) showing intercalation of Cu-O layers (arrowed): (a) digitized image; (b) averaged over 10 multislices; (c) rotation after averaging of (b); (d) multislice calculated image; (e) projection of model; (f) the 3D representation of the faulted structure.

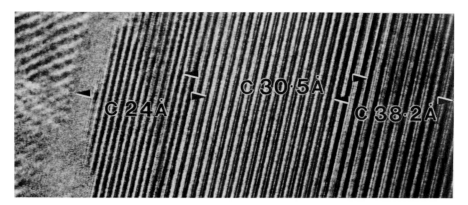

Figure 5.36 Polytypoid structure in orthorhombic Bi-Sr-Ca-Cu-O superconductor showing polytypoids of c-axis lattice parameter = 24, 31, and 38 Å for which T_c is 10, 80, and 110 K, respectively. The lowest-T_c phase is at the grain boundary.

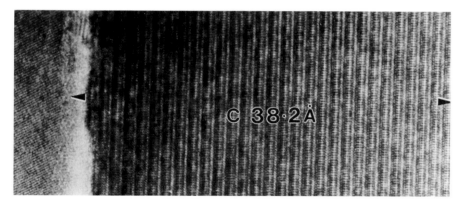

Figure 5.37 As Figure 5.36 for sample sintered with PbO—the low-T_c phase has been eliminated adjacent to the grain boundaries.

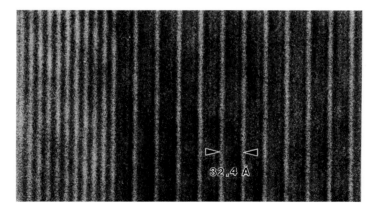

Figure 5.38 Polytypoids in Mg-Si-Al-O-N ceramic due to composition changes near the boundary.

171

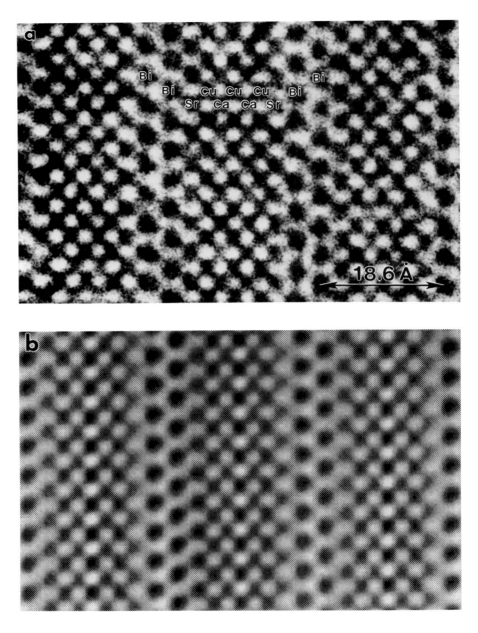

Figure 5.39 Atomic resolution image of the 38 Å polytypoid of Bi-Sr-Ca-Cu-O (2223 phase) with a c-axis lattice parameter of 37.2Å and $T_c = 110$ K.; (b) this image processed using SEMPER software: Notice the central Cu-O rows of high contrast.

oxygen, and nitrogen) (Figure 5.38), as known from the phase diagrams. Unfortunately, as yet, such diagrams are unknown for most ceramic superconductors. Finally, as an example of very-high-resolution imaging Figure 5.39(a) shows the 38 Å polytypoid of Bi-Sr-Ca-Cu-O with the cation columns resolved. Figure 5.39 shows the same image after processing using the SEMPER computer program. Additional high-resolution images of high-T_c superconductors are illustrated in Chapter 7.

ACKNOWLEDGMENTS

I sincerely thank the over 100 graduates and research colleagues with whom I have worked since coming to Berkeley in 1960, without whose talents and challenges this article would not have been possible. The selection of examples has been more of a problem of what to omit rather than what to include, and I trust any omissions will not produce offence.

This work is supported by the Director, Office of Energy Research, Office of Basic Energy Sciences, Materials Sciences Division of the U.S. Department of Energy under Contract No. DE-AC03-76SF00098. Structural ceramics research was supported by the National Science Foundation under Grant DMR-80-23461.

REFERENCES

Chescoe, D., and Goodhew, P. J. (1984). *The Operation of the Transmission Electron Microscope*. Oxford University Press, Oxford

Edington, J. W. (1974–77). *Practical Electron Microscopy in Materials Science (Monographs 1–5)*. Philips Electronic Instruments, Mahwah, N.J.

Hirsch, P. B., Howie, A., Nicholson, R. B., Pashley, D. W., and Whelan, M. J. (1977). *Electron Microscopy of Thin Crystals*. Krieger, New York.

Marton, L. (1968). *Early History of the Electron Microscope*. San Francisco Press, San Francisco.

Thomas, G., and Goringe, M. J. (1979). *Transmission Electron Microscopy of Materials*. Wiley-Interscience, New York.

Williams, D. B. (1984). *Practical Analytical Electron Microscopy in Materials Science*. Philips Electronic Instruments, Mahwah, N.J.

6

Electron Diffraction Images

D. B. WILLIAMS AND K. S. VECCHIO

Several classical imaging processes, such as transmission light and electron microscopy, make use of a lens—the objective lens—to gather a fraction of the radiation scattered by the sample. The sample acts as the object of the lens, which, as shown in Figure 6.1, recreates an image of the sample in the image plane of the lens (CD). Subsequent lenses magnify this image and project it onto an appropriate imaging surface, such as the retina (for a light microscope), a fluorescent screen (in an electron microscope), or any photographic/recording medium. Between the objective lens and the image plane is the focal plane of the lens (AB). Within this plane the radiation from the sample is focused to positions that depend upon the initial angle of scatter, as also shown in Figure 6.1. Thus the intensity distribution of radiation in the focal plane of the lens constitutes a scatter pattern. In this chapter, we will illustrate the use of such scatter-pattern images taken in the focal plane of the transmission electron microscope (TEM). Although these images are not "images" of the sample in the strict sense of the word, they are important indicators of the nature of the sample. These scatter patterns are also termed *diffraction patterns* because the phenomenon of diffraction (which means the radiation is scattered coherently, i.e., all the waves are in step) is very dependent on the crystalline nature and orientation of the sample. Thus, these diffraction patterns may be thought of as images that reflect the crystal nature of the sample. In the TEM, the area of the sample contributing to the diffraction pattern is determined by placing an aperture in the image plane of the objective lens, which creates a virtual aperture in the plane of the specimen. The resulting diffraction patterns are usually termed *selected area diffraction* (SAD) patterns, and typically come from areas of the sample greater than about one micrometer in diameter.

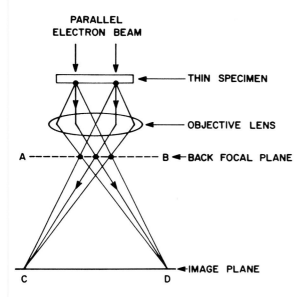

Figure 6.1 Schematic diagram of the creation of a diffraction pattern in the focal plane (AB) of the objective lens in the transmission electron microscope. Undeviated electrons from the illuminated area of a thin sample are focused on the optic axis, and scattered electrons are focused off the axis. The diffraction pattern contains all the information used to reconstitute the image of the sample in plane CD.

Because visible light is not coherently scattered by the structure of materials viewed in the reflection or transmission light microscope, the distribution of light in the focal plane is not of much interest to the optical microscopist. The light is either transmitted (forward scattered) on the optic axis of the microscope (so-called undeviated radiation) or incoherently scattered off the optic axis. Thus it is possible to insert an aperture in the focal plane to select either the undeviated or the scattered light, forming so-called bright-field or dark-field images, respectively (see Chapter 1). However, the scatter pattern itself is never observed in the light microscope.

In contrast, electrons with energies above a few thousand electron volts have wavelengths less than typical interatomic distances and therefore can be scattered coherently by the regular arrays of atoms within the sample. Thus electron diffraction patterns obtained in the TEM are most useful for obtaining information about atomic arrangements, and we will consider only these kinds of diffraction images. It is worth noting in passing that high-energy x-rays have similar properties to high-energy electrons and, indeed, x-ray diffraction and x-ray microscopy are independent disciplines on their own. Where possible, analogs between electron and x-ray diffraction will be described, and the interested reader is referred to other books on x-ray phenomena such as Cohen (1966) and Cullity (1978) for a more complete discussion of x-ray diffraction.

In the transmission electron microscope the diffraction pattern is often related to the conventional bright-field and dark-field images, and, indeed, we cannot fully understand one without reference to the other. As discussed in Chapters 5 and 7 in this book, the diffraction pattern is usually observed prior to deciding whether or not to record the image and is an integral part of the images discussed therein. Thus, this chapter is both complementary and strongly related to these other two chapters.

For those who are interested in the theory of the diffraction process, it is worthwhile summarizing, briefly, the physics of diffraction. However, from the point of view of appreciating the significance of the patterns and interpreting them in a qualitative manner, knowledge of the theory is not essential, and one may choose to skip the next section, where a few basic equations will be used.

THEORY OF ELECTRON DIFFRACTION

Considering crystalline materials only, electron diffraction occurs when the path difference between incident and diffracted electron waves is an integral number of wavelengths ($n \lambda$). For a crystal with unit cell lattice parameters a, b, and c, this condition is described by the Laue equations, which describe the path differences in three dimensions:

$$h \lambda = a (\cos \alpha_1 - \cos \alpha_2) \tag{6.1}$$

$$k \lambda = b (\cos \beta_1 - \cos \beta_2) \tag{6.2}$$

$$l \lambda = c (\cos \gamma_1 - \cos \gamma_2), \tag{6.3}$$

where h, k, and l are integers that correspond to the Miller indices for a specific diffracting plane. The cosines of the angles define the incident ($\alpha_1, \beta_1, \gamma_1$) and diffracted ($\alpha_2, \beta_2, \gamma_2$) beam directions with respect to the rows of scattering atoms with spacings of a, b, and c, respectively, as shown in one dimension in Figure 6.2(a) [in this diagram AB–CD is the path difference given by Equation (6.1)]. Equations (6.1)–(6.3) each describe a cone of coherently scattered electrons emanating from each row of scattering atoms. In three dimensions coherent scattering only occurs along the direction in which all three cones coincide. Thus, depending on the crystal structure, coherent scattering will only occur in specific directions. A more straightforward and mathematically equivalent description has been given by Bragg, who used the (physically incorrect) analogy of electron reflection [see Figure 6.2(b)] to deduce that coherent scatter from atomic planes of spacing d occurs when:

$$n \lambda = 2d \sin \theta, \tag{6.4}$$

where θ is the angle of incidence and reflection, and n is an integer. Thus the path difference between electrons reflected from different planes [distance ABC in Figure 6.2(b)] is again an integral number of wavelengths if diffraction is to occur. It is usual to regard higher-order diffracted beams ($n = 2,3, \ldots$) from a specific (hkl)

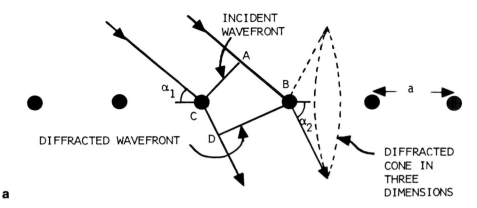

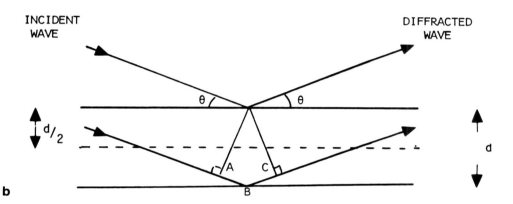

Figure 6.2 (a) The coherent scattering of electron waves by a row of atoms as described by the Laue equations. (b) The reflection of electron waves by parallel atomic planes of spacing d as described by the Bragg equation. Reflection of electrons from the dotted plane spacing $d/2$ would result in destructive interference and no diffracted intensity (assuming all planes are identical).

plane of spacing d, as equivalent to first-order diffraction from parallel planes of indices ($nh\ nk\ nl$) and spacing d/n.

The diffracted intensity in any particular direction is a function of the atomic species that is scattering, and of the atomic distribution in the crystal. These factors are known as the atomic scattering factor [$f(\theta)$] and the structure factor [$F(\theta)$] respectively. The value of $f(\theta)$ is given by:

$$f(\theta) = [(m_0 e^2)/2h^2]\ (\lambda/\sin\theta)^2\ (Z - f_x),\qquad(6.5)$$

where m_o is the rest mass of the electron, e is the electronic charge, h is Planck's constant, Z is the atomic number, and f_x is the atomic scattering factor for x-rays. Equation (6.5) is a combination of Rutherford scattering effects from the atomic nucleus and scattering from the electron cloud. From this equation it can be seen that diffracted intensity is strongly peaked in the forward direction (small θ), and that higher atomic number (Z) atoms scatter more strongly.

The amplitude scattered by a unit cell $F(\theta)$ can be calculated by summing the values of $f(\theta)$ for each of n atoms in the unit cell, taking into account the phase difference between each scattering point. Thus, the scattering from a particular (hkl) plane is given by:

$$F(\theta) = \sum_n f_n(\theta) \exp[-2\pi i(hx_n + ky_n + lz_n)] \qquad (6.6)$$

where x_n, y_n, z_n are the atomic coordinates of the nth atom, and the exponential term represents the phase difference. The actual intensity is given by $|F(\theta)|^2$. Clearly for different crystal structures different intensities will arise from the same (hkl) plane, and under specific geometrical conditions $F(\theta) = 0$ and the scattered intensity will be zero. For example, in Figure 6.2(b), the dotted line represents a plane of atoms with spacing $d/2$, from which scattered electrons would destructively interfere with electrons diffracted from the planes of spacing d, giving zero diffracted intensity. Such so-called systematic absences are characteristic of specific structures [e.g., when $h + k + l$ is odd, $F(\theta) = 0$ in body-centered-cubic structures]. Similarly, if more than one atom type is present (e.g., in alloys), then differences in $f(\theta)$ between different atoms can give rise to large changes in diffracted intensity. For example, some intensity may occur where systematic absences are expected if only one element were present in the lattice.

Indexing Diffraction Patterns

To determine which (hkl) plane is responsible for a specific diffracted beam, it is usual to employ the construction of a reciprocal lattice of the specimen, in which a set of parallel (hkl) atomic planes is represented by points located a distance $1/d_{hkl}$ from the origin. These points can be considered to represent the diffracted intensity for the following reason. If Bragg's law, Equation (6.4), is rewritten as:

$$\sin \theta = 1/(d_{hkl})/(2/\lambda) \qquad (6.7)$$

and if a sphere of diameter $2/\lambda$ (termed the Ewald sphere) is inscribed within the reciprocal lattice, intercepting the origin, then any points in the lattice that intersect the surface of the sphere will satisfy the Bragg equation. Hence the planes that are represented by these points will be diffracting strongly, as shown in two dimensions in Figure 6.3(a). Considering the relative dimensions of d_{hkl} and λ, it can be seen that for x rays, where $\lambda \sim 0.2$ nm and $2/\lambda \sim 10$ nm^{-1}, the chance of many diffraction maxima occurring from a crystal where $d^{-1}(\sim 3$ nm$^{-1})$ is very small. This explains why it is necessary to use white x radiation (wide range of λ) or to oscillate, rotate, or powder the specimen (thereby creating many variations of d and

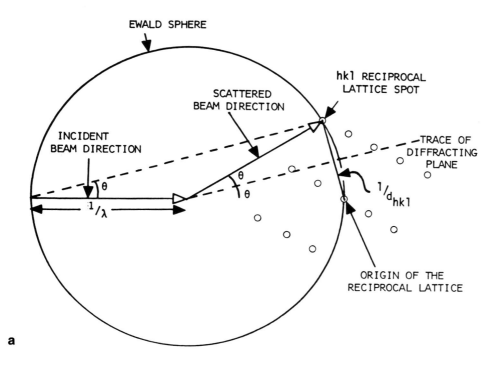

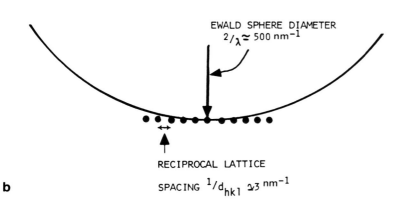

Figure 6.3 (a) The Ewald-sphere–reciprocal-lattice construction, showing how Bragg's law is only satisfied by atomic planes whose reciprocal lattice spots lie on the surface of the sphere. If the specimen or beam orientation is changed, as can be done routinely in the TEM, then different planes will diffract and the diffraction pattern will change. (b) Relative magnitudes of the Ewald sphere and reciprocal lattice in the TEM, showing that the surface of the sphere is almost planar with respect to the reciprocal lattice.

θ) in order to obtain a sufficient number of diffraction spots to analyze the structure. For 100 keV electrons, however, λ is 0.0037 nm and 2/λ is 540 nm^{-1}, and thus the surface of the Ewald sphere is almost planar in comparison with the array of reciprocal lattice spots [Figure 6.3(b)]. As such, an electron diffraction pattern from a thin sample will contain many diffraction maxima corresponding to the interception of the Ewald sphere with many points in the reciprocal lattice. Numerous examples of such patterns are shown in this chapter.

Rather than always indexing all spots in all diffraction patterns, which would be a tedious exercise, it is a common practice to orient the specimen such that a low-index zone of planes is diffracting, and then to compare the observed pattern with standard ones, which are available in most TEM textbooks (e.g., Edington, 1975; Hirsch et al., 1977).

The fact that it is possible to obtain diffraction from several planes in a zone at once is also due to the effect of the specimen shape on the diffracted intensity distribution. The diffraction spot is only a mathematical point if the specimen is infinite in all directions. For example, a TEM specimen is effectively infinite (~3 mm) in the plane of the specimen, and very thin (<0.5 μm) parallel to the electron beam. As a result, the diffracted intensity can be represented in reciprocal space as a rod stretched parallel to the electron beam rather than as a point. Therefore, over a range of angles, the Ewald sphere will still intercept the rod, and diffracted intensity will still exist. This is equivalent to saying that the Laue condition is relaxed in one dimension in the TEM owing to the specimen shape; that is, Equations (6.1) and (6.2) remain, but Equation (6.3) is eliminated. For this reason, accurate structural analysis of unknown specimens is very difficult in conventional TEM diffraction, and x-ray diffraction is the preferred approach. However, as we will discuss later in this chapter, the technique of convergent beam electron diffraction can be employed for very accurate structural analysis.

CONVENTIONAL DIFFRACTION

From the discussion of the theory of diffraction, it should be clear that the most important information in an electron diffraction pattern is the structural nature of the sample that is scattering the electrons. The structure of materials in the solid state may be grossly divided into amorphous and crystalline. The amorphous state may be most readily described as a "frozen liquid," in which the atoms occupy almost random positions. Many glasses have amorphous structures, as do some metals and alloys when cooled exceptionally quickly from the liquid state. Because an amorphous structure is approximately random, the process of diffraction (i.e., coherent scatter), which relies on the scattering centers (i.e., atoms in this case) being regularly positioned in space, cannot occur. Therefore, amorphous materials, strictly speaking, do not generate diffraction patterns, but only scatter patterns. However, the term *diffraction* is generally used even when referring to amorphous samples, and Figure 6.4 shows a "diffraction" pattern from a thin film of amorphous carbon.

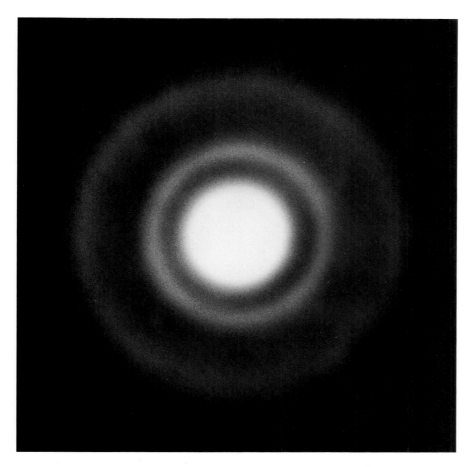

Figure 6.4 Scatter pattern from amorphous carbon film, showing an intense central region of undeviated electrons and two diffuse rings of elastically scattered electrons. Note the decrease in intensity at higher scattering angles.

The amorphous carbon was created simply by striking an arc between two carbon electrodes. The arc generates a sufficiently high temperature to vaporize the carbon, and the carbon atoms are then collected on a cold surface, where they condense to an amorphous solid. The diffraction pattern then consists of a bright central region containing all those electrons that passed straight through the thin film and a diffuse ring of intensity around the central maximum. The uniform nature of the ring means the electrons were scattered radially in a uniform manner, which would be expected from a random structure. However the diffuse scatter does show a clear maximum at some distance around the central spot. Remember that distance corresponds to angle of scatter (θ) that is related by Bragg's Law to some interatomic spacing (d). Thus, this implies that the atomic spacing in amorphous carbon is not

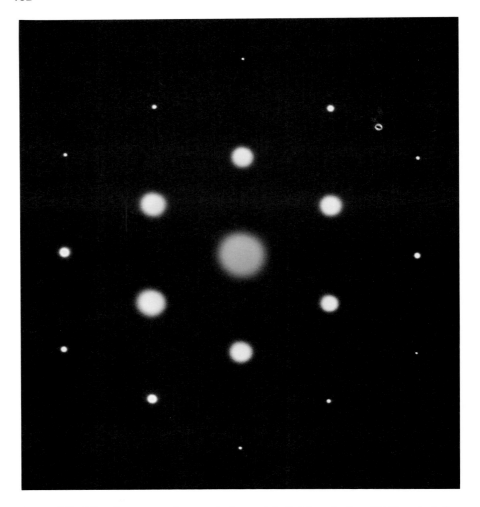

Figure 6.5 Diffraction pattern from a single crystal of silicon in the <111> orientation. Several diffraction maxima are observed, and at least two orders of diffraction maxima are visible. The slightly different intensities in the diffraction maxima are an indication that the sample is in fact slightly misoriented from <111>.

truly random. In fact, most atoms in an amorphous structure are spaced by a distance called the *first-nearest-neighbor distance,* and this can be deduced from such a pattern as Figure 6.4. The concept of "first nearest neighbor" merely reflects the fact that, if you throw a bunch of atoms (or any round objects, for that matter) together randomly, there will be a minimum distance of approach between any two, and this distance will be the most common distance separating adjacent atoms.

If the atoms in a material are regularly spaced in all three dimensions, then the material is crystalline. Under these circumstances the electrons are only scattered coherently in certain directions, giving rise to a true diffraction pattern. Figure 6.5 is a diffraction pattern from a thin single crystal of silicon, oriented with the (111) planes normal to the electron beam. The pattern in this figure can be used to determine the crystal orientation and an approximate value of the spacing of the planes of atoms. Figure 6.5 was from a single crystal, which means that in the area of the sample illuminated by the electrons ($\sim 10^{-4}$ cm across) all the atoms and their planes were similarly oriented. In some materials, however, the single-crystal regions are much smaller than 10^{-4} cm, and so many randomly oriented crystals are illuminated by the electron beam. Under these circumstances many different single-crystal patterns superimpose themselves (much like placing a pin through the middle of Figure 6.5 and spinning it around), and the result is a "ring pattern," as shown in Figure 6.6. The sample that gave rise to Figure 6.6 was evaporated gold, which condenses onto a substrate as very small crystals ~ 10 nm in diameter, so many thousands of crystals contribute to the pattern. Such a ring pattern is the electron equivalent of the x-ray Debye–Scherrer powder pattern.

Figures 6.4–6.6 illustrate diffraction images from the major structural forms of materials: amorphous, crystalline, and microcrystalline. If the crystals become small enough (less than a few nanometers), it becomes difficult to say whether the structure is microcrystalline (or nanocrystalline) or amorphous, and the transition between the two is a matter of some terminological discussion in the scientific literature. Another type of structure has been discovered recently, termed *quasicrystalline,* in which the atoms are neither randomly nor regularly spaced, but positioned in approximately pentagonal arrays, which cannot be arranged to fill space with a periodic lattice. These structures also give characteristic diffraction patterns, which we will discuss later (see Figure 6.18).

We have already seen the relationship between the electron diffraction pattern and TEM images. If you look at the ray diagram in Figure 6.1, you can see that all the electrons that constitute the bright-field image pass through the central spot in the pattern, and all electrons that constitute a specific dark-field image must also pass through one of the off-axis (diffracted) spots. Therefore, in a pattern such as Figure 6.5, all the various images must be in effect "stored" in each spot. To show that this is so, all we need to do is defocus the imaging lenses of the microscope slightly so that, instead of focusing on the back focal plane AB in Figure 6.1, they focus on a plane above or below AB. Then, instead of seeing a focused point on the microscope viewing screen, we see an expanded point or disc. Within each disc is a bright-field or dark-field image of the sample, as shown in Figure 6.7, which is called a *multiple dark-field image.* As you can see, areas such as the lamellar particles that appear dark in the intense central disc appear bright in the discs on either side, illustrating again the complementary nature of bright and dark-field images seen in Chapter 5 (e.g. Figure 5.14) and also Chapter 1 (e.g., Figures 1.11, 1.12, 1.14, and 1.16). Such multiple dark-field images serve to show directly the relationship between directions in the image and crystallographic directions in the sample. In this particular example, the row of particles is aligned almost parallel to the [001] direction, a major axis in the aluminum (cubic) crystal system.

Figure 6.6 Ring diffraction pattern from polycrystalline gold indexed to show which atomic planes (*hkl*) are contributing to each ring. Compare the sharply defined rings in this pattern with the diffuse rings in Figure 6.4 from the amorphous sample.

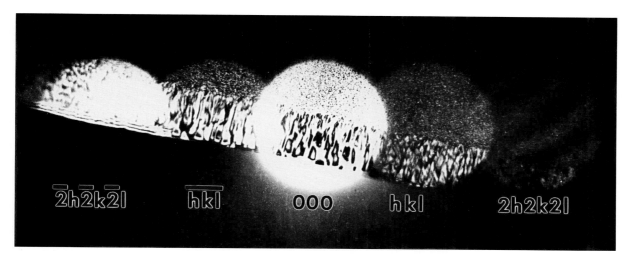

Figure 6.7 Multiple dark-field image of lamellar precipitates formed at a grain boundary in Al–10.7 at. % Li. The information in the 000 disc is a bright-field image, and dark-field images of the precipitates and matrix are visible in the *hkl* maxima.

Other Information in SAD Patterns

From Figures 6.4–6.6, it is obvious that a diffraction pattern can distinguish the various structures of materials in the grossest sense. Furthermore, by indexing the pattern [i.e., determining which rows of atoms (crystal planes) scatter electrons into which spot] it is possible to deduce the crystal orientation, and relate it to the TEM image. Because the TEM is impossible to calibrate very accurately (due to variations in lens currents, thermal effects, and magnetic hysteresis), we do not use these patterns to give accurate measurements of the crystal dimensions. However, accurate crystal measurements can be achieved by nonconventional or so-called convergent beam electron diffraction, which is discussed later in this chapter, or by x-ray diffraction methods, which do not use lenses.

However, the relatively simple information about the sample that we have just described is not all that is available in the diffraction patterns. If you look again at Figure 6.5, which is an SAD pattern from a thin piece of silicon (probably less than ~50 nm thick) in the <111> orientation, we see a set of intensity maxima showing six-fold symmetry, which is the characteristic pattern for a cubic material in the (111) zone axis orientation. If we take a pattern in the same orientation, but from a much thicker sample (Figure 6.8), the spot pattern is replaced by an array of lines called *Kikuchi lines,* which are discussed briefly in Chapter 5 (see Figure 5.29). These arise because the thick sample scatters electrons by inelastic processes. These inelastically scattered electrons that have lost energy are still diffracted by the atom planes, but in a different manner. As you can see in Figure 6.8, there is no central

Figure 6.8 Kikuchi pattern from a thick sample of silicon. The pairs of parallel dark–bright Kikuchi lines move with sample tilt, and their six-fold symmetrical distribution in this pattern is a good indicator that the sample is very close to the <111> orientation (compare with Figure 6.5).

bright spot, which means the sample is so thick that no electrons traveled straight through; they were all scattered one way or another.

These Kikuchi lines have the useful property that they shift if the crystal is tilted. The eye is acutely sensitive to changes in symmetry, and so we can use Kikuchi patterns to measure very accurately the orientation of the crystal (to much better than 1°). The pattern in Figure 6.8 is very symmetrical, indicating that the silicon is very close to the <111> crystal orientation. If it were more than a fraction of a degree off this orientation, the Kikuchi pattern would become asymmetric, while the spot pattern in Figure 6.5 would remain essentially unchanged, at least until the

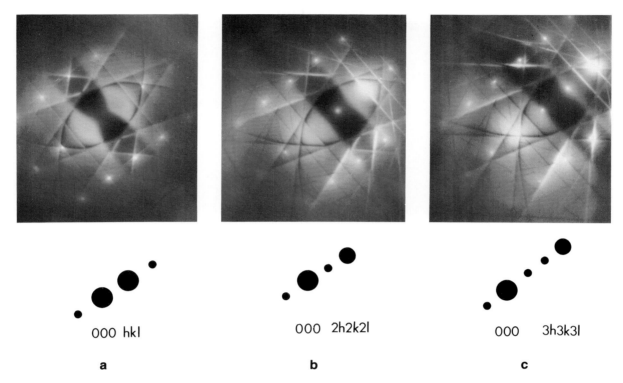

000 hkl   000 2h2k2l   000 3h3k3l

a   b   c

Figure 6.9 Sequence of Kikuchi spot patterns showing that, as the sample is tilted slightly, the lines move, but the spots remain stationary. The spots only change in intensity, reaching a maximum when the bright Kikuchi lines intersect them, which occurs at the exact Bragg condition for (a) *hkl*, (b) 2*h*2*k*2*l*, and (c) 3*h*3*k*3*l*, respectively.

misorientation was much greater than 1°, at which point the intensities of the diffracted spots would become slightly unequal.

Thus a spot pattern is an inaccurate measure of crystal orientation, while a Kikuchi pattern is a much better measure. This point is further demonstrated in the series of patterns in Figure 6.9, which is taken from a sample of pure aluminum, well away from a major crystal axis. In Figure 6.9(*a*), both diffraction spots and Kikuchi lines are present, and the inner pair of the parallel Kikuchi lines goes through two of the diffraction spots. Under these circumstances, it can be shown that the planes diffracting into the spot labeled *hkl* are at the exact Bragg condition (and so the spot is brighter). By tilting the sample slightly, the Kikuchi lines move, but the spots remain in the same position, changing only in intensity. In Figure 6.9(*b*), we have tilted the sample so the (2*h* 2*k* 2*l*) planes are at the exact Bragg condition, and in Figure 6.9(*c*) the (3*h* 3*k* 3*l*) spots are at the exact Bragg condition (within the accuracy of Kikuchi line measurements). It can be seen from Figures 6.9(*b*) and (*c*) that, although the specimen orientation has changed compared with

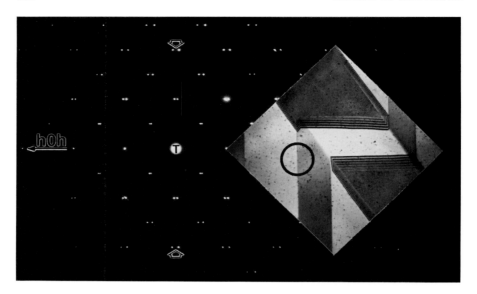

Figure 6.10 Composite SAD pattern from a coherent mirror twin boundary in $CaZrO_3$. The black circle in the micrograph indicates the position of the selector aperture. There is an unsplit $\{h0h\}$ systematic row of spots perpendicular to the interface. Spot splitting (arrowed) parallel to the unsplit row is characteristic of the mirror-twin nature of the interface. Courtesy V. P. Dravid.

Figure 6.9(*a*), the (*hkl*) reflection remains in the same position and has only changed in intensity. Using Kikuchi lines in this way, we can carefully orient any crystal in the microscope such that only one chosen set of atomic planes (*hkl*) is exactly at the Bragg condition. Under these circumstances we can get TEM images that are very easy to understand in terms of the crystallographic (diffraction) contrast that they contain (see Chapter 5).

Sometimes in diffraction patterns we see that, instead of discrete spots, we get spots that are split into two, as shown in Figure 6.10. This splitting implies that we have more than one crystal contributing to the pattern, and that both crystals are closely related in terms of their orientations. Figure 6.10 was obtained from a sample of $CaZrO_3$ containing coherent twins in which two different, but very specific, orientations co-exist. When the diffraction pattern is taken from both twins simultaneously, spot splitting occurs in the direction normal to the twin plane. No splitting occurs along the single row of reflections normal to the twin plane that also passes through the origin of the pattern. (This effect is a signature of coherent mirror twins.) The angle between these twin-related orientations can be determined by measuring the angle between a split pair of reflections and the transmitted beam. The occurrence of these coherent twin boundaries in $CaZrO_3$ has been recently shown to be a direct consequence of a displacive phase transition from the higher-symmetry space group ($Pm\bar{3}m$) to a lower-symmetry space group ($Pcmn$) (Dravid et al., 1989). The symmetry elements lost during this phase transition manifest themselves as mirror twins along {101} planes. We will discuss these space-group symmetry concepts later in the chapter.

Extra spots can also arise in diffraction patterns for a number of other reasons. Figure 6.11 shows an enlargement of the fundamental diffraction spots from an artificial "superlattice." Such a superlattice consists of very thin layers of two slightly different materials deposited alternately on each other, giving a very regular set of parallel crystal layers. This particular sample consists of alternate layers of AlGaP and GaP. These individual layers, although consisting of perhaps ten or so crystal planes, are themselves small enough to scatter electrons. This coarse artificial superlattice gives rise to extra spots around each fundamental reflection, termed *satellite spots*. The insets show highly magnified images of the 222 fundamental reflection in a [110] diffraction pattern. The fundamental spot is surrounded by a pair of less intense superlattice spots, which are aligned in the direction normal to the (200) planes. Similar effects are observed in materials that contain regular composition fluctuations such as occur during spinodal decomposition, or in crystals in which regular arrays of defects such as dislocations or antiphase boundaries occur.

As we have already discussed for SAD, the diffraction spot is a mathematical point only if the sample shape is infinite and homogeneous in all three dimensions. Of course a TEM sample is not infinite and is, in fact, deliberately thinned in one dimension so the electron beam can penetrate it. This shape effect relaxes the Bragg equation, permitting many planes to diffract at once, and accounts for the presence of many spots in a diffraction pattern. In a similar manner, if there are diffracting regions in the sample with a very thin planar shape, then they too will not diffract uniformly but rather will produce a streak of scattered electrons perpendicular to the face of the plate. Such "shape-factor" effects are visible in Figure 6.12, which was obtained from a complex aluminum alloy containing many different phases (hence the many faint spots in addition to the strong fundamental reflections from the aluminum matrix). In the inset image a set of thin, planar precipitates is visible, and examination of the diffraction pattern confirms that the two sets of streaks are normal to the faces of the two sets of plates. In fact there is a third set of plates in the plane of the sample that also generates streaks. But these streaks are perpendicular to the diffraction pattern and so are intercepted by the photographic plate, which records a cross section of the streak as one set of the faint spots in the pattern.

In summary, conventional electron diffraction patterns give us a gross sense of the sample structure (amorphous or crystalline), and in crystalline samples we can get approximate crystal orientations from spot patterns, more accurate orientations from Kikuchi patterns, and information about other regular crystal defects that can diffract in addition to the basic atomic lattice and shape effects. Remember that all of this information comes from an area not much less than 1 μm^2 in size due to limitations of the SAD technique. If the sample is badly deformed, we lose much of the information about the sample. Similarly, if there are discrete crystal regions less than $\sim$1 μm in size, we cannot get a pattern from one of these crystals alone. Under these circumstances we have to use convergent beam electron diffraction, which makes use of a small converging beam of electrons to select a region of the sample anywhere from $\sim$1 μm down to $\sim$ 1 nm in diameter. This technique, which we will

Figure 6.11 Artificial superlattice consisting of alternate Ga$_x$Al$_{-x}$P and GaP layers grown on a buffer of (100) GaP. The 222 diffraction maximum from the sample is shown in the inset, and either side of the fundamental reflection are two less intense superlattice spots aligned in the <200> direction. (A second-order superlattice reflection is also faintly visible on one side of the fundamental reflection). These superlattice spots arise because of the regular nature of the superlattice. Measurement of the spacing between the fundamental and superlattice spots gives a superlattice wavelength of ~17 nm, which agrees reasonably with the measured spacing in the image (~19 nm). Courtesy C. Hills.

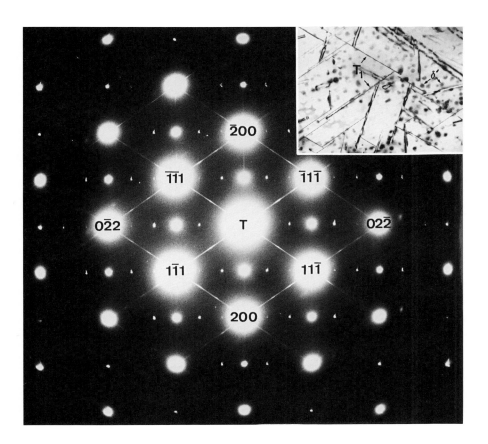

Figure 6.12 Diffraction pattern from an Al-Li-Cu alloy, showing shape effects. The intense fundamental spots come from the aluminum matrix, normal to <011>. The area from which the pattern was obtained is shown in the inset. Small precipitates of Al_3Li are visible in the background and plate-like precipitates of Al_2CuLi can be clearly seen. These precipitates give rise to the faint diffraction spots and the streaks normal to the plane of the Al_2CuLi plates.

now discuss, is a most exciting development in electron diffraction, giving us patterns from more localized regions of the sample than SAD, a wealth of new information about the crystal dimensions and symmetry, as well as some particularly striking images.

CONVERGENT BEAM DIFFRACTION

The technique of SAD, while giving us useful information about the sample, has two severe limitations. First, as we have just mentioned, the smallest area from which an SADP can be obtained is ~0.5 μm, which is large compared with the dimensions of many crystalline features of interest to materials scientists. Second, an SAD pattern contains only rather imprecise two-dimensional crystallographic information. The technique of convergent beam electron diffraction (CBED) overcomes both of these limitations. In this section we will describe the concepts and terminology of CBED and illustrate its application to many materials. As with many sophisticated analytical techniques, CBED uses many obscure definitions and acronyms, which we will attempt to clarify.

CBED can provide three-dimensional structure information with a spatial resolution on the order of 5 nm or less. In order to achieve the very high spatial resolution of CBED, a small convergent probe must be used, and such probes have only recently been obtainable with the advent of the probe-forming analytical electron microscopes (AEMs). In SAD, a parallel beam of illumination at the specimen plane is employed, and discrete diffraction spots result in the back focal plane of the objective lens (see Figure 6.13). In CBED, a convergent beam is focused on the specimen, and diffraction discs are formed in the back focal plane (again, see Figure 6.13). Even more important than the high spatial resolution achievable with CBED is the fact that these diffraction discs contain an intensity distribution that, when properly interpreted, can provide three-dimensional crystallographic information.

These two points (i.e., high spatial resolution and three-dimensional crystallographic information) are illustrated in Figure 6.14, which is a CBED pattern from pure copper, obtained with an electron probe size of ~50 nm. Details of the symmetry in such a pattern (such as the obvious mirror planes and rotational symmetry) will be discussed later in this section, but it is sufficient to note at this stage the incredible amount of detail in terms of dark and bright lines, rings of spots, etc., that is visible in such patterns. The various terms that are used to describe the features in the CBED pattern are also noted on Figure 6.14.

The central region of Figure 6.14 is termed the *zero-order Laue zone* or ZOLZ. The ZOLZ pattern is essentially equivalent to an SAD pattern, except that the diffracted beams appear as discs in the pattern rather than spots. The individual discs overlap slightly in the central region of the ZOLZ and are more clearly visible in the inset. Around the ZOLZ region, at higher angles of scatter (remember that distance in a diffraction pattern is related to the angle of scatter), the intensity drops

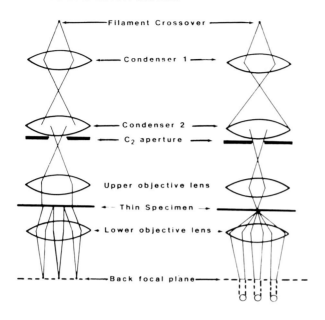

Filament Crossover

Condenser 1

Condenser 2

C_2 aperture

Upper objective lens

Thin Specimen

Lower objective lens

Back focal plane

**Selected Area
Electron Diffraction**

**Convergent Beam
Electron Diffraction**

Figure 6.13 Schematic diagram showing the difference between the electron ray paths for conventional SAD patterns and CBED patterns. The angular disc size is limited by the second condenser aperture and the angle of the nearest Bragg reflections. One advantage of the CBED technique is that the specimen area required to form the diffraction pattern is of the same order of magnitude as the probe size.

off until a ring of bright lines is observed. This ring is the first of a series of possible *higher-order Laue zones,* or HOLZ, rings. This ring arises from high-angle scatter from crystal planes that are not parallel to the electron beam. If you read the section on the theory of diffraction, then the rings can also be explained in terms of the interception of the Ewald sphere with the reciprocal lattice. The first higher layer to intercept the sphere is called the *first-order Laue zone* (FOLZ). The ring radius is dependent upon the atomic stacking in the crystal, the electron wavelength, and the pattern magnification (i.e., camera length of the microscope). Figure 6.15 shows a schematic representation of the distribution of reflections from the ZOLZ and HOLZ layers in a CBED pattern. Additional higher-order Laue layers can sometimes be observed. However, these reflections are generally less intense due to weak electron scattering that occurs at high scattering angles. Measurement of the HOLZ ring radius can be used to deduce the third dimension of the crystal parallel to the beam direction. Raghavan et al. (1984) have established analytical expressions for calculating the spacing between atomic planes parallel to the beam. These ex-

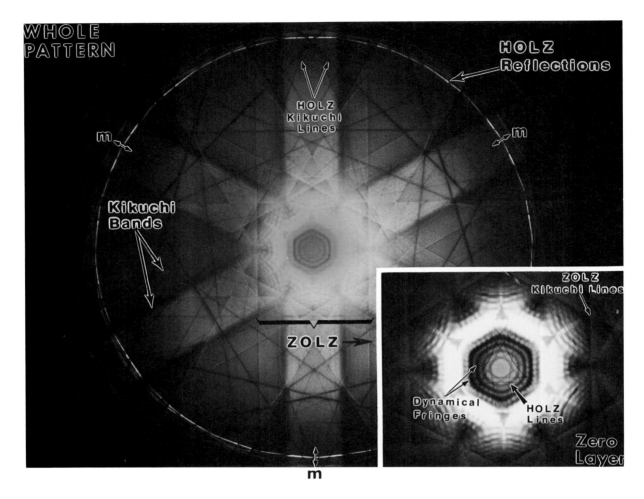

Figure 6.14 CBED pattern from pure copper aligned along <111>, illustrating the information available in CBED patterns.

pressions can be used to determine lattice constants, lattice type (e.g., cubic, tetragonal, hexagonal, etc.), and also the type of lattice centering (e.g., primitive, body centered, face centered, or base centered). In addition, Raghavan et al. point out that measuring the diameters of HOLZ rings can provide a reliable means of verifying the orientation of the CBED pattern. Thus another unique aspect of CBED patterns is that from a single two-dimensional pattern it is possible to obtain three-dimensional information about the sample.

Within the CBED pattern are a number of sharp and diffuse lines. The bright bands of intensity across the pattern are Kikuchi lines formed by inelastic scatter from the ZOLZ planes, and the sharp dark and bright lines are HOLZ lines. These

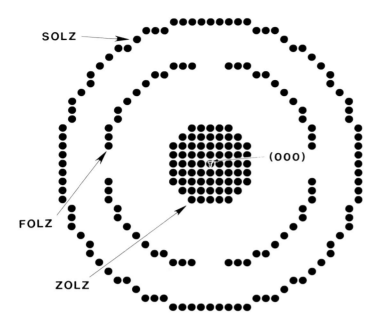

Figure 6.15 Schematic diagram of the distribution of diffraction maxima in a CBED pattern.

HOLZ lines are the elastic analogs of Kikuchi lines. For each bright line constituting the ring of HOLZ reflections, there is a corresponding parallel dark line in the central disc, shown magnified in the inset in Figure 6.14. These HOLZ lines can be very useful, as we shall see.

Figure 6.16 shows a CBED pattern taken down the <111> crystal axis of a thin sample of pure silicon. A straightforward comparison of this figure with the silicon <111> SAD pattern (Figure 6.5) shows the enormous increase in information between SAD and CBED patterns. Again the basic similarity between the spatial distribution of diffraction discs in the ZOLZ in Figure 6.16, and the distribution of spots in the SAD pattern in Figure 6.5 is obvious. There are six diffraction maxima symmetrically placed around the central disc. This six-fold symmetry is characteristic of all <111> SAD patterns from cubic materials and reflects the six-fold (hexad) symmetry that you would see in a two-dimensional array of atoms in the {111} close-packed plane. In three dimensions, however, the true symmetry of the crystal is three-fold rotation (triad) symmetry. This symmetry is visible in the central disc of the CBED pattern in Figure 6.16, which is also shown magnified in Figure 6.17. The criss-cross pattern of dark HOLZ lines shows three-fold (triad) symmetry, and careful examination of the other dark (HOLZ Kikuchi) lines between the discs in Figure 6.16 also reveals the correct, three-dimensional, three-fold symmetry. Compare this symmetry with the six-fold symmetry in the Kikuchi pattern in Figure 6.8, and again you can see that all SAD patterns contain essentially two-dimensional information, and CBED patterns contain three-dimensional information.

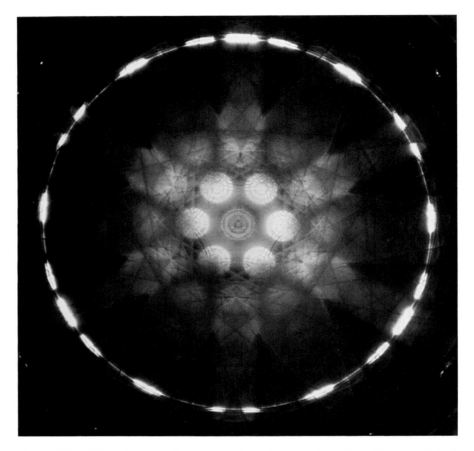

Figure 6.16 <111> CBED pattern from silicon. Compare the information with that obtained in the SAD pattern (Figure 6.5).

Even without knowing anything about why this difference between SAD and CBED patterns exist, it is possible to use such information in a qualitative but very constructive manner. Figure 6.18 shows a comparison between an SAD and a CBED pattern from the sample shown in the inset. The sample shows a five-fold star external shape (so-called icosahedral) symmetry and is an example of a possible quasicrystalline material. This sample is an intermetallic with the approximate composition Al_6Li_3Cu. Such five-fold symmetric phases were only recently discovered (Schechtman and Blech, 1985) and caused considerable controversy since classical crystallography teaches that pentagons and icosahedra cannot fill two- and three-dimensional space, respectively, and therefore such shapes are forbidden. The explanation for the five-fold symmetry requires a quasicrystalline structure to be postulated, involving pentagonal and other shapes (known as Penrose tiling), which together fill space, with a quasiperiodic spacing of atoms. Confirmation of this

Figure 6.17 Expanded central ZOLZ region of the CBED pattern in Figure 6.16, showing the three-fold symmetry of the HOLZ line array in the central disc. Compare this symmetry with the six-fold symmetry visible in the silicon <111> Kikuchi pattern in Figure 6.8, and in the ZOLZ discs in Figure 6.16 and this figure.

apparent symmetry is shown in the SAD pattern next to the inset. Around the central spot (T) the diffraction maxima are arranged in circles containing ten regularly spaced spots (i.e., five-fold symmetry plus a mirror plane).

However, in a CBED pattern taken from the same sample, which is also shown in Figure 6.18, the Kikuchi bands are not observed to show regular spacing around the pattern. Each Kikuchi band is composed of two sub-bands (arrowed), and the intensity profile is different across each pair. Also, the HOLZ discs are not regular in their intensity distribution. The conclusion from such a pattern is that *no* symmetry exists because the rotation of the pattern around the center does not result in an identical pattern at any other position. An explanation for this phenomenon (Vecchio and Williams, 1988) is that the gross five-fold symmetry present in the microstructure is created by multiple twinning of tiny (~30 nm) crystals. A similar

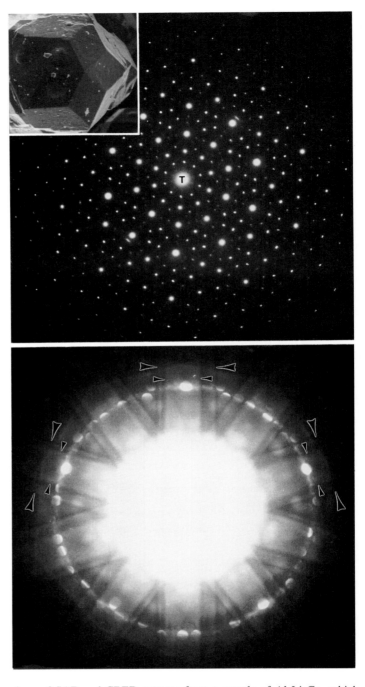

Figure 6.18 Comparison of SAD and CBED patterns from a sample of Al_6Li_3Cu, which shows external icosahedral symmetry (see inset SEM image). The SAD pattern exhibits a five-fold symmetric array of diffraction maxima consistent with the external symmetry, but the CBED pattern shows that pairs of Kikuchi lines are asymmetric (see arrows), which indicates a breakdown of any symmetry in this orientation.

explanation for quasicrystals was advanced by Pauling (1986) when the controversy first erupted, and this evidence supports his approach. Nevertheless, the debate continues.

Symmetry Determination From CBED Patterns

In this section we will describe the steps necessary to obtain full crystal symmetry information about the sample. This kind of detail will mainly be of interest to the practicing electron microscopist, and so you may wish to move on to the last section of the chapter at this stage.

The development of CBED for symmetry determination in crystals has evolved from the early work of Friedel (1913) and Von Laue (1948) on the kinematical theory of x-ray diffraction. One of the fundamental aspects of Friedel's work was the development of an understanding of diffracted beam intensities in x-ray diffraction and is known as Friedel's Law. This law states that the intensity of some reflection *hkl* is equivalent to the intensity of its paired reflection $\overline{hkl}$. As such, Friedel's Law acts automatically to introduce a center of symmetry into most x-ray diffraction patterns from single crystals. For example, the presence of a mirror plane in a crystal, parallel to the crystal axes *a* and *b*, causes the intensity of all reflections of type *hkl* to be equivalent to the intensity of the corresponding $\overline{hkl}$ reflection. Therefore, the presence of a mirror plane cannot be readily distinguished from a two-fold rotational axis parallel to the mirror plane, under kinematical diffraction conditions. Similarly, the presence of a four-fold rotation axis in a crystal, parallel to the *c* axis, results in $I_{hkl} = I_{\bar{h}kl} = I_{\bar{h}\bar{k}l} = I_{h\bar{k}l}$, where $I_{(hkl)}$ denotes the diffracted intensity of a reflection of type *hkl*. This diffraction pattern cannot be distinguished from one containing two independent mirror planes perpendicular to the rotation axis. The fact that Friedel's Law holds in x-ray diffraction severely limits the use of this technique for space-group determination, since crystals that do not possess true centers of symmetry (noncentrosymmetric crystals) appear in x-ray diffraction to possess a center of symmetry, and thus cannot be readily distinguished from those crystals that are centrosymmetric. As such, the 32 crystal point groups are reduced to the 11 Laue groups in x-ray diffraction, which severely hampers point-group and space-group analysis.

One of the most important revelations that has come out of the dynamical theory of electron diffraction is that Friedel's Law no longer holds for noncentrosymmetric crystals (Goodman and Lehmpfuhl, 1968). Simply stated, the breakdown of Friedel's Law occurs because the mass associated with electrons makes electron radiation more sensitive to changes in the phase of scattering centers within crystals (i.e., differences in structure factors). This sensitivity of electron diffraction to different symmetry elements is particularly useful in convergent beam electron diffraction because the intensity distributions within individual reflections can be readily compared in CBED patterns. Therefore, one of the main advantages of CBED is its ability to distinguish between centrosymmetric and noncentrosymmetric crystals. That is to say, the 32 crystal point groups are not reduced to the 11 Laue groups, as occurs in x-ray diffraction. Another important concept from Goodman and Lehmpfuhl's work regards the projected structure of the crystal that can be ascertained from the projection diffraction symmetry in the zero-layer pattern itself. The term *projection diffraction* refers to the transmitted beam plus the diffracted beams in the zero layer, ignoring contributions from upper Laue layers. Projection diffraction symmetry is a two-dimensional diffraction phenomenon and refers to the symmetry of the diffraction discs, their relative positions in the two-dimensional pattern of the zero layer, as well as the diffuse intensity within these discs. The diffuse contrast within these discs arises from the dynamical interaction within the zero layer of the crystal and is referred to as zero-order information. This zero-order information is sensitive to crystal thickness and changes dramatically with changing thickness of the specimen (e.g., note the circular dynamical fringes in the transmitted disc shown in Figure 6.17). This sensitivity of the diffuse intensity to specimen thickness can be used to measure foil thickness with a reasonable degree of accuracy (Kelly et al., 1975; Allen, 1981). The symmetry of the projection diffraction corresponds to the projected two-dimensional symmetry of the crystal along the particular zone axis of interest. The projection diffraction symmetry at any orientation must belong to one of the ten two-dimensional diffraction point groups.

Goodman and Lehmpfuhl pointed out that the zero-layer pattern must contain a center of symmetry even when the crystal itself does not. As such, the zero-layer pattern provides no true structural information. However, the value of the zero layer lies in its ability to reveal the higher-order Laue zone effects, which may have the result of destroying symmetries in a diffraction pattern. The breakdown of Friedel's Law is therefore most evident in asymmetries introduced into the diffraction pattern as a whole. These asymmetries are generally observed in the HOLZ lines or HOLZ reflections in the CBED pattern. Thus, the whole pattern, particularly HOLZ effects, provides the means by which the noncentrosymmetric nature of certain crystals can be revealed. When the upper-layer effects are combined with the centrosymmetric zero-layer pattern, the complete structural information can be deduced.

Whole-pattern symmetry refers to the relative positions of the HOLZ reflections, as well as the symmetry of the HOLZ Kikuchi lines observed in the pattern at low camera length (see Figure 6.14). The whole-pattern symmetry reveals the true crystal symmetry and must correspond to one of the 31 diffraction groups (see the next section). The term *bright-field symmetry* refers to the symmetry of the transmitted disc only. This can either be two-dimensional diffuse intensity within the disc or the three-dimensional symmetry of the HOLZ lines present within the transmitted disc. Both of these symmetries are deduced separately by ignoring the contributions of one form of intensity while examining the other.

Goodman (1975) derived symmetry rules for relating the CBED pattern symmetries to the three-dimensional symmetry of the crystal itself. Several definitions or explanations of symmetries were given by Goodman that are particularly important. The first refers to the symmetry of the intensity distribution within individual diffraction discs of the CBED pattern. Each disc contains a two-dimensional intensity distribution of the crystal, commonly called the *rocking curve* of the crystal, in which the angles of the incident electrons are transformed into *x-y* coordinates within the disc. These discs are then distributed in the same geometrical relationship as the reciprocal lattice points of the crystal. In describing the pattern symmetry as a whole, we must consider both the geometry of the intensity within each disc, as well as the relative positions of the discs themselves. Since the transmitted disc and its center are unique (i.e., the pattern has an origin at the center of the transmitted disc), it might be supposed that we only need to consider the point-group symmetry of the diffracted discs. While this is true for the whole-pattern symmetry, it is useful to consider the symmetries within each disc, and this includes geometrical relationships between a given pair of reflections, *hkl* and $\overline{hkl}$ (i.e., $\pm G$ reflections). The term *G reflection* refers to the situation in which the image of the second condenser aperture coincides with the exact Bragg position of a particular *hkl* reflection. In this situation, the symmetry of individual Bragg reflections can be identified, and the relation between $\pm G$ reflections compared. Investigating the intensity distribution at particular Bragg reflections allows certain symmetry elements to be identified. For example: a two-fold screw axis produces a mirror along a central line of an *hkl* reflection to which it is perpendicular, and a central mirror or glide plane produces a center of symmetry at the central point of each *hkl* reflection. Having discussed the principles behind CBED symmetry determination, we will now describe the methods to carry it out.

Point-Group Determination

One of the most important contributions to the use of CBED for symmetry determination is the work of Buxton et al. (1976). Buxton et al. used both group theory and graphical constructions to deduce the relationships between CBED pattern symmetries and crystal point groups, and they established tables by which diffraction groups could be related to the specimen point groups themselves. These tables (shown as Tables 2 and 3 in Buxton et al. 1976) provide the means by which we can interpret the symmetries in CBED patterns to deduce crystal point groups.

The first analysis to be made from the CBED pattern is to determine the two-dimensional symmetry of the ZOLZ. This is done by determining the symmetry of the zero-order reflections with respect to each other and with respect to the center of the pattern, as well as the symmetry of the bright-field disc itself. The symmetry of the diffuse intensity and the HOLZ lines within these reflections must also be considered. For a given pattern only certain symmetries can be observed in either the bright-field disc or the whole pattern. A pattern may contain symmetries: 1, *m*, 2, 2*mm*, 3, 3*m*, 4, 4*mm*, 6, or 6*mm*. These symbols refer to the observable symmetry in the whole pattern or the bright-field disc, and four examples of different pattern symmetries are shown schematically in Figure 6.19. In this figure, symmetry "2" refers to a two-fold rotation axis; that means the pattern has symmetry when rotated through 180°. Symmetry "2*mm*" refers to a two-fold rotation symmetry with two independent mirror planes. Symmetry "3*m*" indicates three-fold rotation symmetry with one mirror plane; that is to say the pattern has rotational symmetry every 120° with one mirror plane present at each rotational position. Symmetry "4*mm*" indicates a four-fold rotational symmetry with two independent mirror

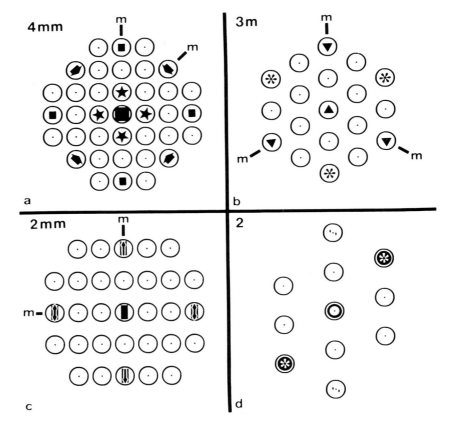

Figure 6.19 Schematic diagram illustrating four examples of different symmetries that can occur in CBED patterns.

planes. In any given pattern only the ten symmetries listed above can be observed and, as an example, Figure 6.20 shows that 6*mm* symmetry exists for both the bright-field disc and the whole pattern.

Table 2 in Buxton et al. (1976) contains lists of the diffraction groups, projection diffraction groups, and the corresponding symmetries (including bright-field, whole-pattern, dark-field, and $\pm G$ symmetries). The second analysis to be conducted on the CBED pattern is to determine the two-dimensional symmetry of the whole pattern. By combining the information from the whole-pattern and the bright-field symmetry, the diffraction group can be identified using Buxton's Table 2.

The Steeds method (1979) for point-group determination uses several different zone axis patterns, and analysis is conducted on only the projection diffraction symmetry (two-dimensional ZOLZ symmetry), transmitted beam symmetry (including HOLZ lines), and the whole-pattern symmetry (which includes HOLZ reflections and HOLZ Kikuchi lines, i.e., three-dimensional symmetries). All of these symmetries can usually be observed in one properly exposed pattern (see for

m

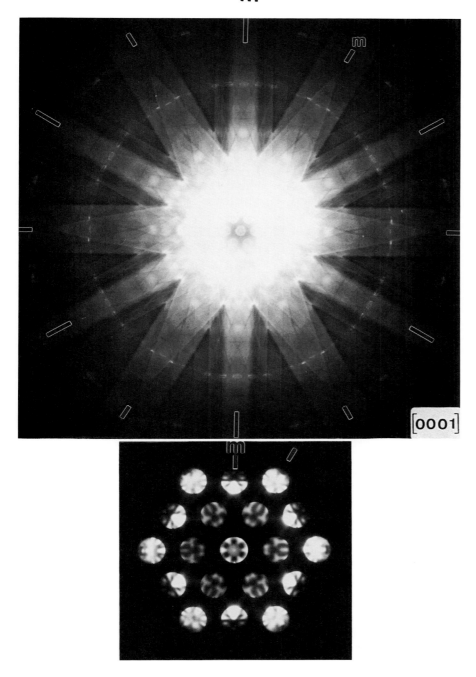

Figure 6.20 Low-magnification [0001] CBED pattern from the hexagonal phase Al$_2$CuLi, showing the 6mm diffraction group symmetry that is obvious down this axis. From this diffraction symmetry, only certain point-group symmetries are possible. The lower image is the central region of the upper pattern, correctly exposed, obtained with a small C2 aperture.

Figure 6.21 Large-angle (LA)CBED pattern from <111> silicon. In the LACBED technique, no C2 aperture is employed, and the field of view is limited by the selector aperture. In this figure the dark-field LACBED reflections are shown (exposed separately) around the wide angular view of the 000 transmitted disc.

204

example Figure 6.16). Because the symmetries at different orientations from a single crystal must be related, analysis of the symmetries present at several different orientations reduces the ambiguity of the analysis. As such, this technique provides a very simple and straightforward method of determining crystal point groups without the difficulty of interpreting dark-field and $\pm G$ symmetries.

The same type of analysis is conducted for each of the other highest-symmetry orientations in the crystal. When possible diffraction groups for each orientation are determined, the crystal point group can be deduced using Table 3 in Buxton et al. (1976). Table 3 contains the 31 diffraction groups and 32 crystal point groups, and the diffraction groups consistent with particular crystal point groups can be deduced. For each crystal orientation and diffraction group, one or more different crystal point groups are possible. The crystal point group that is consistent with all the diffraction groups observed is the true point group of the crystal.

Using a technique known as *large-angle CBED*, it is possible to obtain spectacular patterns showing the bright- and dark-field pattern symmetries over a very wide angle of convergence. The electron beam is overfocused so that cross-over occurs above the specimen plane and the specimen receives defocused illumination. Thus, a considerably larger specimen area is required compared to conventional CBED. The main advantage of the LACBED technique is that a very large angular view of any particular reflection can be obtained. However, separate exposures are still required to record all of the necessary information from a particular zone axis. An example of such a pattern is shown in Figure 6.21. To the experienced TEM operator these patterns are obviously related to bend centers in diffraction contrast images, and indeed they are formed in an analogous, but reciprocal, manner.

An alternative analysis procedure to determine point-group symmetry uses patterns known as simultaneous zone axis (ZA), $\pm G$, and BF patterns, an example of which is given in Figure 6.22. In this technique (Tanaka and Terauchi, 1986), condenser and objective lens focus conditions are satisfied at the specimen, and the specimen area is selected by the selector aperture. The second condenser aperture is then removed to provide maximum convergence of the electron beam, and the specimen height is raised so that the specimen receives defocused illumination. By careful adjustment of condenser and objective lens setting and specimen height, a diffraction pattern can be obtained in the back focal plane of the objective lens, in which the exact Bragg condition is satisfied for the center of the individual discs. The main advantage of this technique is that the zone axis (ZA) whole-pattern, bright-field (BF), and individual diffraction discs ($\pm G$) can be recorded simultaneously. The primary disadvantage is that fairly large defect-free specimen areas of uniform thickness are required. This method mainly aims to find three-dimensional symmetry elements from a single pattern.

Analysis of Lattice Centering

The second step in the CBED analysis for structural information is to determine the specific lattice type of the crystal. That is to say, determine whether the cell is

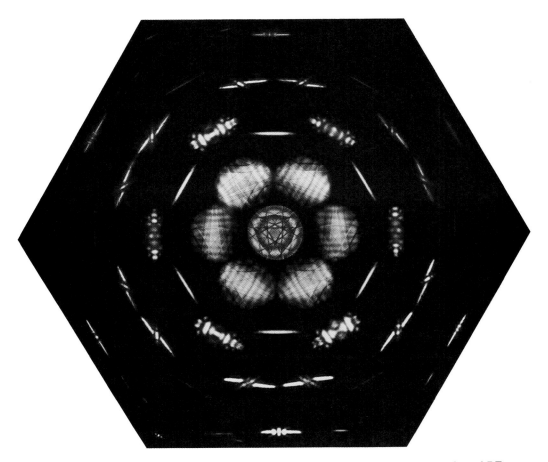

Figure 6.22 A special kind of CBED pattern known as the simultaneous ZA, $\pm G$, and BF technique can also be utilized for symmetry determination. In this figure the pattern is obtained from <111> silicon.

primitive, body centered, face centered, or base centered. This is done by indexing the diffraction maxima in the HOLZ ring and determining how the higher-order Laue zones overlap with the zero-order Laue zone. Figure 6.23 shows schematically the overlap between different Laue layers in the [001] orientation for three different lattices. By overlaying the HOLZ pattern and the zero-layer pattern, the stacking of the layers can be determined. When the higher-order Laue zones overlap directly with the zero layer, the structure is primitive, and the lattice stacking is atom upon atom directly. If HOLZ stacking is not directly coincident with the zero layer, then the structure is nonprimitive. In the latter case, the lattice type can only be determined by exact indexing of the zero-layer pattern, and determining the displacement vector between the zero- and higher-order Laue layers. For example, a face-cen-

[001] cubic projections

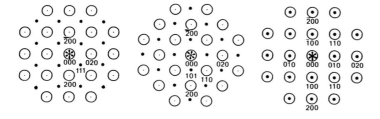

Face-centered   **Body-centered**   **Primitive**

✳ Transmitted beam
⊙ Zero-order Laue zone
• First-order Laue zone

Figure 6.23 Schematic diagram illustrating the overlap between the ZOLZ and FOLZ in the [001] orientation of three different types of cubic crystals.

tered-cubic (fcc) structure, observed along the [001] crystal direction, requires a [111]-type displacement vector to bring the FOLZ coincident with the ZOLZ. On the other hand, due to differences in atomic stacking between fcc and body-centered-cubic (bcc) structures, the displacement vector for bcc structures would be [011] for the same orientation and Laue layers.

Space-Group Determination

Having determined the crystal point group and lattice type, several different space groups are possible. The third step in the CBED analysis is to analyze the reflections that are kinematically forbidden for each possible space group, and thus identify the specific space group of the crystal. Due to the dynamical nature of electron diffraction, reflections that are forbidden by kinematical diffraction often occur in CBED patterns by double diffraction. Kinematically forbidden reflections can occur by double diffraction either due to the crystal centering or to additional symmetry elements (such as glide planes or screw axes). In CBED we are interested in the kinematically forbidden reflections that occur due to the additional symmetry elements. When two or more equivalent double-diffraction paths exist in a given orientation, the kinematically forbidden reflection that occurs will have a central line of zero intensity passing through the disc. The kinematically forbidden reflections, which generally have multiple double-diffraction routes, usually lie along systematic rows of reflections, as shown in Figure 6.24. These so-called dynamical

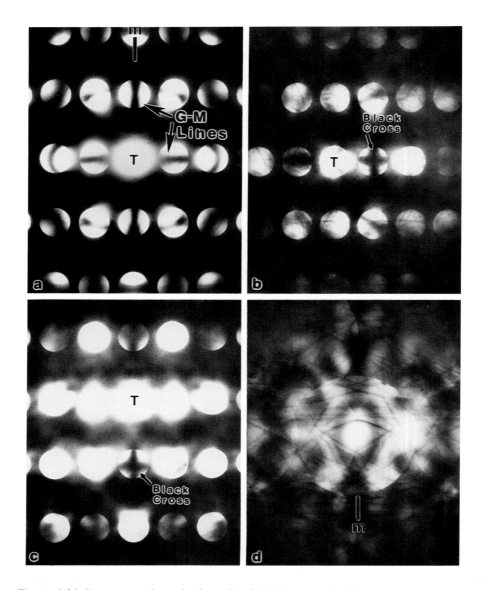

Figure 6.24 Space-group determination using CBED is greatly facilitated by observing the presence and orientation of Gjønnes–Moodie (GM) lines in certain kinematically forbidden reflections with respect to the bright-field mirror (3D-HOLZ line symmetry).

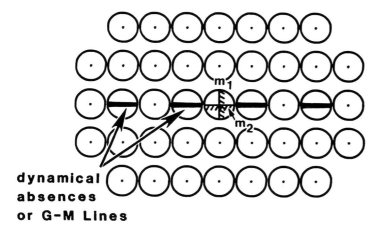

dynamical
absences
or G-M Lines

Figure 6.25 Schematic diagram illustrating the relationship between dynamical absences (or GM lines) and the bright-field mirrors in a CBED pattern.

absences occur in these reflections because the diffracted beams from two equivalent paths undergo complete destructive interference along the central line of the disc to which the beams are perpendicular. The existence of a dynamical absence in a kinematically forbidden reflection indicates that the electron beam is aligned either parallel to a glide plane or perpendicular to a screw axis in the crystal (Gjønnes and Moodie, 1965). Steeds and Vincent (1983) established tables based on the earlier work of Gjønnes and Moodie that described the relationship between the dynamical absences (also referred to as Gjønnes-Moodie or G-M lines) and the number of symmetry elements that can be responsible for those dynamical absences. These standard tables are used as references for interpreting the presence of screw axes and glide planes in space-group determinations using three-dimensional CBED effects.

The presence of these kinematically forbidden reflections implies that there is either a glide plane parallel to the beam direction or a screw axis perpendicular to the beam direction. In order to ascertain whether one or both of these symmetry elements are present, an analysis of the orientation of the G-M line with respect to the bright-field mirrors within the zone axis must be conducted (see Figure 6.25). Once the additional symmetry elements in each orientation are identified (i.e., glide planes and screw axes), the space group can be identified with the aid of Volume 4 of The *International Tables of X-Ray Crystallography* (Hahn, 1983) in conjunction with the rules for forbidden reflections. We can illustrate the determination of translational symmetry elements by going back and observing Figure 6.24 in more detail.

$CaZrO_3$ possesses the *mmm* point-group symmetry of the orthorhombic system. Within the *mmm* point group the only translational symmetry elements allowed are two-fold screw/rotation axes and {100} glide/mirror planes. Determination of which translational symmetry elements are present can be deduced by analyzing CBED patterns along <UV0>-type zone axes. <UV0>-type zone-axis patterns, such as [210] within the *mmm* point group contain only a single bright-field/whole-pattern mirror. Figure 6.24(*a*) shows two orthogonal pairs of G-M lines in alternate reflections along the (00*l*), $l \neq 2n$, and (*hk*0), $h + k \neq 2n$, systematic rows. Figure 6.24(*d*) shows the bright-field disc that displays single mirror symmetry, with (*hk*0), $h + k \neq 2n$, G–M lines being parallel to the bright-field mirror as a consequence of a (001) *c*-glide plane. On the other hand, (001), $l \neq 2n$ G–M lines, which are normal to the bright-field mirror, indicate the presence of a two-fold screw axis along [001]. Confirmation that these dark bands are in fact G-M lines is given by the observation of "black crosses" when the Bragg reflection condition is satisfied for the forbidden reflections. For example, Figure 6.24(*b*) shows a black cross in (001), and Figure 6.24(*c*) shows a black cross in (120). A combination of the *mmm* point group and the *c*-glide plane and the two-fold screw axis along with other zone axis patterns permits the space group of $CaZrO_3$ to be determined as *Pcmn*.

HOLZ Lines for Lattice Parameter Determination

The final step in the CBED analysis is to determine the lattice constants of the crystal. A reasonably accurate measurement of the lattice constants (~2 parts in 100) can be obtained by indexing the reflections in the ZOLZ and/or by measurement of the HOLZ ring diameter. However, the best method is to use the positions of HOLZ lines in the transmitted disc that are very sensitive to changes in lattice parameter. An order of magnitude more accurate result (~2 parts in 1000) can be obtained by computer simulation of the position of the HOLZ lines using different lattice constants (Ecob et al., 1981). The lattice constants that produce the best matching with the experimentally observed HOLZ line positions can be identified as the lattice constants of the crystal.

As discussed earlier, HOLZ defect lines are sharp lines within the transmitted and diffraction discs of the zero layer, and arise from the three-dimensional diffraction between higher-order Laue layers and the zero-order layer within the reciprocal lattice (Jones et al., 1977). One of the most powerful aspects of CBED analysis is the use of HOLZ lines for absolute lattice parameter determinations, and even more accurate determination of relative changes in lattice constants (Ecob et al., 1982). The theory for the origin of HOLZ lines is fairly complicated, and the reader is referred to Jones et al. (1977) for a detailed description. For the purpose of this chapter only a short, simplified discussion of the origin of HOLZ lines will be given. Simply stated, the elastic scattering of electrons from planes within the higher-order Laue layers results in diffracted waves that possess a nonzero component in the incident beam direction. Therefore, these diffracted waves are rediffracted back into the zero layer. Since the HOLZ lines arise from elastic scattering with reflections that occur in upper Laue zones, well away from the origin [i.e., the (000) beam in the zero-order Laue zone], the "**g** vectors" between the origin and the HOLZ reflections are much larger than those for reflections within the zero-order layer. As a result, HOLZ lines appear as narrow lines within the diffracted and

Plate 1.1 Tint etching of the low-carbon sheet steel (Figure 1.3) with aqueous 10% $Na_2S_2O_3$ plus 3% $K_2S_2O_5$ colors the ferrite grains.

Plate 1.2 CDA 510 phosphor bronze (Figure 1.4) etched with Klemm's I tint etchant.

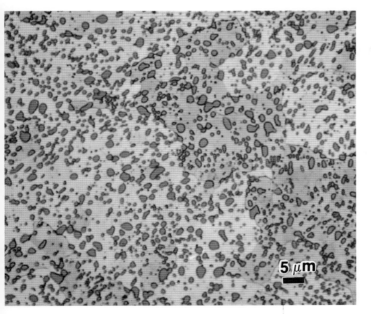

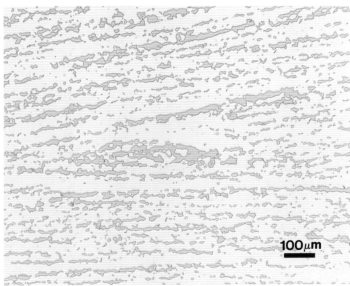

Plate 1.3 Microstructure of AISI W2 carbon tool steel (Figure 1.5) etched with Beraha's sodium molybdate tint etch.

Plate 1.4 Microstructure of α-β brass (Figure 1.6) tint etched with Klemm's I to color the β phase.

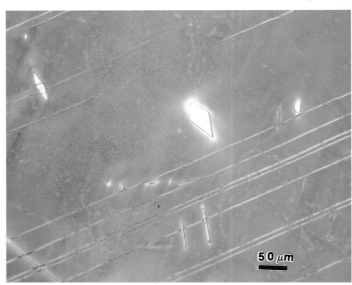

Plate 1.5 Microstructure of 7-Mo PLUS® showing the ferrite colored preferentially by tint etching with aqueous 10% HCl plus 1% $K_2S_2O_5$.

Plate 1.6 Microstructure of Coahuila, a hexahedrite meteorite, tint etched with aqueous 10% $Na_2S_2O_3$ and 3% $K_2S_2O_5$, which colors the kamacite (ferrite). The long diagonal lines are mechanical twins; the rod- and prismatic-shaped particles are rhabdite (Fe-Ni phosphide).

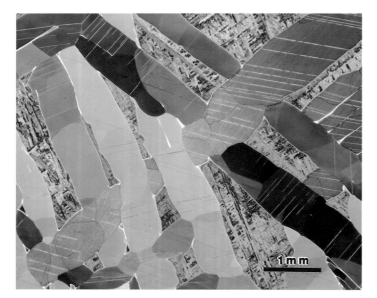

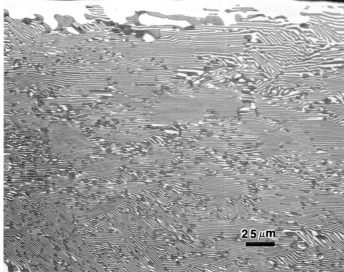

Plate 1.7 Microstructure of Gibeon, a fine octahedrite meteorite, tint etched (as in Plate 1.6) to color the kamacite grains. Note the mechanical twins (short parallel lines) within the elongated kamacite grains. The dark cross-hatched constituent is "finger" plessite, a mixture of kamacite and taenite (austenite). The white films between the kamacite grains are taenite.

Plate 1.8 A high-magnification view of "pearlitic" plessite in Odessa, a coarse octahedrite meteorite, tint etched (as in Plate 1.6). The kamacite in the plessite is colored, while the taenite remains unaffected. The outer rim around the plessite (at top) is taenite highly enriched in nickel.

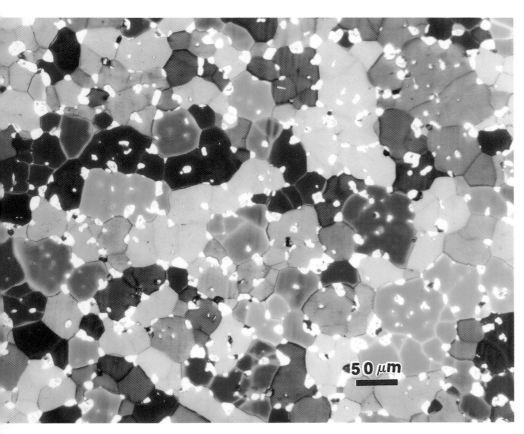

Plate 1.9 Microstructure of the Washington County ataxite meteorite, tint etched (as in Plate 1.6) to color the equiaxed kamacite grains which are free of mechanical twins. The white particles are taenite.

Plate 1.10 Microstructure of beryllium using cross-polarized light (Ahrens prism polarizer plus Berek compensator, as in Figure 1.8(d).

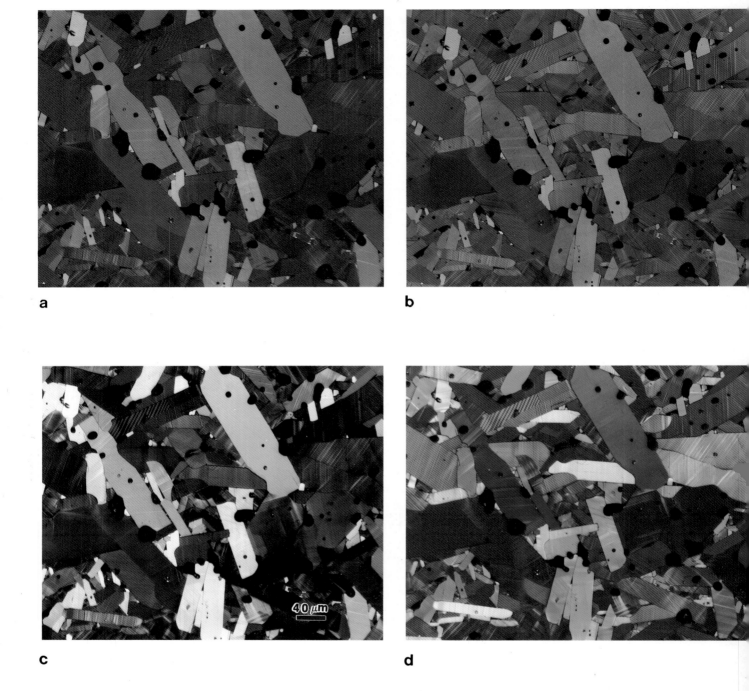

Plate 1.11 Microstructure of a $YBa_2Cu_3O_{7-x}$ superconducting oxide viewed with crossed Polaroid polarizer and analyzer: (a) crossed polarizers only, (b) with the addition of a full wave plate, (c) with the addition of a Berek prism compensator, and (d) with the addition of the full-wave plate and the Berek prism compensator.

Plate 1.12 Higher-magnification view of the superconducting oxide, shown in Plate 1.11, using an Ahrens prism polarizer, a Polaroid filter analyzer, and a Berek prism compensator.

Plate 1.13 Microstructure of ruthenium in cross-polarized light (Foster calcite prism polarizer plus full-wave plate).

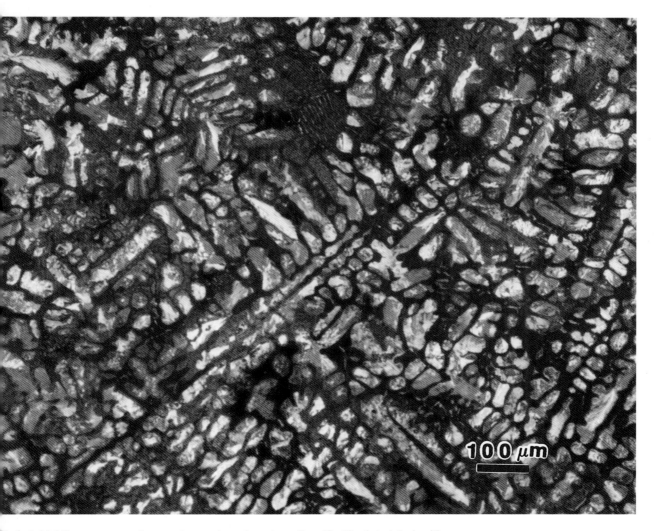

Plate 1.14 Microstructure of a gravity sand casting zinc alloy, Zn-12, tint etched with Klemm's I reagent and viewed with cross-polarized light plus a sensitive tint plate.

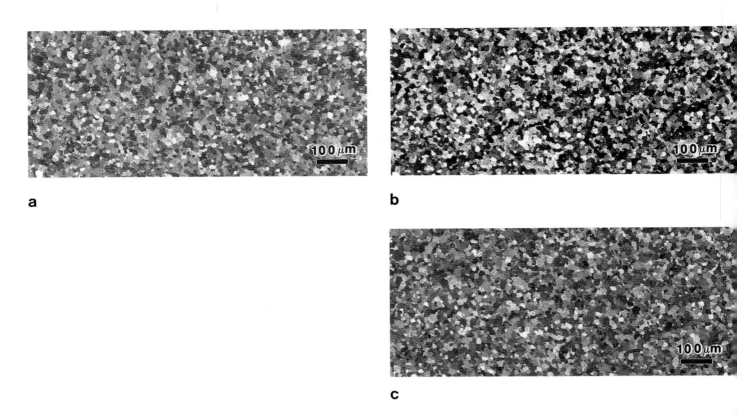

Plate 1.15 Microstructure of α-phase Ti–5% Al–2.5% Sn etched with aqueous 0.5% H[using cross-polarized light (Polaroid filters) and a full-wave plate showing three differer settings of the full-wave plate. (a) Front, (b) middle, and (c) rear.

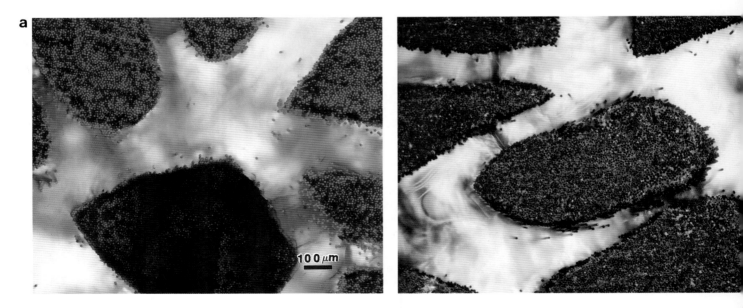

Plate 1.16 Microstructure of a graphite-fiber-reinforced polysulfone composite viewed wit (a) cross-polarized light plus a sensitive tint plate, and (b) dark-field illumination (differen areas are shown).

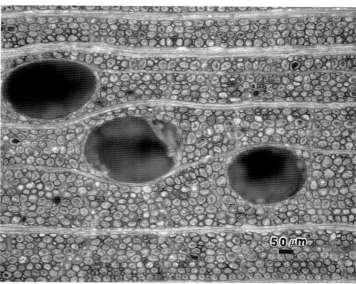

Plate 1.17 Nodular graphite in annealed ductile iron using cross-polarized light plus sensitive tint [as in Figure 1.10(b)].

Plate 1.18 Microstructure of polished walnut viewed with dark-field illumination.

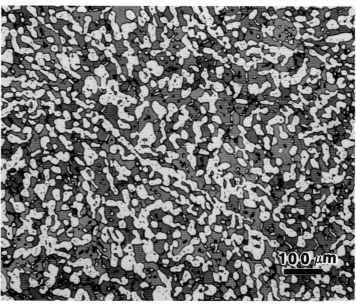

Plate 1.19 Microstructure of Waspaloy specimen tint etched with a modified Beraha's solution (as in Figure 1.13).

Plate 1.20 Microstructure of AISI 312 (Figure 1.15) specimen tint etched with a modified Beraha's solution to color the ferrite.

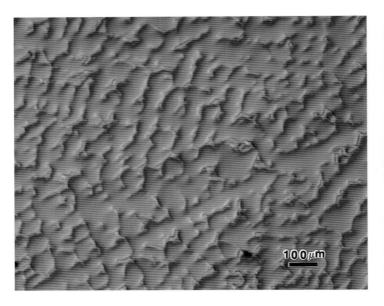

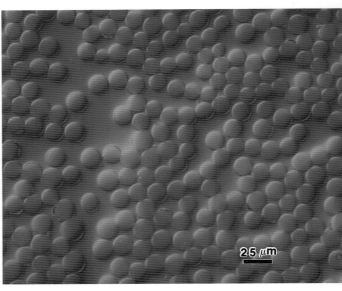

Plate 1.21 Microstructure of an as-cast Cu–37% Zn alloy in the as-polished condition viewed with differential interference contrast showing a "cored" dendritic structure.

Plate 1.22 Microstructure of an as-polished fiberglass reinforced plasti◌ viewed with differential interference contrast illumination.

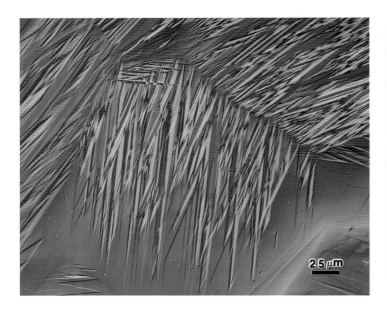

Plate 1.23 Microstructure of an as-polished Cu–26% Zn–5% Al shape memory alloy viewed with differential interference contrast illumination showing β_1 martensite.

Plate 1.24 "Butterfly" martensite formed on the polished surface of an Fe–22% Ni–3% Cr alloy cooled to −100°F (−73°C) viewed withou◌ etching using differential interference contrast.

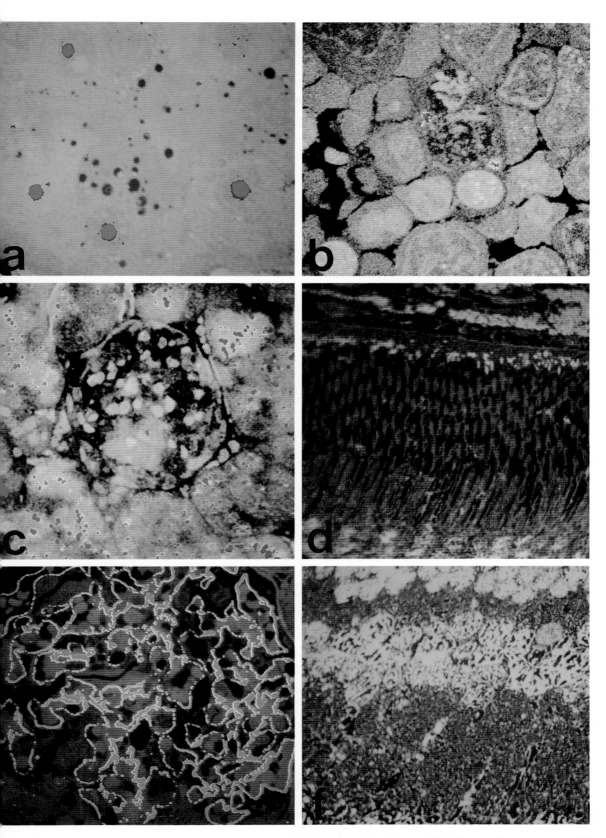

Plate 3.1 Ion microprobe chemical imaging of biological materials. Unless otherwise specified, the samples were prepared by standard electron microscopy embedding techniques. (a) Cross-sectional view of a proximal tubule in the kidney of a rat that had been injected with a solution of beryllium sulfate. CN, green distribution; Be precipitates in cellular nuclei, blue. 40 μm full scale (Levi-Setti et al., 1988a). (b) Be inclusions (blue) in a section of rat bone marrow. CN distribution, describing tissue morphology, yellow. 40 μm full scale. (c) Indium inclusions (blue) in the cytoplasm of tubule cells surrounding a glomerulus in the rat kidney. CN, yellow. 160 μm full scale. (d) Cross section of cryogenically prepared, resin embedded rat retina. CN, red; K, pink. 60 μm full scale. (Burns et al., 1988). (e) Silver deposition (blue) in a silver-poisoned rat kidney glomerulus. CN distribution, red. 160 μm full scale. (f) Another view of the rat retina. CN, blue; Ca, white. 60 μm full scale.

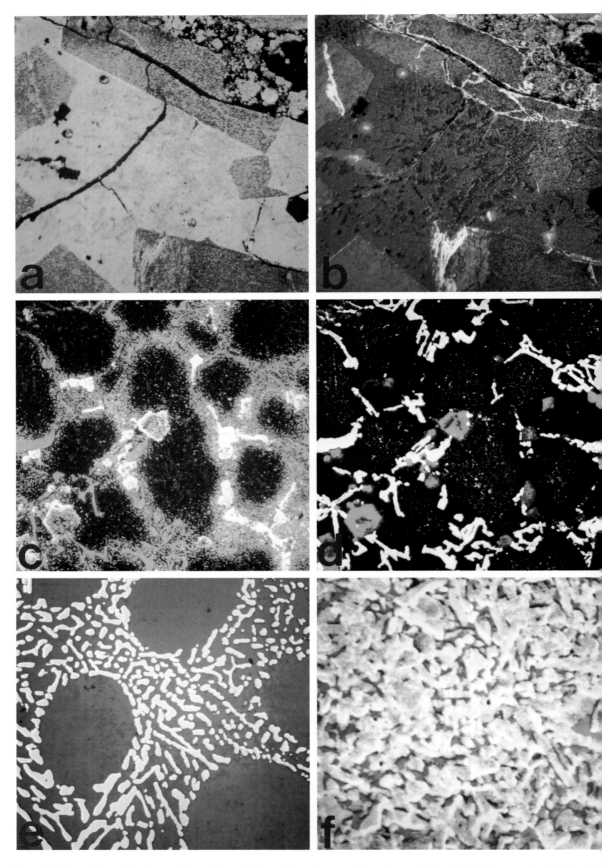

Plate 3.2 Ion microprobe elemental mapping of objects from the materials sciences. (a) and (b): Micrographs of a chondr[?] the basic subunit of stoney meteorites. Si distribution, green; Al, red; Mg, blue. 160 μm full scale. (Levi-Setti et al., 198[?] (c) Elemental composition of a polished section of an Al-based casting alloy. Cu, green; Si, blue. 80 μm full scale. (Levi-S[?] et al., 1988c). (d) Additional elemental maps of the same area of the alloy pictured in (c). Si, yellow; Cs, blue. (e) Analysi[?] another Al-based casting alloy. Al, blue; Si, yellow. 80 μm full scale. (Levi-Setti et al., 1988b). (f) Yttria-sintered sili[?] nitride ceramic. Implanted, resputtered Ga, white, Y, blue. 20 μm full scale. (Chabala et al., 1987b).

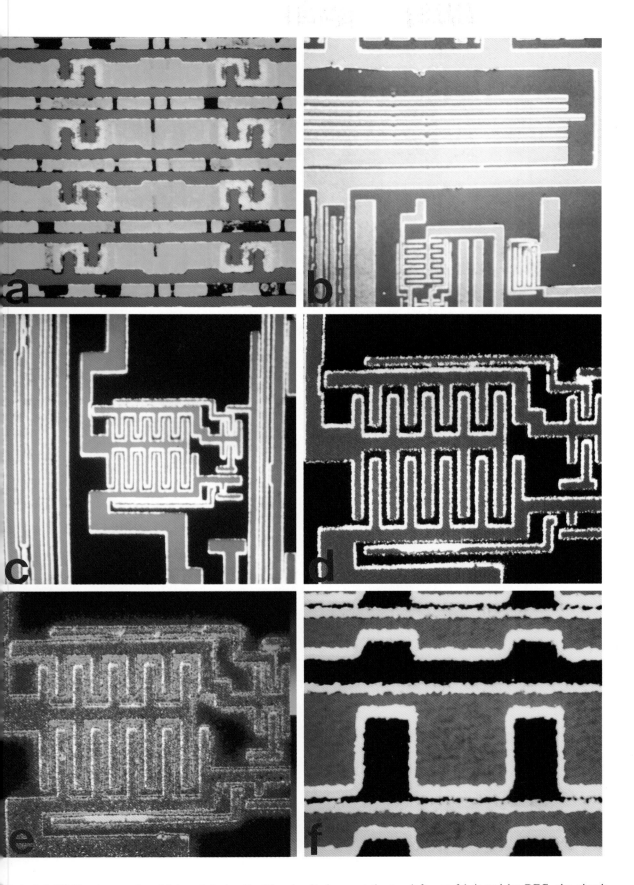

Plate 3.3 SIMS micrographs of integrated circuits. The circuit shown at the top left was fabricated by DEC, the circuit pictured in the rest of the plate was constructed at IBM. (a) Si, red; Al, green. 160 μm full scale. (b) Si, blue; O (from aluminum oxide), yellow. 80 μm full scale. (c) Al, blue; Ti, white. 40 μm full scale. (d) Al, red; Si, blue. 20 μm full scale. (e) Pseudocolor display of Al distribution. The Al wires are shown in red, the sputtered halo is blue. 20 μm full scale. (f) Al, red; Ti, yellow. 10 μm full scale.

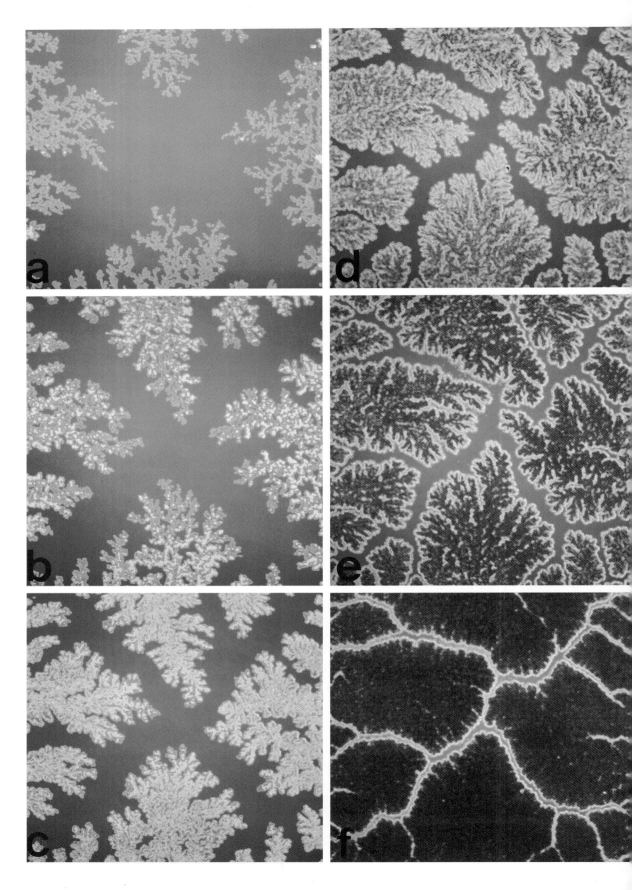

Plate 3.4 Series of micrographs monitoring the growth of gallium oxide (green-yellow-red, in order of increasing intensity) on the clean surface of liquid gallium (which appears blue). This time sequence begins with (a) and progresses to (f). The oxide extends in a fractal pattern from the edges, thickening (appearing redder), until it eventually covers the entire surface (f). 4 μm full scale.

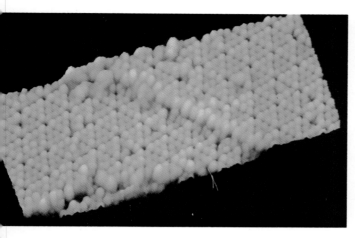

Plate 8.1 Three-dimensional view of the reconstruction of the clean Si(111)(7 × 7) surface across an atomic step, showing the characteristic 12 adatoms per unit cell and a corrugation of 2 Å. Image-processing techniques (Wilson and Chiang, 1988) have been used to enhance the observation of the adatoms on the terraces bordering the step in this ~290 × 170 Å² image of filled states of the sample. From Chiang and Wilson, 1987).

Plate 8.2 Three-dimensional STM image of a region of Si(111)(7 × 7) character (top) next to a domain of ($\sqrt{3} \times \sqrt{3}$)R 30° Ag/Si(111) structure (bottom). The image has been contrast enhanced by statistical differencing (from Wilson and Chiang, 1988) to allow the simultaneous display of the 2 Å corrugation of the (7 × 7) with the 0.2 Å corrugation of the $\sqrt{3}$ structure. The ~60 × 50 Å² image of empty states of the sample was obtained for tip bias of −2.0 V and tunneling current of 2.0 nA. From Wilson and Chiang (1987c).

Plate 8.3 Three-dimensional view of STM tunneling-current image of ~25 × 25 Å² region of a Au(111) thin film measured in air, with an atomic spacing of 2.8 ± 0.3 Å. The current modulation is ~10%, with a dc level of 2 nA and a tip bias of +50 mV. The image has been corrected for thermal drift and piezoelectric creep. From Hallmark et al. (1987).

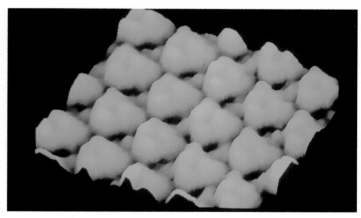

Plate 8.4 At bias $|V_T| < 0.5$ V, a nominally three-fold structure with a depression at the center is observed. The lobes appear to be localized between, rather than over, the underlying metal atoms. The three-dimensional view emphasizes the ringlike appearance. From Ohtani et al. (1988).

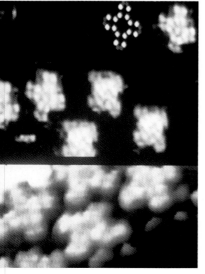

Plate 8.5 (a) High-resolution image of Cu-phth molecules on Cu(100) at submonolayer coverage for (−0.15 V, 2 nA). Fine structure has been emphasized by baseline subtraction, and a gray-scale representation of the HOMO, evaluated 2 Å above the molecular plane, has been embedded in the image. (b) High-resolution image near 1 monolayer coverage with $V_T = -0.07$ V, $i_T = 6$ nA. From Lippel et al. (1989).

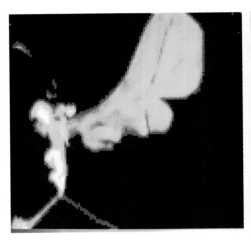

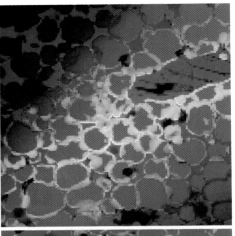
a

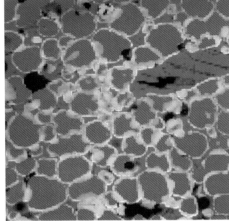
b

Plate 10.1 Quantitative electron probe compositional map for the same region as Figure 10.1. The concentration data from 0 to 10 wt % zinc are depicted with a thermal color scale, and structures from 0.1 to 10 wt % Zn are simultaneously visible in this image. The image field width is 100 μm.

Plate 10.2 Comparison of electron probe compositional maps of a multicomponent ceramic $(MgO–V_2O_5–CoO)$. (a) No correction for defocusing of the wavelength-dispersive spectrometers. (b) Correction applied (Marinenko et al., 1987). Image field width, 500 μm. Sample courtesy of John Blendell, National Institute of Standards and Technology.

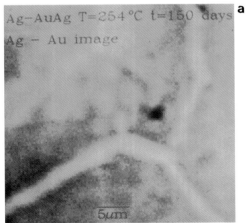
a

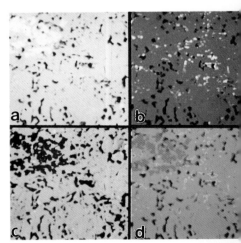

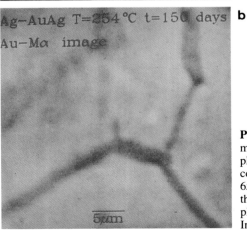
b

Plate 10.3 Detection of a small concentration modulation against a general background. Sample: Ag-Au. (a) Ag compositional map, with contrast expansion; Ag enrichment 5% above a 65% average background. (b) Au decrease in the corresponding Au compositional map. Sample courtesy of Daniel Butrymowicz, National Institute of Standards and Technology.

Plate 10.4 Use of the primary color overlay technique. Sample: $YBa_2Cu_3O_{7-x}$ high-T_c ceramic superconductor. Gray-scale images are shown for (a) Ba; (b) Cu; and (c) Y. The primary color superposition of these three images is shown in (d), with the color assignment: Ba, red; Cu, green; and Y, blue. Image field width, 223 μm. Sample courtesy of John Blendell, National Institute of Standards and Technology.

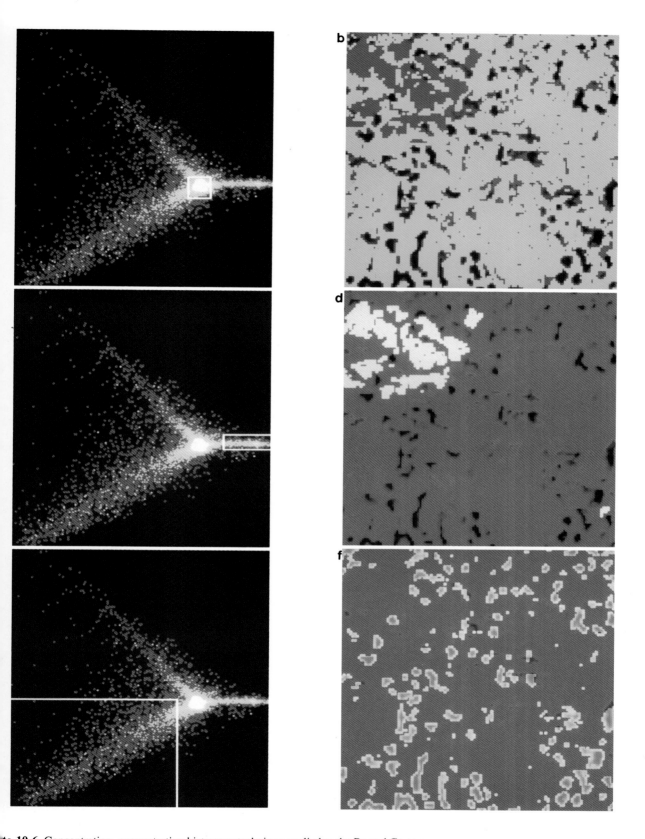

te 10.6 Concentration–concentration histogram technique applied to the Ba and Cu com-
itional maps of Plate 10.4. (a) CCH with Ba on the horizontal axis (0–74%) and Cu on the
tical axis (0–74%). (b) Color superposition of the pixels selected by the TRACEBACK
ction in (a) depicted in green upon the barium image (red), revealing the majority phase.
Selection of the zone in the CCH with constant Cu and variable Ba. (d) Corresponding
els located by TRACEBACK, revealing a minority phase. (e) Selection of the zone in the
H with both Cu and Ba varying. (f) Corresponding pixels located by TRACEBACK, high-
ting the edges of the voids. Image field width, 223 μm. Sample courtesy of John
ndell, National Institute of Standards and Technology.

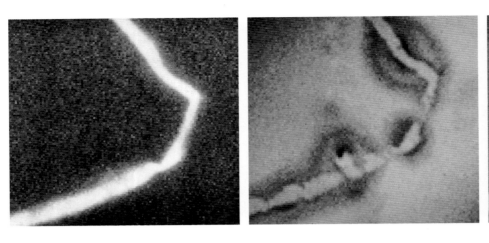

a   b   c

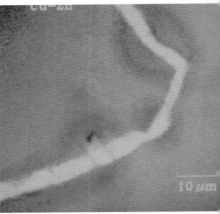

Plate 10.5 Primary color overlay with prior enhancement. Sample: Cu-Zn system showi diffusion-induced grain boundary migration. (a) Zn map, with contrast expansion so that 2 Zn is white; (b) Cu map, with 90% Cu being black; (c) color superposition of these imag with Zn red and Cu green. Sample courtesy of Daniel Butrymowicz, National Institute Standards and Technology.

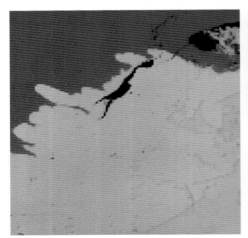

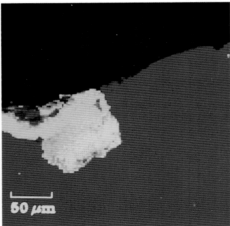

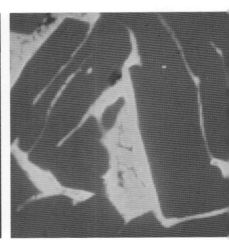

Plate 10.7 Application of electron probe quantitative compositional mapping to examine the interaction zone of a steel screw and an aluminum wire in a failed residential electrical circuit. The compositional maps for Al (green), Fe (red), and Zn (blue) are superimposed and show the zone of intermetallic compound formation in the secondary color yellow. Image field width, 500 μm.

Plate 10.8 Application of electron probe quantitative compositional mapping to examine the cross section of a fracture interface. Color superposition of compositional maps for Fe (red) and S (green). The penetration of sulfur below the original fracture surface is shown as the secondary color, yellow. Note also the banding in the sulfur zone apparent in the color variation.

Plate 10.9 Application of electron probe qu titative compositional mapping to examine inclusion in an explosively formed GaAs c pound. Maps for As (red), Cr (green), and (blue) are superimposed and reveal the prese of a matrix of GaAs (magenta) with inclusi of an As-Cr compound (yellow) from a conta nant. Image field width, 100 μm. Sample c tesy Professor O. T. Inal, New Mexico Insti of Mining and Technology.

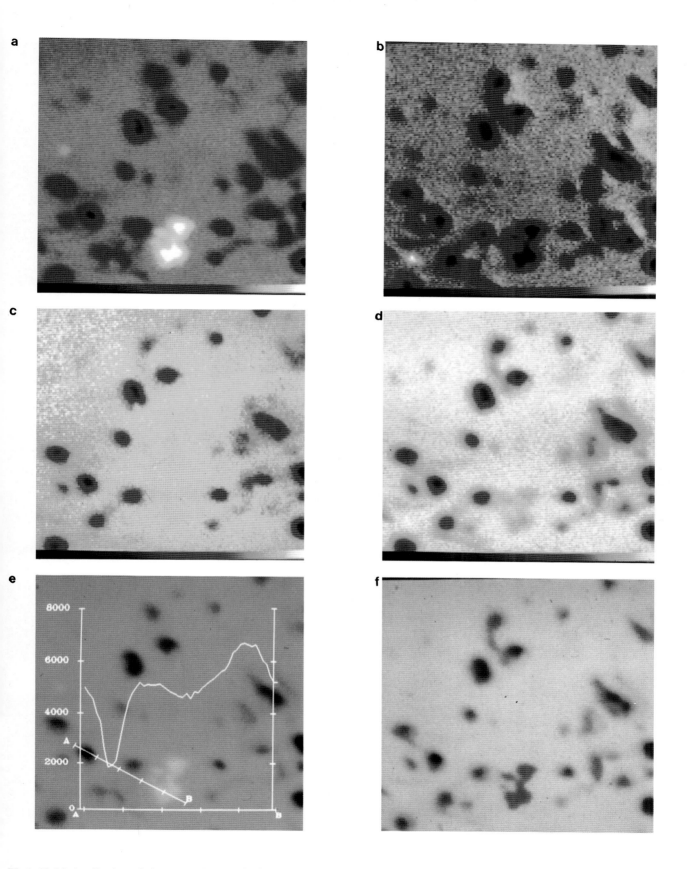

Plate 10.11 Application of electron probe quantitative compositional mapping to examine a high-T_c ceramic superconductor, $YBa_2Cu_3O_{7-x}$. Three different forms of image processing and presentation are shown on the same image data. Thermal scale presentation of single-element compositional maps: (a) Ba; (b) Cu; (c) Y; and (d) O (calculated by assumed stoichiometry). (e) Quantitative numerical data superimposed on the image for the Ba map. (f) Primary color overlay with Ba, red; Y, green; Cu, blue. Image field width, 47 μm. Sample courtesy of John Blendell, National Institute of Standards and Technology (U.S.).

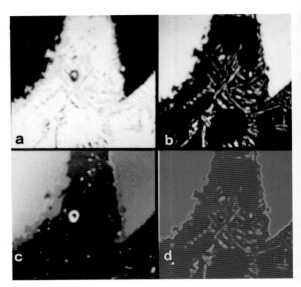

Plate 10.10 Application of electron probe quantitative compositional mapping to examine an Al-SiC ceramic: (a) Al map. (b) Si map. (c) C map (calculated by difference). (d) Primary color overlay with Al, red; Si, green; C, blue. Note Al-Si precipitate (yellow) and Al-C precipitate (magenta). Image field width, 114 μm. Sample courtesy of Carol Handwerker, National Institute of Standards and Technology.

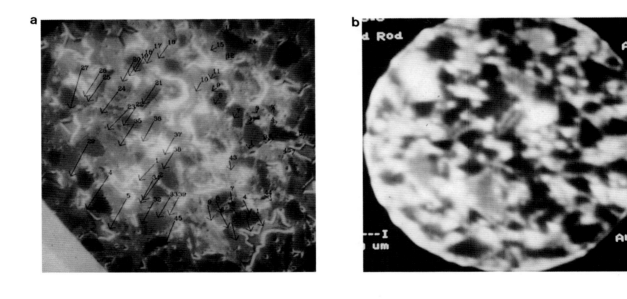

Plate 10.12 Application of secondary ion mass spectrometry compositional mapping to study the microstructure of reaction-bonded silicon carbide. (a) Scanning electron micrograph prepared with a combined SE/BSE detector corresponding to SIMS image; manually chosen correspondence vectors for image warping are shown; (b) SIMS ion intensity image for aluminum from this same area; image diameter, 150 μm. (c) Superposition of warped SIMS image on SE/BSE SEM image showing correspondence. Sample courtesy of Trevor Page, University of Newcastle.

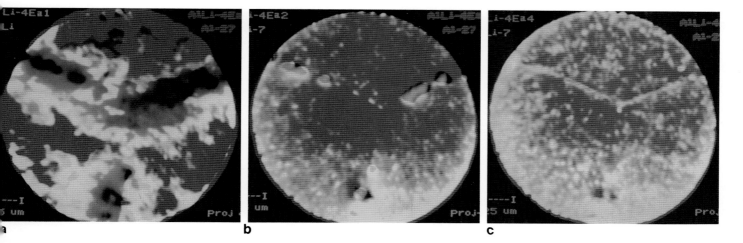

Plate 10.13 Application of secondary ion mass spectrometry compositional mapping to study the microstructure of aluminum–lithium alloys. (a)–(c) A series of Al and Li ion intensity maps, taken as a function of ion sputtering time and viewed after superposition by primary color overlay with Al, red, and Li, green, reveals dramatic changes in the apparent distribution of Li. Preparation of the specimen by mechanical polishing in water caused leaching of Li from the interior to the surface. Image diameter, 150 μm. Sample courtesy of David Williams, Lehigh University.

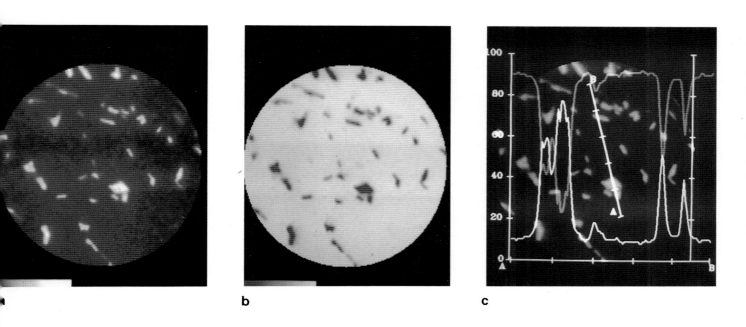

Plate 10.14 Application of secondary ion mass spectrometry compositional mapping with full quantitative corrections to yield concentrations. (a) Li compositional map. (b) Corresponding Al compositional map. (c) Li compositional map with concentration plotted for the line trace along the vector *AB*. Image diameter, 150 μm. Sample courtesy of David Williams, Lehigh University.

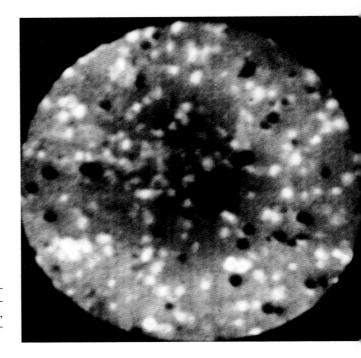

Plate 10.15 Application of secondary ion mass spectrometry compositional mapping to the study of the microstructure of $YBa_2Cu_3O_{7-x}$. Primary color overlay with Cu, red; O, green; and Y, blue. Image diameter, 150 μm. Sample courtesy of John Blendell, National Institute of Standards and Technology.

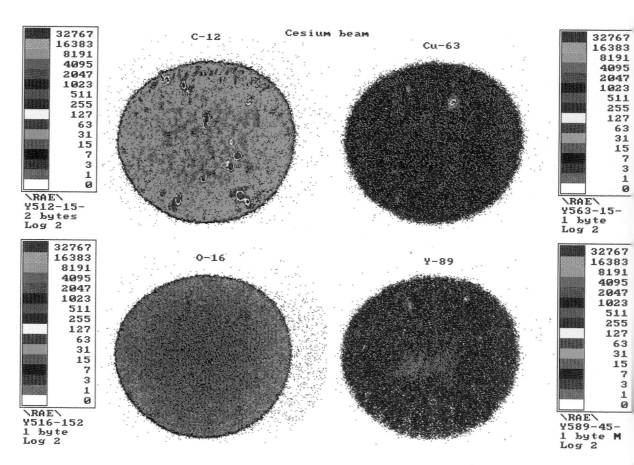

Plate 10.16 Application of secondary ion mass spectrometry compositional mapping to the study of the microstructure of $YBa_2Cu_3O_{7-x}$. Primary color overlay with O, red; C, green; and Cu, blue. Image diameter, 150 μm. Sample courtesy of John Blendell, National Institute of Standards and Technology.

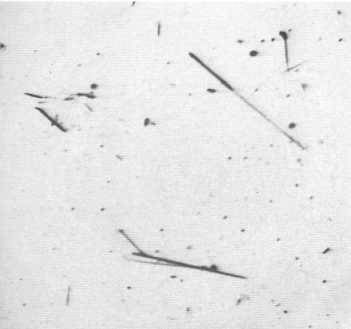

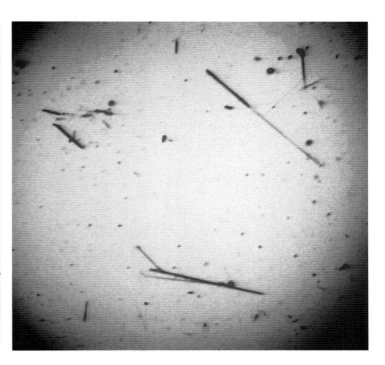

Plate 11.1 A transmission light micrograph of asbestos fibers with low contrast. Field width, 100 μm.

Plate 11.2 Plate 11.1 with contrast enhanced, with uneven illumination and the edges of the vignette showing a contouring effect. Field width, 100 μm.

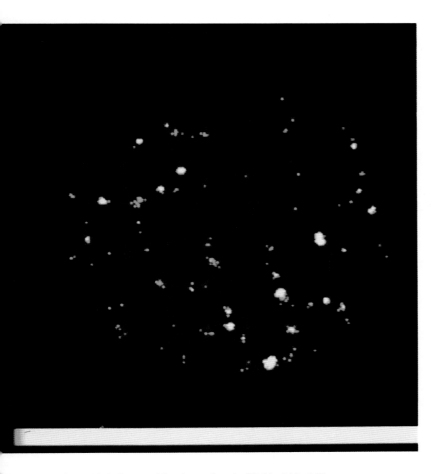

Plate 11.3 Figure 11.1 shown with a thermal scale. Field width, 150 μm.

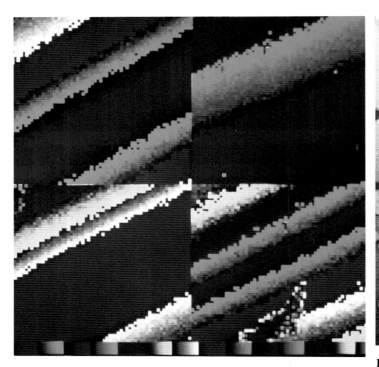

Plate 11.4 Figure 11.2 shown with a pseudocolor scale designed to locate the symmetry of the defocusing effect. Field width for each map, 250 μm.

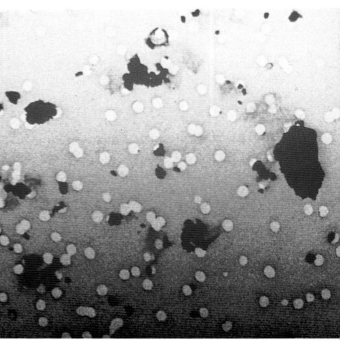

Plate 11.5 Figure 11.3 after histogram equalization. Background deta enhanced. Filter pores (nominal 0.4 μm) and uneven background i lumination are visible. Field width, 20 μm.

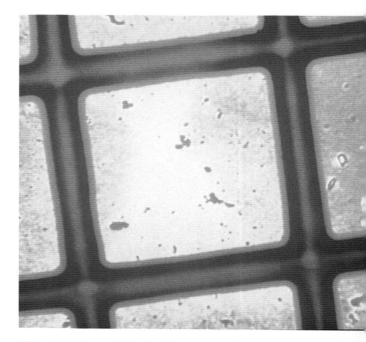

Plate 11.6 Figure 11.4 after histogram equalization. Inhomogeneities the carbon film and stray reflection from the grid are visible. Field widt 150 μm.

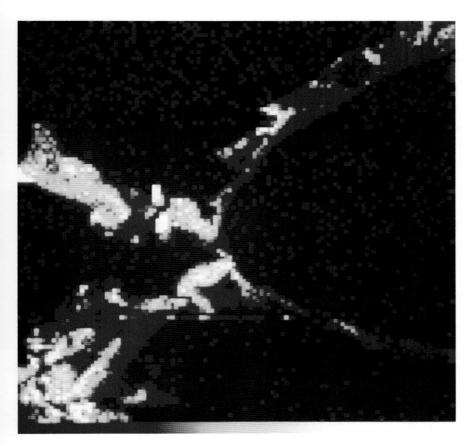

Plate 11.7 Figure 11.6 rendered with a thermal scale. Field width, 250 μm.

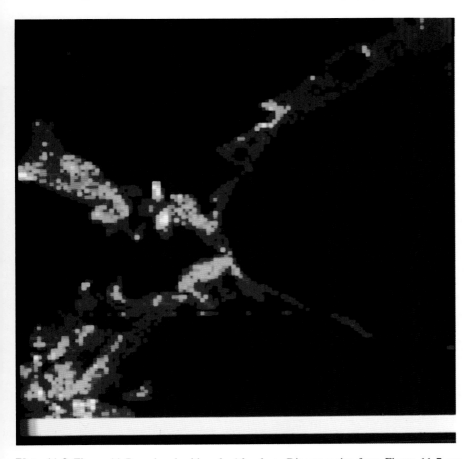

Plate 11.8 Figure 11.7 rendered with only 16 colors. Discrepancies from Figure 11.7 are more noticeable. Field width, 250 μm.

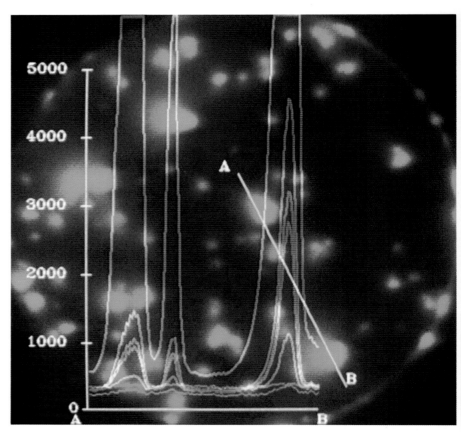

Plate 11.9 ^{138}Ba image taken with an ion microscope. Intensity profiles for line A–B, all five isotopes, magenta. Background profiles, green. Field width, 150 μm.

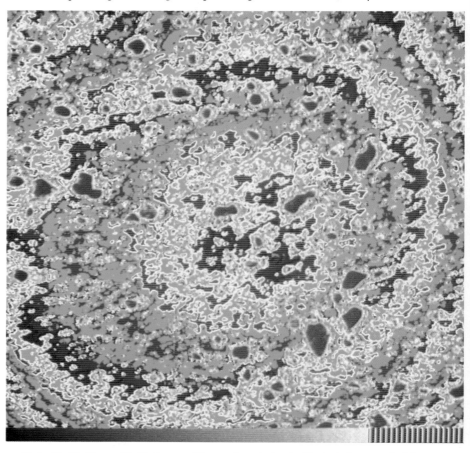

Plate 11.10 Figure 11.15 shown with pseudo-coloring to illustrate large-scale inhomogeneities in the background (uneven illumination). Field width, 100 μm.

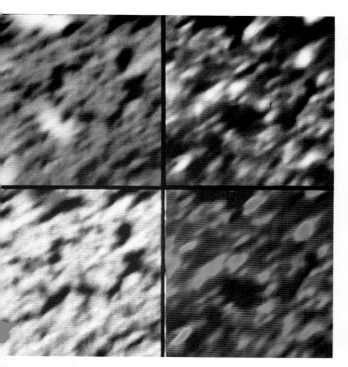

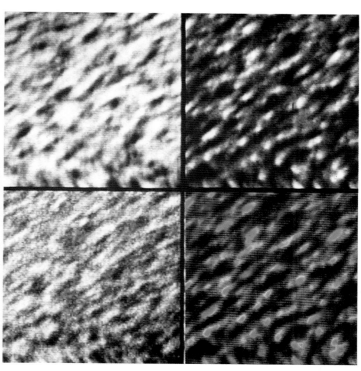

te 11.11 Electron probe x-ray maps of a thin film superconducting ceramic. Cu, Ba, Y and color overlay: upper left, upper right, lower left, lower right. Field width for each map, 50 μm.

Plate 11.12 Maps of superconductor, same as Figure 11.11, different sample. Field width for each map, 50 μm.

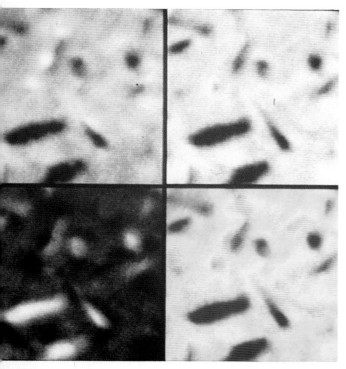

te 11.13 Maps of superconductor, same as Figure 11.11, different ple. Field width for each map, 50 μm.

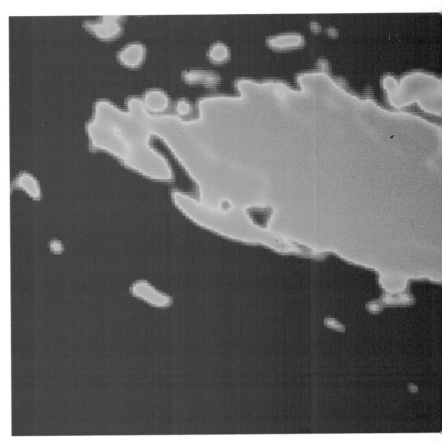

Plate 11.14 Color overlay of Figures 11.17 and 11.18. Field width, 50 μ

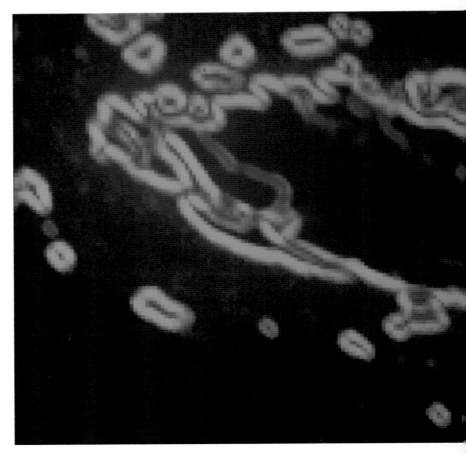

Plate 11.15 Color overlay of Figures 11.19 and 11.20 to highlight the differences. Fi width, 50 μm.

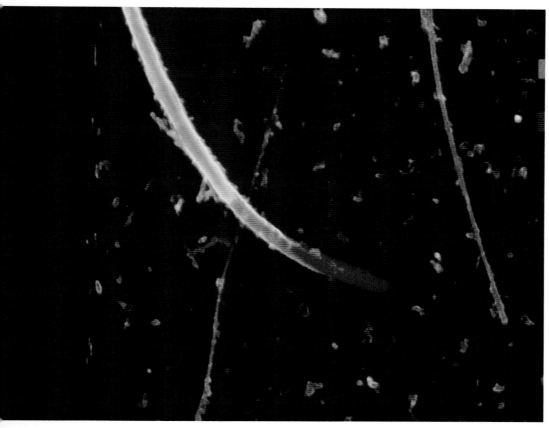

Plate 11.16 Color overlay of Figures 11.21 and 11.22 showing the portion of the end of the large fiber missing from SEM image (red) and showing enhanced edges in the SEM image (green). Field width, 50 μm.

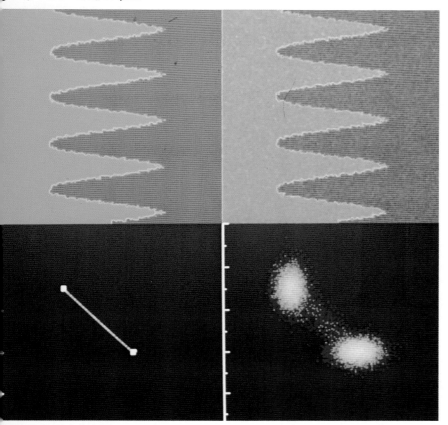

Plate 11.17 A computer simulation of electron probe maps and matching CCHs (see text). Simulated map width, 128 pixels; CCH resolution, 256 × 256 pixels. CCH concentration scales are arbitrary.

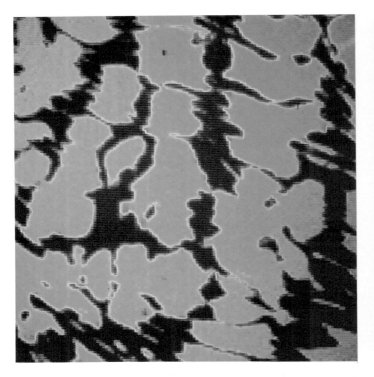

Plate 11.18 A color overlay of Cu (red) and Ti (green) electron probe maps of a Cu-Ti alloy. Field width, 120 μm.

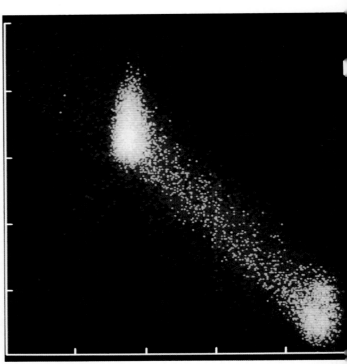

Plate 11.19 Essentially the same CCH as Figure 11.23, but shown with a thermal scale. CCH resolution, 256 × 256 pixels. Abscissa, Cu, 28–5 wt. %. Ordinate, Ti, 43–73 wt. %.

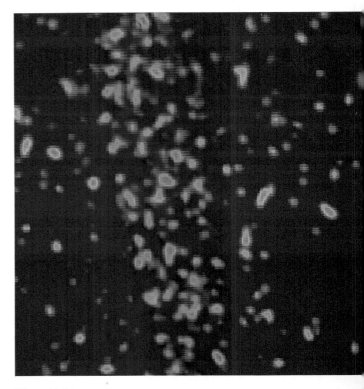

Plate 11.20 A color overlay of Ba, Cu, and Y maps of Figure 11.24 Field width, 114 μm.

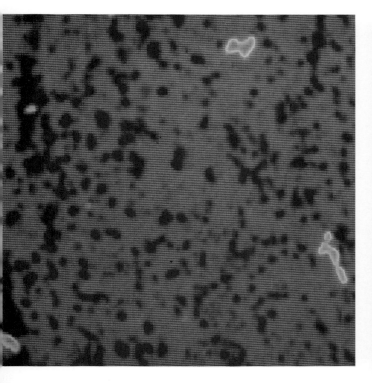

Plate 11.21 A color overlay of Ba, Cu, and Y maps of Figure 11.25. Field width, 114 μm.

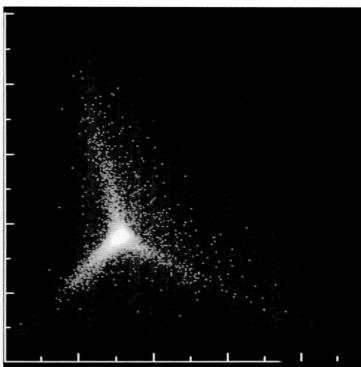

Plate 11.22 Cu-Y CCH for maps in Figure 11.24 and Plate 11.20. Abscissa, Cu, 8–71 wt. %. Ordinate, Y, 2–34 wt. %.

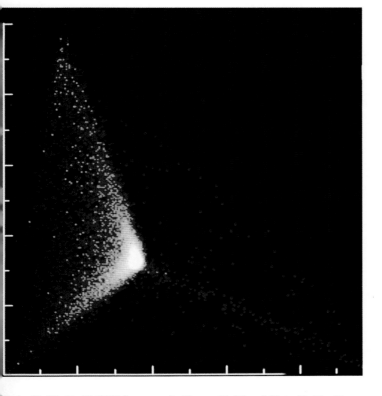

Plate 11.23 Cu-Y CCH for maps in Figure 11.25 and Plate 11.21. Abscissa, Cu, 3–77 wt. %. Ordinate, Y, 1–42 wt. %.

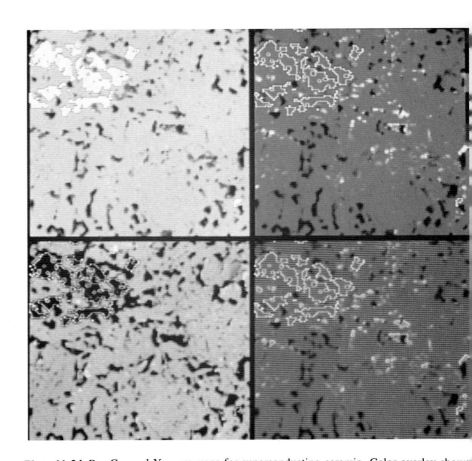

Plate 11.24 Ba, Cu, and Y x-ray maps for superconducting ceramic. Color overlay shows high Ba region outlined (see text). Field width for each map, 223 μm.

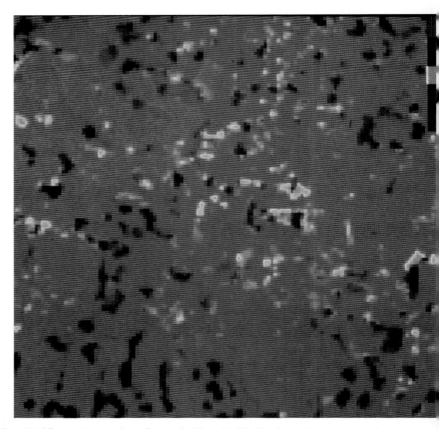

Plate 11.25 A color overlay of maps in Plate 11.24. Small outlined region to lower right in Plate 11.24 could be missed in this image. Field width, 223 μm.

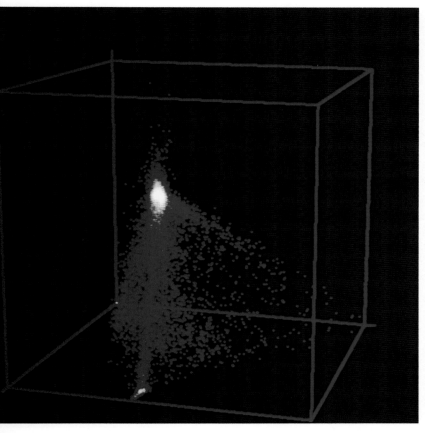

e 11.26 CCCH of maps in Plate 11.24. Axis toward viewer, Ba, 7400 counts per pixel scale. Horizontal axis, Cu, 7400 counts per pixel full scale. Vertical axis, Y, 2100 counts pixel full scale.

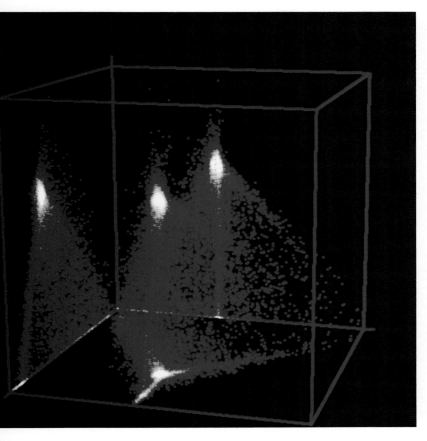

e 11.27 The same CCCH as in Plate 11.26 with projections on sides of cube.

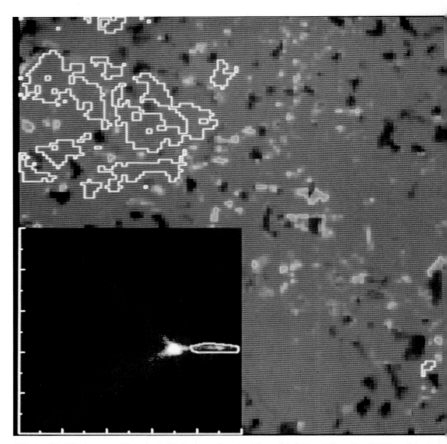

Plate 11.28 A magnification of the overlay in Plate 11.24 and Ba-Cu CCH, with the lined arm corresponding to outlined regions in the overlay. CCH abscissa—Ba, ordinat Cu, axes have same scales as Plate 11.26.

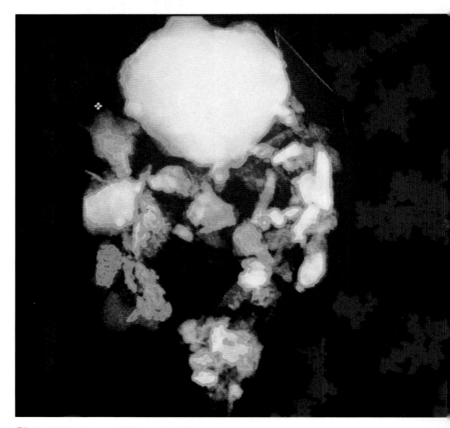

Plate 11.29 Areas of Figure 11.26 found by the blob-splitting algorithm (see text) grou as parts of particles (green) and background (red). Field width, 20 μm.

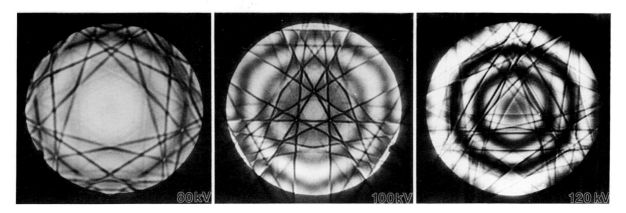

Figure 6.26 The variation in HOLZ line positions with accelerating voltage, illustrating the effect of the change in electron wavelength. This variation can be used to calibrate the accelerating voltage of the microscope. (The sample is copper in the <111> orientation.)

transmitted discs of the zero layer and are essentially equivalent to Kossel lines in x-ray diffraction patterns. Examples of the HOLZ lines can be seen in Figures 6.16–6.17 and 6.21–6.22. These HOLZ defect lines are very sensitive to lattice parameter changes because small changes in lattice parameter or electron wavelength produce large changes in the size of the **g** vectors responsible for the HOLZ reflections. As such, HOLZ lines can be used to measure lattice parameters (e.g., Jones et al., 1977), relative changes in lattice strain (e.g., Kaufman et al., 1986), precipitate/matrix mismatches (e.g., Ecob et al., 1982), and small symmetry changes associated with phase transformations (e.g., Porter et al., 1981). Sharp lines are also observed outside the diffraction discs as a result of inelastic scattering. These are conventional Kikuchi lines and are known as HOLZ Kikuchi lines. The HOLZ Kikuchi lines are continuous with the HOLZ lines within the discs but are generally less sharp (see Figure 6.17).

For determining absolute lattice parameters a simulation technique is employed in which the experimental HOLZ line pattern is compared with a computer-simulated pattern. The position of the HOLZ lines in the simulation is derived from kinematical diffraction theory and generally does not incorporate dynamical effects that may sometimes be important (Britton and Stobbs, 1987). Generally a standard specimen of known lattice parameter is used to establish the exact electron wavelength to be used in subsequent simulations. It is first necessary to establish the electron wavelength for the patterns, since the HOLZ lines shift with changes in accelerating voltage (see Figure 6.26). The lattice parameter is then varied until good matching is achieved between the simulated and experimental patterns (see Figures 6.27 and 6.28). Theoretically, an accuracy of 2 parts in 10,000 should be achievable. However, in practice an accuracy of 2 parts in 1000 is generally obtained.

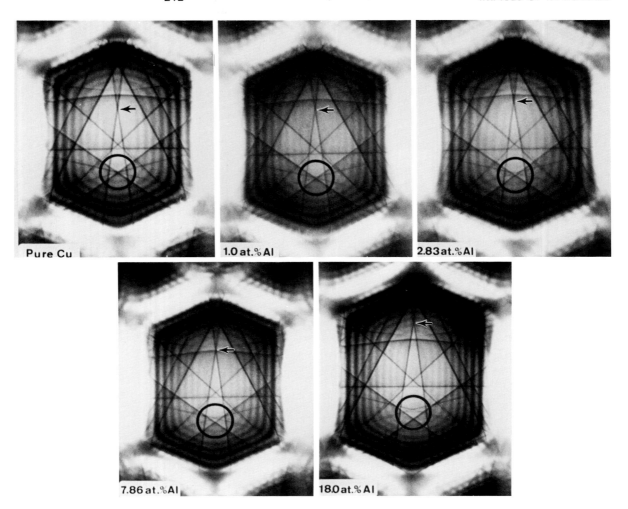

Figure 6.27 HOLZ line patterns obtained from several different copper-aluminum alloys in the <114> orientation at fixed accelerating voltage. The HOLZ line shifts are due to changes in lattice parameter, which result from changes in composition.

Another unique aspect of CBED patterns is shown in Figure 6.29. We can use HOLZ/CBED patterns to deduce whether a crystal shows right or left handedness, since under these circumstances the pattern symmetries vary depending on which way the beam enters the sample. In point groups in which only rotation axes are allowed, the crystals belonging to these point groups can be either right handed or left handed. The handedness of a crystal can be readily determined using CBED. If a sample is enantiomorphous (i.e., it has no symmetry elements that change "hand" such as inversion centers or mirrors), then the patterns are different when the beam

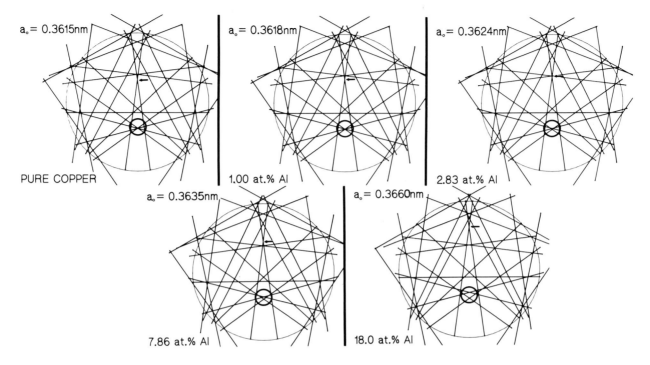

Figure 6.28 Computer simulation of the HOLZ line patterns in Figure 6.27, showing the lattice parameter that corresponds to each pattern in Figure 6.27.

enters the sample from opposite directions. In crystals that do not possess a handedness, CBED patterns obtained with illumination from the top or bottom surface of the crystal are identical. However, in right/left-handed crystals (e.g., quartz) the CBED patterns obtained in this way are no longer identical but are related to each other by a two-fold rotation axis. This is clearly the case in Figure 6.29, which is from enantiomorphous quartz (space group $P3_121$ or $P3_221$). An example of an image from an enantiomorphous material is illustrated in Figure 5.14 in Chapter 5.

Finally, it is possible to obtain some direct correlation between CBED symmetry information and the image of the sample. This technique, termed *convergent beam imaging* (CBIM) is a real space analog of the multiple dark-field technique shown in Figure 6.7. The CBIM technique involves adjustment of the condenser and/or objective lens conditions so as to produce either a defocused image of the image plane or a defocused image of the back focal plane of the objective lens. The CBIM image in Figure 6.30 is a defocused image of the back focal plane, created by slightly underfocusing the probe with the objective lens. In Figure 6.30, the change in symmetry across a two-phase interface between NiO and CaO can be seen, as well as the diffraction contrast associated with the interface. A map of HOLZ lines from both the CaO and NiO crystals is visible, superimposed on the bright-field

TOP

BOTTOM

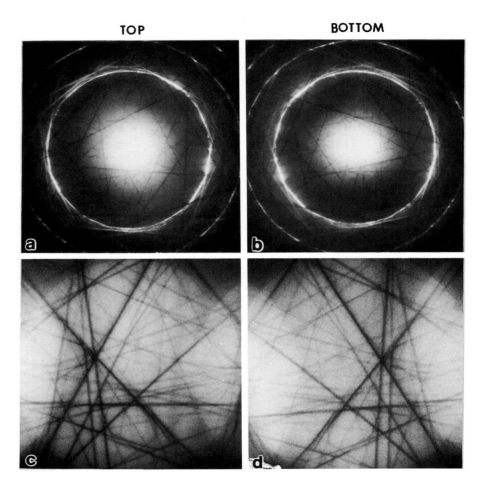

Figure 6.29 (a) and (c) The whole-pattern and bright-field disc, respectively, from a quartz specimen illuminated from the top surface. (b) and (d) The same patterns as (a) and (c), except that the sample was inverted to illuminate the bottom surface. Note the mirror symmetries between (a) and (b), as well as between (c) and (d).

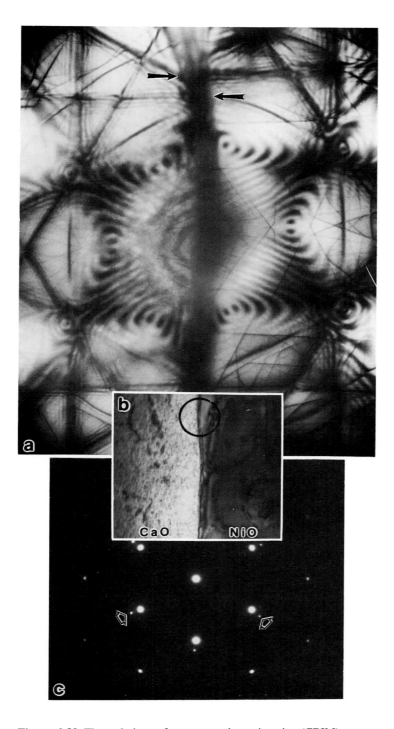

Figure 6.30 The technique of convergent beam imaging (CBIM) creates a mixture of both image and diffraction pattern in one. By adjustment of condenser and objective lenses, either the image or the diffraction pattern can be focused, but not both at the same time. (a) A CBIM pattern, using the LACBED technique, of a NiO-CaO eutectic interface. (b) In this NiO-CaO directionally solidified eutectic all direction and planes in both phases are parallel to each other, as evident in the SADP. (c) The CBIM pattern in (a) displays the parallelism of (220) mirror planes that are normal to the planar interface, and the lack of continuity of HOLZ lines and Kikuchi bands indicates the lattice mismatch across the interface.

Figure 6.31 Channeling pattern from a bulk sample of <111> silicon, showing a distorted six-fold symmetrical star, similar to the Kikuchi pattern in Figure 6.8. The distortion arises because the sample is tilted in the SEM stage relative to the backscattered electron detector.

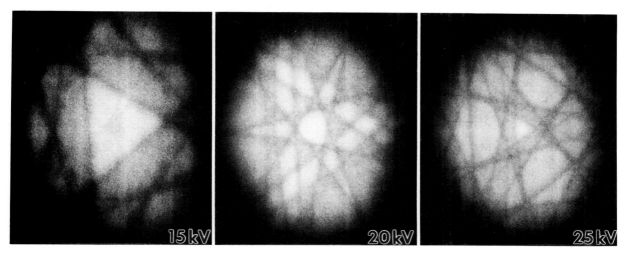

Figure 6.32 Fine structure in the central region of the channeling pattern in Figure 6.31. The structure shows the true three-fold symmetry of the <111> axis and the channeling lines shift with accelerating voltage in a manner similar to the HOLZ line behavior shown in Figure 6.26.

image of the interface. The mismatch of the HOLZ lines across the interface is indicative of the difference in lattice parameter between the two phases. The insets show a TEM image of the interface and an SAD pattern across the interface, showing the orientation relationship between the phases, which is $(111)_{NiO}$ // $(111)_{CaO}$.

SCANNING BEAM DIFFRACTION

We end the chapter with some examples of analogous diffraction phenomena obtained in scanning electron beam instruments. The phenomenon of electron channeling is the only way to obtain crystallographic information from a bulk crystal in the SEM. This technique is described in detail in Chapter 2, and channeling patterns are only shown here for purposes of comparison.

Figure 6.31 is a channeling pattern from <111> silicon, and the dark lines represent regions of the sample close to the Bragg orientation. Direct comparison with the Kikuchi pattern in Figure 6.7 is illuminating. What is more surprising is that the fine structure in the central region of the channeling pattern is a direct result of HOLZ scatter. Therefore the channeling pattern symmetry in this central region shows the true three-fold symmetry characteristic of HOLZ line CBED patterns from a thin area of the same sample (Figure 6.17). Furthermore (see Figure 6.32),

Figure 6.33 A series of Eades double-rocking zone-axis patterns obtained from a thin sample of aluminum in the [001] orientation. (a) The bright-field image; (b) the 200 dark-field image; and (c) an energy-filtered version of (a) in which the removal of energy-loss electrons removes the chromatic aberration effects, thus sharpening the image. Courtesy Philips Electronic Instruments.

218

changing the accelerating voltage causes the channeling pattern to change in exactly the same manner as the HOLZ lines (Figure 6.26).

Finally, similar scanning–diffraction phenomena are obtainable in scanning–transmission instruments using a so-called "double-rocking technique" (Eades, 1980), which involves rocking the beam over a range of angles in two dimensions off the optic axis, both above and below the sample. The resultant patterns are very striking (Figure 6.33) and can be viewed in bright field [Figure 6.33(a)] or dark field [Figure 6.33(b)]. In addition, the images can be sharpened by sending the electrons through an energy filter prior to displaying the pattern on the scanning microscope screen [Figure 6.33(c)], thereby increasing the signal-to-noise ratio of the image by reducing chromatic aberration effects.

SUMMARY

We have shown electron diffraction patterns from a variety of materials in a variety of states. Samples of metals, ceramics, semiconductors, composite structures, and heterostructures have been used to show the similarities and differences in the diffraction images produced by these materials. The various states observed have been amorphous, quasicrystalline, microcrystalline, and crystalline. We have demonstrated that the diffraction images in the back focal plane of the TEM objective lens contain information about orientation relationships, sample shape, two- or three-dimensional symmetry, lattice parameter, crystal class, point group, and space group. But at all times these patterns can be related directly to other TEM or SEM images of the sample. This is the power of "diffraction imaging."

REFERENCES

Allen, S. M. (1981). *Phil. Mag. A* **43,** 325.

Britton, E. G., and Stobbs, W. M. (1987). *Ultramicroscopy* **21,** 1.

Buxton, B. F., Eades, J. A. Steeds, J. W., and Rackham, G. M. (1976). *Phil. Trans. Roy. Soc.* **281** (A1301), 171.

Cohen, J. B. (1966). *Diffraction Methods in Materials Science,* Macmillan, N.Y.

Cullity, B. D. (1978). *Elements of X-Ray Diffraction,* 2nd ed. Addison-Wesley, Reading, Mass.

Dravid, V. P., Sung, C. M., Notis, M. R., and Lyman, C. E. (1989). *Acta Cryst.* **B45,** 218.

Eades, J. A. (1980). *Ultramicroscopy* **5,** 71.

Ecob, R. C., Ricks, R. A., and Porter, A. J. (1982). *Scripta Met.* **16,** 1085.

———, Shaw, M. P., Porter, A. J., and Ralph, B. (1981). *Phil. Mag. A* **44,** 1117.

Edington, J. W. (1975). *Practical Electron Microscopy in Materials Science,* Macmillan, London.

Friedel, M. G. (1913). *C. R. Acad. Sci. Paris.* **157,** 1533.

Gjønnes, J., and Moodie, A. P. (1965). *Acta Cryst.* **19,** 65.

Goodman, P. (1975). *Acta Cryst.* **A31,** 804.

———, and Lempfuhl, G. (1968). *Acta Cryst.* **A24,** 339.

Hahn, T. (ed.) (1983). *International Tables for X-Ray Crystallography 4.* Reidel, Dordrecht, Holland.

Hirsch, P. B., Howie, A., Nicholson, R. B., Pashley, D. W., and Whelan, M. J. (1977). *Electron Microscopy of Thin Crystals.* Krieger, N.Y.

Jones, P. M., Rackham, G. M., and Steeds, J. W. (1977). *Proc. Roy. Soc.* **A354,** 197.

Kaufman, M. J., Konitzer, D. G., Shull, R. D., and Fraser, H. L. (1986). *Scripta Met.* **20,** 103.

Kelly, P. M., Jostons, A., Blake, R. G., and Napier, J. G. (1975). *Phys. Stat. Sol.* **A31,** 771.

Pauling, L. (1986). *Phys. Rev. Lett.* **58,** 365.

Porter, A. J., Shaw, M. P., Ecob, R. C., and Ralph, B. (1981). *Phil. Mag.* **A44,** 1135.

Raghavan, M., Scanlon, J. C., and Steeds, J. W. (1984). *Met. Trans.* **15A,** 1299.

Schechtman, D., and Blech, I. A. (1985). *Met. Trans.* **16A,** 1005.

Steeds, J. W. (1979). *Introduction to Analytical Electron Microscopy.* J. J. Hren, J. I. Goldstein, and D. C. Joy (eds.), Plenum, N.Y., 387.

———, and Vincent, R. J. (1983). *J. Appl. Cryst.* **16,** 317.

Tanaka, M., and Terauchi, M. (1986). *JEOL News* **24E-3,** 62.

Vecchio, K. S., and Williams, D. B. (1988). *Phil. Mag.* **B57,** 535.

Von Laue, M. (1948). *Materiewellen und Ihre Interferenzen.* Akae Verlag. Leipzig, 257.

7

Atomic-Resolution Microscopy

R. GRONSKY

The resolving power of any optical instrument depends primarily upon the wavelength of the radiation used to illuminate the object. As shown in many other chapters of this book, when all other factors are equal, the smaller the wavelength, the smaller the detail that can be resolved. With conventional visible-light optics, resolution is limited to about 500 nm, the average wavelength of light in the visible region of the electromagnetic spectrum. However, electrons accelerated to high energies have an associated wavelength that is much less than one-tenth of a nanometer, and should therefore be capable (all other factors being equal) of easily resolving atoms separated by a few tenths of a nanometer, as they normally are in most materials. Unfortunately, "all other factors" are not equal in the electron microscope.

For one thing, the *relative* effects of lens aberrations become much more severe at this level of resolution, and when the lens is a magnetic field and not a piece of ground glass, correction measures are much more tedious. Another factor is the electronic stability of the microscope, which could introduce a number of sources of chromatic aberration. Electrons accelerated to different energies due to a voltage fluctuation have different wavelengths and are therefore focused differently. Last but not least is the mechanical stability needed to resolve atoms; remember that if the sample or the camera moves during a photographic exposure by *atomic dimensions,* the image is blurred!

Nevertheless there have been tremendous advancements made in recent years that have removed these obstacles and made atomic resolution possible. Aberrations have been corrected by an order of magnitude, electronic fluctuations are controlled to within the part-per-million range, and mechanical stability is sufficient to use exposure times of many seconds if required. Coupled to the use of microprocessors

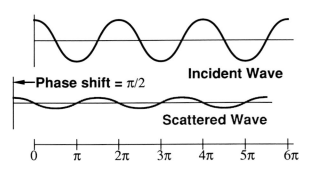

Figure 7.1 Schematic of two waves representing the incident (or forward-scattered) beam and the Bragg-diffracted beam in the transmission electron microscope. In this case the sample is very thin; consequently the Bragg-scattered wave has very small amplitude relative to the incident wave but has been delayed in phase by $\pi/2$ radians.

for digital control of the instrument, and computers for image simulation and processing (see Chapters 10–12), scientists are now imaging atoms routinely. The present chapter highlights this special triumph of modern transmission electron microscopy.

PHASE-CONTRAST IMAGING

The transmission electron microscope produces images using a variety of conventional operating modes, as described in Chapter 5. To image materials with *atomic resolution* in the transmission electron microscope, it is necessary to employ a different imaging procedure. *Phase-contrast imaging* is the name of this imaging technique, which exploits the differences in phase among the various electron beams scattered by the sample to produce contrast. In its simplest modification, this technique is illustrated by considering just two beams that pass through the specimen, one scattered in the forward direction, one scattered by diffraction through a Bragg angle. The Bragg-scattered beam will be of very low intensity relative to the forward-scattered beam for those samples yielding the best images: thin samples that behave as "weak phase objects." As a consequence of this model, the forward-scattered beam is negligibly attenuated relative to the incident beam. The propagation of these two beams might then be visualized as the two waves shown in Figure 7.1. Note that for each point along the horizontal axis representing their direction of travel, the two waves are shown to be out of phase by $\pi/2$ radians or 90°. This simply means that everywhere the forward-scattered wave has maximum amplitude, the Bragg-scattered wave has zero amplitude, and vice versa, as is observed.

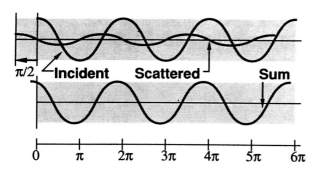

Figure 7.2 Schematic showing the unmodified addition of the two waves in Figure 7.1. With a $\pi/2$ phase difference between the two waves, there is no interference, and therefore no change in amplitude between the "sum" wave and the incident wave. Compare the amplitudes with reference to the shaded background marker.

Now the principle of phase-contrast imaging becomes clear when it is understood that Bragg scattering is caused by the atoms in the sample. To see those atoms, it is necessary to establish contrast between those Bragg-scattered waves that contain information about the atoms and the forward-scattered or transmitted wave that contains only "background" information. The contrast in the image then represents the contrast between regions of the sample where there are atoms and regions where there are no atoms; for example, atoms would appear as black dots against a white background (or white dots against a black background, depending upon the relative amplitudes of the "signal" and "background"). Unfortunately, it is not a routine matter to obtain contrast. Returning to the case of the two waves differing in phase by 90° as a result of a diffraction event (Figure 7.1), note that, if these are simply added with their native 90° phase difference as shown in Figure 7.2, the resultant wave at the image plane looks just like the transmitted wave, shifted slightly in phase. Consequently, because there is no change in amplitude between the "sum" wave and the transmitted wave, there is no contrast in the image.

To obtain contrast it is essential that the "sum" wave differ in amplitude from the incident or forward-scattered wave that serves as "background." The best cases occur when the Bragg-scattered wave is either perfectly "in phase" with the forward-scattered wave, or perfectly "out of phase" with the forward-scattered wave. These are illustrated schematically in Figures 7.3 and 7.4. Note that, when both waves are perfectly in phase (Figure 7.3), their amplitudes add, yielding an image with bright or negative contrast. Under these conditions, the atoms responsible for the Bragg-scattering event would appear bright against a dark background in the image. However, when both waves are perfectly out of phase (Figure 7.4), their amplitudes subtract, yielding an image with dark or positive contrast. Under these conditions, the atoms responsible for the Bragg-scattering events would appear dark

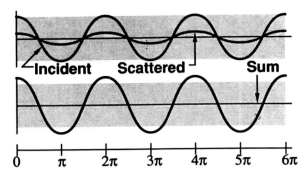

Figure 7.3 Schematic showing the addition of the two waves in Figure 7.1 with an additional phase shift such that the total phase difference between the waves is zero or 2π. Total constructive interference results, yielding a "sum" wave that has greater amplitude than the incident wave; consequently, the scattering species (atoms) will look bright against a darker background.

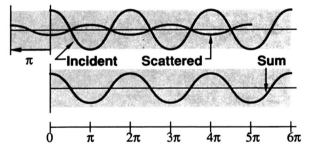

Figure 7.4 Schematic showing the addition of the two waves in Figure 7.1 with an additional phase shift such that the total phase difference between the waves is π. Total destructive interference results, yielding a "sum" wave that has lower amplitude than the incident wave. This causes the scattering species to appear dark against the brighter background of the incident wave.

224

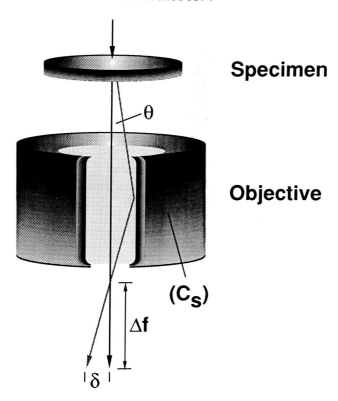

Figure 7.5 Ray diagram of the objective assembly in a transmission electron microscope. The beam scattered by the specimen through angle θ suffers a lateral displacement (δ) from the focal point of an on-axis beam due to the combined effects of spherical aberration (described by the spherical aberration coefficient C_s) and defocus (Δf) of the objective lens.

against a bright background. Now it remains to be seen how the operator of a microscope can control these contrast conditions.

The optics of an electron microscope introduce other phase shifts into the image-forming beams. These are the direct result of the path differences traveled by off-axis (e.g., Bragg-scattered) beams relative to the transmitted beam that travels directly down the optic axis. Once scattered, the trajectory of an off-axis beam is influenced by the objective lens, as illustrated schematically in Figure 7.5. Spherical aberration leads to a lateral displacement of the focal point by $C_s\theta^3$, where C_s is the spherical aberration coefficient of the lens and θ is the scattering angle, while the defocus (Δf) of the lens causes an independent displacement of $\Delta f\theta$. The total displacement $\delta(\theta)$ is therefore the sum of these two terms:

$$\delta(\theta) = C_s\theta^3 + \Delta f\theta, \tag{7.1}$$

as is schematically shown in Figure 7.5. This displacement can be integrated over all scattering angles to obtain the difference in total path length $D(\theta)$ traveled by an off-axis beam relative to the axial beam.

$$D(\theta) = \int_0^\theta \delta(\theta)\, d\theta \qquad (7.2)$$

or

$$D(\theta) = C_s \frac{\theta^4}{4} + \Delta f \frac{\theta^2}{2}. \qquad (7.3)$$

Substituting (λu) for θ from Bragg's law, where $\mathbf{u}$ is a reciprocal lattice vector:

$$D(\mathbf{u}) = C_s \frac{\lambda^4 u^4}{4} + \Delta f \frac{\lambda^2 u^2}{2}, \qquad (7.4)$$

and noting that the phase difference between beams is $2\pi/\lambda$ times their path difference, the final expression for the difference in phase $\phi(u)$ between a beam located at $\mathbf{u}$ in reciprocal space and the beam at $\mathbf{u} = 0$ is:

$$\phi(\mathbf{u}) = \frac{2\pi}{\lambda}\left(C_s \frac{\lambda^4 u^4}{4} + \Delta f \frac{\lambda^2 u^2}{2} \right). \qquad (7.5)$$

It should now be clear where the operator of the microscope has control over the phases of the beams contributing to the image: by changing focus!

Phase-contrast imaging in the transmission electron microscope therefore amounts to selecting the proper *focus* setting that establishes *contrast* from the atoms in the thin sample. The proper focus setting compensates in part for both the phase shift due to diffraction and the phase shift due to spherical aberration. As illustrated in Figures 7.3 and 7.4, when the total phase shift $\phi(\mathbf{u})$ between the transmitted and scattered beams is an even integer multiple of π, atoms appear bright, and when $\phi(\mathbf{u})$ is an odd integer multiple of π, atoms appear dark. This means that the operator also has control over the "color" of the atoms in the image, but also suggests a potential problem. Because different beams located at different values of $\mathbf{u}$ may have different values of $\phi(\mathbf{u})$, the image could get hopelessly confused. Consider for example the case of bright atoms against a dark background intermixed with dark atoms against a bright background in a single image. Naive interpretation, that is, without knowledge of the exact phases of the various beams contributing to the image, of such a micrograph is dangerous, if not impossible.

It is for this reason that the phase differences of all "beams" or all position vectors $\mathbf{u}$ in the diffraction pattern must be tracked simultaneously. As a convenience, these can be mapped in the form of a "phase-contrast transfer function" $T(\mathbf{u})$, defined for a weak phase object by:

$$T(\mathbf{u}) = 2\sin\phi(u). \qquad (7.6)$$

This formulation clarifies the oscillatory nature of phase contrast in the transmission electron microscope, illustrated in Figure 7.6. When $T(\mathbf{u})$ is positive, image detail appears bright; when $T(\mathbf{u})$ is negative, image detail appears dark; and when $T(\mathbf{u})$ is zero, image contrast goes to zero. In effect, the transfer function is a filter function, describing how contrast is transferred from the object to the image via phase

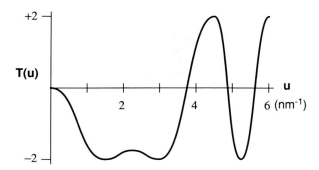

Figure 7.6 Phase-contrast transfer function for a weak phase object, plotted against distance (**u**) in reciprocal space. When superimposed (with radial symmetry) over the electron diffraction pattern, the phase-contrast transfer function gives the relative phases of the different beams scattered by the specimen, and thereby describes how the beams contribute to the contrast in the image.

alteration. Note that the abscissa represents a distance in reciprocal space or, equivalently, in the diffraction pattern. The larger the value of **u**, the smaller the features in the object that are transferred to the image. For the case shown in Figure 7.6, all scattered beams that appear between $u = 1$ (representing 1 nm detail in the object) and $u = 3.3$ nm^{-1} (representing 0.3 nm detail in the object) are transferred to the image with nearly the same phase, and because the value of $T(\mathbf{u})$ is negative, the image will show black dots against a white background at atom positions.

Using this concept of the contrast transfer function, the resolution limit of the microscope can also be conveniently defined: The value of **u** at which $T(\mathbf{u})$ first goes to zero [e.g., 3.8 nm^{-1} (0.26 nm object detail) in Figure 7.6] is a common one. Higher-order beams may contribute to the image, but with errant phase. (Note that a beam at 4.5 nm^{-1} in Figure 7.6 will produce opposite contrast to the lower-order beams that pass through the main transfer interval. Clearly, when the operator knows the transfer function of the microscope, phase contrast images of an "unknown" specimen can be correctly interpreted. Conversely, when the specimen is known, the operator can choose the correct conditions for phase contrast imaging by control of the transfer function.

More sophistication can be used to interpret phase-contrast images by employing computer-simulation techniques. Matching of simulated and experimental images over a range of parameters (e.g., specimen thickness, specimen orientation, microscope focus) is at present the most accurate method of image interpretation, and essential for complex structures, as illustrated by Figure 7.7. This collage of 50 completely different simulated images represents only one sample in one orientation. It graphically reveals just how sensitively image contrast depends upon very small phase changes. Computers are also being used to obtain such images in the first place. With proper dynamic feedback through a video link, the computer can

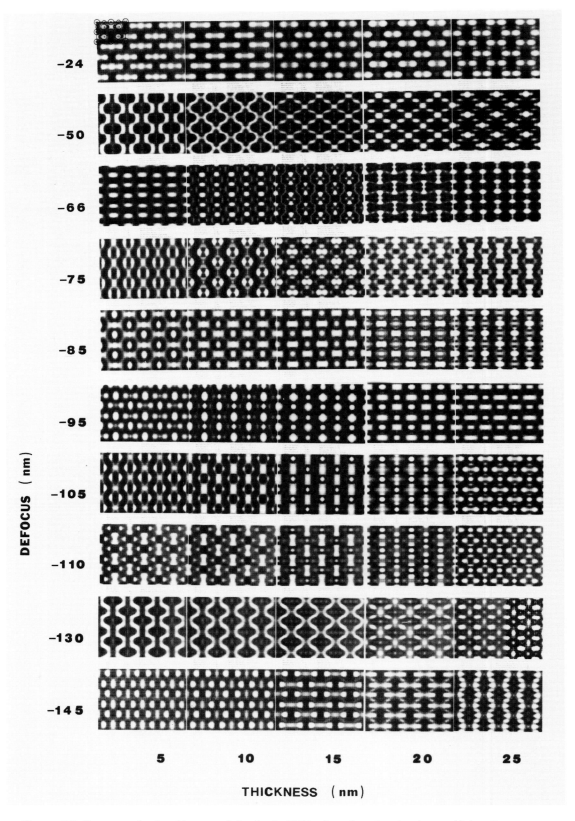

Figure 7.7 Computer-simulated images of alumina in [011] orientation, showing the sensitivity of contrast to small changes in defocus (vertical axis) and thickness (horizontal axis). All 50 images represent the same atomic structure; they just look different because of different phase interference conditions.

directly communicate to the microscopist (or the microscope) the settings needed for optimum phase-contrast imaging.

Although this simplified picture of phase-contrast imaging is basically sound, it omits many of the complications that are often encountered in practice. Thick specimens, or those with a large layer of surface contamination, increase multiple scattering and absorption of the electron beam, preventing phase-contrast imaging. Improperly oriented specimens also decrease the signal in the scattered beams contributing to the image, destroying contrast. Chromatic aberration from fluctuations in accelerating voltage, filament current, and objective lens current scramble the phases of the imaging electron beams, decreasing resolution. These are often modeled as a "damping envelope" that is convolved with the transfer function and has the effect of prematurely driving $T(\mathbf{u})$ to zero for larger values of $\mathbf{u}$. It is essential therefore to employ a stable high-resolution instrument, and choose thin, clean, and flat samples, properly oriented under the beam.

ATOMIC RESOLUTION

With the proper set of conditions, phase contrast reveals atomic structure. Of course it helps to have the best instrumentation, and Figure 7.8 is an example. The atomic-resolution microscope (Gronsky, 1980) is unique in that it was specifically constructed for optimum resolution performance over a range of operating voltages from 400 to 1000 kV, a range of specimen orientations (using a $\pm 40°$ biaxial tilt stage), and the tremors of earthquake country! It is mounted on a 100 ton, pneumatically damped, seismically restrained, vibration isolation system, housed within a three-story air-conditioned tower, utilizing digital feedback compensation of voltage fluctuations, independent water cooling of the objective lens and power supplies, and microprocessor control of all beam drivers, including six separate channels for astigmatism correction alone. Commissioned in 1982, it led the way for a new generation of "intermediate voltage" microscopes now popularly used for high-resolution phase-contrast imaging. At present, it has the unique capability of achieving better than 0.16 nm point-to-point resolution [at the first zero of $T(\mathbf{u})$].

EXAMPLES

Even at lower resolution, phase-contrast imaging directly reveals structural details that are difficult to discern by other techniques. Figure 7.9 shows graphite, the covalently bonded planes of carbon atoms spaced by 0.34 nm, with a full 360° variation in orientation visible within this field of view. Such localized structural information is averaged out by diffraction studies. However, the image shown here readily depicts why graphite is frequently used as a solid lubricant. It is easy to visualize these planes of carbon atoms, weakly bonded to one another by van der

Figure 7.8 The atomic-resolution microscope (ARM) at the National Center for Electron Microscopy, Lawrence Berkeley Laboratory. As of this writing, it holds the resolution performance record of better than 0.16 nm point-to-point over a ±40° tilt range.

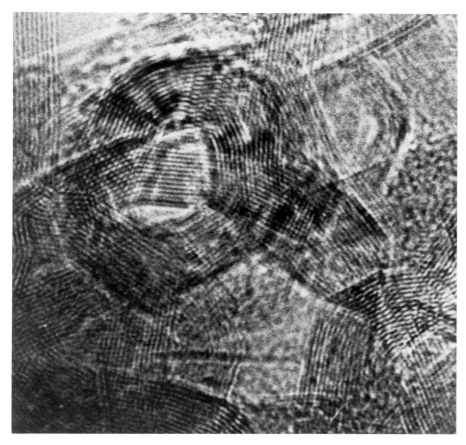

Figure 7.9 Phase-contrast image of graphite lattice planes spaced by 0.34 nm. Note the omnidirectional nature of the stacking of these close-packed planes.

Waals interactions, slipping over one another for nearly any external loading condition.

In another example in which diffraction studies proved inconclusive, Figure 7.10 reveals the intergrowth of three twin variants of tetragonal $YBa_2Cu_3O_{7-\delta}$. Because the c/a ratio is very close to 3 in this nonsuperconducting phase of the well-known high-T_c superconductor, overlapping x-ray peaks confused the interpretation of its complex structure (Zandbergen et al., 1988). In fact, without detailed evaluation, or phase-contrast images like this one, it is easy to mistake this phase for the superconducting one.

The superconducting phase of $YBa_2Cu_3O_{7-\delta}$ is not without its own high-resolution secrets. Passing horizontally just above the center of Figure 7.11 is a planar fault, diagnosed after extensive computer simulation (see overlay on the right) to be a double layer of CuO (Zandbergen et al., 1988). Resulting from a surface-nucleated reaction, the intercalation of this extra layer is an insidious problem that eventually leads to fragmentation and spalling of the surface as long as the material is exposed to a source of oxygen. Obviously the materials will have to be encapsulated in some way before being put into service as a superconductor.

Computer matching also helped in the analysis of twinning in alumina, shown in Figure 7.12. The offset dashed lines in the figure suggest a rigid lateral displacement across the twin plane, confirmed by the model used in the simulations. Although a "coarse" atomic model has been verified, the details of the interfacial configuration, particularly the site occupancy of the oxygen ions, has yet to be completed. It is not too difficult to spot the problem: note the skew rotation of bright ovals in the experimental image compared with the simulated image.

The sides of Figure 7.13 are not parallel for a purpose. Cropped along the columns of dots representing silicon atoms in {111} planes, the image has 17 columns at the bottom edge of the figure and 18 columns at the top of the figure. The extra column ends somewhere in the center of the image at a Shockley partial dislocation; it is the extra plane associated with an extrinsic stacking fault. Even the lattice strain appears in the image, as evidenced by the bowing of the columns around the core of the dislocation. At this scale, the magnitude of the atomic disturbances makes it easy to understand why such defects lead to the breakdown of electronic devices.

Metals present a bigger challenge for atomic resolution because they are the most densely packed materials. Body-centered-cubic (bcc) molybdenum is shown in Figure 7.14, an image of a grain boundary running nearly vertically through the center of the micrograph. It is possible to determine all atom positions completely at this internal junction of two grains; all atom positions are imaged (Penisson et al., 1982). However, the bcc structure is still an "open" structure compared to the "close-packed" face-centered-cubic (fcc) arrangement of atoms in aluminum. For comparison, Figure 7.15 is an atomic-resolution image of a grain boundary in aluminum, the boundary and all atom positions unmistakably visible in white dot contrast (Dahmen et al., 1988). Because grain boundaries are present (and often problematic) in all engineering alloys, images such as these represent significant opportunities for alloy diagnostics at the atomic level.

The last image (Figure 7.16) takes some study. It is a phase-contrast image of rapidly solidified Al-Mn, captured somewhere between the familiar amorphous arrangement of atoms in the liquid state and the familiar crystalline structure of the solid state (Gronsky et al., 1985). It is clear that there are rows of dots that continue in linear chains for measurable distances, suggesting lattice planes such as those imaged in the previous figures. However, a careful measurement of the spacings of any parallel set of these "planes" reveals that there are in fact two spacings, and that they are not periodic. Furthermore, the angle between the parallel sets is consistently 72°, indicative of a five-fold rotational axis normal to the observation plane. It is also possible to find, throughout the image, pentagonal arrangements of both white and black dots, indicating a high degree of rotational symmetry. Nevertheless, the lack of translational symmetry is the key to the identity of this phase as a "quasicrystal."

Figure 7.10 Phase-contrast image of $YBa_2Cu_3O_{7-\delta}$, showing three twin variants of the high-temperature tetragonal phase. The darker bands in the lower right-hand corner are the c planes, spaced by 0.117 nm.

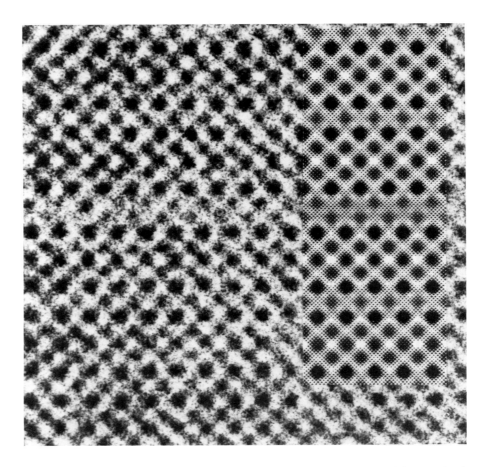

Figure 7.11 Atomic-resolution images of $YBa_2Cu_3O_{7-\delta}$, experimental (backdrop) and calculated (inset). The defect running horizontally through the image at just above center is a double layer of CuO at the location where a single layer should be. Excellent agreement between simulated and experimental images confirms the atomic structural model used in the calculations.

234

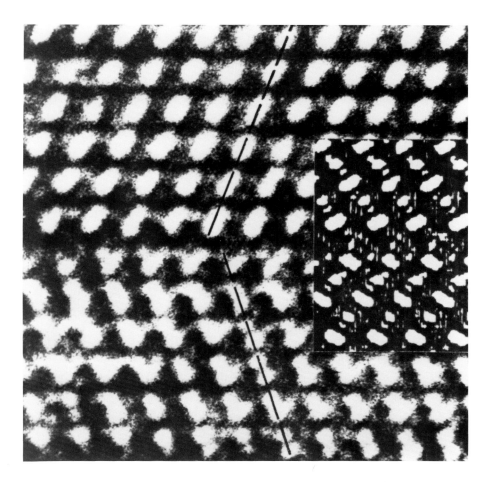

Figure 7.12 Phase-contrast image of a twin boundary (horizontal plane at center of image) in Al_2O_3. Here the inset simulated image shows a greater rotation of image detail across the twin plane than observed in the experimental image, but the amount of rigid translation (dashed line) is confirmed.

Figure 7.13 Phase-contrast image showing termination of an extrinsic stacking fault in silicon. The extra plane enters from the top of the image and ends in the center of the image at the position of a Shockley partial dislocation. Note the pronounced bending of the lattice planes around the defect. The closest separation distance between white dots in this image is 0.32 nm.

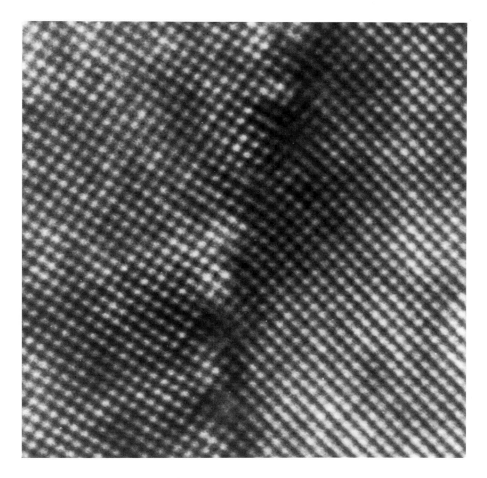

Figure 7.14 Atomic-resolution image of a grain boundary (vertical band) in molybdenum, a body-centered-cubic metal. The closest separation distance between white dots in this image if 0.23 nm.

Figure 7.15 Atomic-resolution image of a grain boundary (vertical band) in aluminum, a face-centered-cubic metal. The closest separation distance between white dots in this image is 0.2 nm.

Figure 7.16 Phase-contrast image of quasicrystalline Al-Mn, showing five-fold symmetrical distribution of white dots (closest separation distance 0.24 nm) representing atomic positions.

SUMMARY

Compared with other atomic imaging methods such as field ion microscopy (Chapter 9) and scanning tunneling microscopy (Chapter 8), the technique described here is a true "microscopy" because, in the traditional sense, it uses lenses. It therefore also uses the equations of optics, and it draws on a long and familiar history of image simulation and image interpretation. Nevertheless, it is also a "destructive" technique, requiring that the sample be made thin enough to transmit electrons. It is a hardware-intensive technique, requiring that the transmission electron microscope used for atomic resolution imaging be of exceptional design and condition. And it is a precise technique, requiring that the operator tune in a specific focus setting to within a few nanometers at worst. Help is on the way, however, with the incessant development of new computer-based instrumentation, new and "friendly" algorithms for image simulation, and on-line, real-time computer processing. The outlook for future images of materials at atomic resolution is excellent.

ACKNOWLEDGMENTS

The author is grateful for the invaluable assistance of his co-workers at the National Center for Electron Microscopy, Dr. Crispin Hetherington, Dr. Ulrich Dahmen, Dr. Michael O'Keefe, and Dr. Roar Kilaas, and for the excellent technical support provided by Chris Nelson, John Turner, and Doreen Ah Tye. This work, and the National Center for Electron Microscopy, are supported by the Director, Office of Energy Research, Office of Basic Energy Sciences, Materials Sciences Division of the U.S. Department of Energy under Contract No. DE-AC03-76SF00098.

REFERENCES

Dahmen, U., Douin, J., Hetherington, C. J. D., and Westmacott, K. H. (1988). In *Mat. Res. Soc. Symp. Proc.*, Vol. **139**, Materials Research Society, Pittsburgh, Penn., p. 87.

Gronsky, R. (1980). In *Proc. 38th Ann. Meeting Electron Mic. Soc. Amer.*, G. W. Bailey (ed.), Claitor's, Baton Rouge, p. 2.

Gronsky, R., Krishnan, K. M., and Tanner, L. E. (1985). In *Proc. 43rd Ann. Mtg. Elect. Mic. Soc. Amer.*, G. W. Bailey (ed.), San Francisco Press, San Francisco, p. 34.

Penisson, J. M., Gronsky, R., and Brosse, J. (1982). *Scripta Met.* **16**, 1239.

Zandbergen, H. W., Gronsky, R., Chu, M. Y., DeJonghe, L. C., Holland, G. F., and Stacy, A. (1988). In *Mat. Res. Soc. Symp. Proc.*, Vol. **99**, Materials Research Society, Pittsburgh, Penn., p. 553.

8

Surface Imaging by Scanning Tunneling Microscopy

S. CHIANG AND R. J. WILSON

The knowledge of the structure of a surface is extremely important to the fundamental understanding of the chemical reactions that occur there. Low-energy electron diffraction (LEED), atom diffraction, and ion channeling are some of the common experimental methods for studying surface structure. These methods are all indirect because calculations must be performed for model structures in order to fit the data. Periodic surface structures are also required in order to measure structures using some diffraction techniques. Other methods of measuring surface structure, such as photoemission and vibrational spectra, may be even less direct and more difficult to interpret. The measurement of the real-space image of a surface with atomic resolution is clearly very desirable for improved understanding of surface structure.

Real-space atomic resolution of surfaces is possible for special types of samples by well-known microscopic techniques. The advent of the high-energy transmission electron microscope (TEM) has recently permitted the real-space imaging of individual atoms in very thin samples with high mass contrast (see Chapters 5 and 7). The field ion microscope (FIM) is also able to image individual atoms at the edges of each layer of a very sharp tip using very high electric fields, as discussed in Chapter 9. While these microscopes have atomic-scale lateral resolution, comparable vertical resolution is not readily attainable.

The scanning tunneling microscope (STM) is an instrument that was developed by Binnig, Rohrer, Gerber, and Weibel (1982a,b) at the IBM Zurich Research Laboratory in 1982. Their work attracted attention because it provided for the first time a method of measuring the surface topography of electrically conducting materials in three dimensions with resolution as high as 0.01 Å vertically and ~2 Å laterally. Their early work applied the technique to such diverse systems as the famous Si(111)(7 × 7) reconstruction (Binnig et al., 1983a), reconstructed Au(110) (Binnig et al., 1983b) and Au(100) (Binnig et al., 1984) surfaces, and chemisorbed atoms (Baro et al., 1984). For their achievements, Binnig and Rohrer shared half of the 1986 Nobel Prize in Physics (Binnig and Rohrer, 1987). Several review articles present more detailed information on STM research (Binnig and Rohrer, 1985a,b; Hansma and Tersoff, 1987).

PRINCIPLES OF OPERATION OF STM

The tunneling of electrons from one conductor to another through an insulator is a purely quantum-mechanical phenomenon that has been known for over 60 years (Fowler and Nordheim, 1928). Although classically the electrons involved do not have enough energy to surmount the potential energy barrier of the insulator, quantum mechanically they have a small probability of reaching the other side of the barrier if the height and width of the barrier are sufficiently small. If two conductors are placed within a few angstroms of each other, the exponentially decaying electronic wave functions of the two conductors can overlap sufficiently that the electrons then have a probability of "tunneling" from one conductor to the other. If a potential is applied between the two conductors, the resulting tunneling current can be measured. For two flat, parallel electrodes, this tunneling current I_T has an exponential dependence on the distance s between the two conductors (Fowler and Nordheim, 1928):

$$I_T \alpha (V_T/s) \exp(-A\phi^{\frac{1}{2}}s)$$

Here $A \approx 1.025$ $(eV)^{-\frac{1}{2}}$ $Å^{-1}$ for a vacuum gap, ϕ is the average of the two electrode work functions, s is the distance between the electrodes, and V_T is the applied voltage. With work functions of a few eV, I_T changes by an order of magnitude for every angstrom change of s. Therefore, the current will always flow between the closest protrusions on the tip and sample.

In an STM the tip is mounted onto a piezoelectric scanner that allows its movement in three-dimensional space, as shown in Figure 8.1. As the tip is scanned laterally in a raster pattern across the surface, a feedback circuit maintains a constant tunneling current by moving the tip perpendicular to the sample. In the absence of chemical effects causing spatially varying work functions, the tip will trace a path equidistant from the atoms on the surface. Since we know the voltages applied to the piezoelectric elements, we know the path of the tip and can make a three-dimensional topographic map of the surface under study. This method of operation is depicted in Figure 8.1(a).

Alternatively, this tunneling current can be measured directly and used to make an image as the tip is kept at a constant height above the surface, as shown in the schematic diagram in Figure 8.1(b). Detailed theoretical analysis of either type of STM image can be extremely complicated, however, because it involves electronic

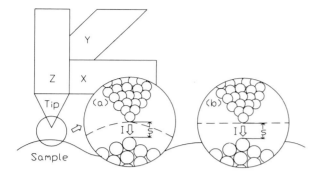

Figure 8.1 Schematic diagram showing operation of a scanning tunneling microscope. The tip is shown mounted onto a piezoelectric tripod with three orthogonal scanners marked *x, y,* and *z.* In the magnified regions (a) and (b), circles represent individual atoms on the tip and the sample, which are separated by distance *s.* The tunneling current *I* from tip to sample is indicated by arrow. The dashed lines show the path of the tip in (a) constant current mode and (b) constant height mode.

wave functions in both the sample and the tip, plus the three-dimensional tunneling process itself.

Spectroscopic information can also be obtained with the STM. In particular, measurements of the local tunneling barrier (which, at infinite distance from the sample, would be equivalent to the work function) can be obtained by modulating Δs while scanning. This is possible since the logarithmic derivative of tunneling current with respect to the electrode separation, from the equation above, is proportional to $\phi^{\frac{1}{2}}$ (Binnig and Rohrer, 1985a).

More detailed spectroscopic information can be obtained from actually measuring *I* versus *V* or *dI/dV* versus *V* curves at various lateral positions on the surface. Such measurements yield information on the surface density of states and electronic structure of the sample (Hamers et al., 1986; Stroscio et al., 1986).

STM INSTRUMENTATION

We are currently operating a state-of-the-art ultrahigh-vacuum (UHV) STM with *in situ* sample and tip transfer to an ultrahigh-vacuum surface analysis system (Chiang et al., 1988a). A schematic diagram of this system is shown in Figure 8.2. The analysis system consists of a VG Escalab, with LEED optics, 500 Å resolution

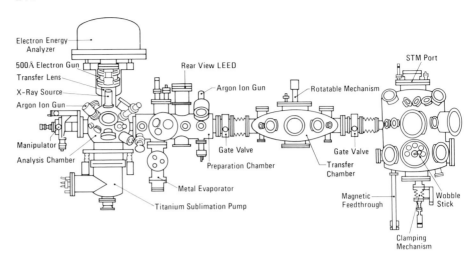

Figure 8.2 Layout of our UHV surface analysis and STM system, with VG Escalab analysis and preparation chambers on the left, and transfer chamber and STM chamber on the right. From Chiang et al. (1988a).

scanning Auger and scanning electron microscopy (SAM/SEM) and x-ray photoemission spectroscopy (XPS). The sample preparation chamber includes facilities for sputtering samples with inert gas ions, heating samples using an electron beam, dosing samples with gas atoms, and evaporating metal or subliming molecular layers onto samples.

The UHV STM itself, shown in the schematic diagram in Figure 8.3, is modeled after the early working designs of Binnig and Rohrer (Binnig et al., 1982a; Binnig and Rohrer, 1985b), with the tip mounted on a piezoelectric tripod scanner, a piezoelectric louse walker for the approach of the sample to the tip, and double spring stage suspension with magnetic eddy current damping. For additional vibration isolation, the STM has its principal elements mounted onto the top plate of a stack of stainless steel plates separated by viton spacers, like the "pocket" STM design (Gerber et al., 1986). The piezoelectric tubes in the tripod scanner have a sensitivity of ~100 Å/V, giving them each a range of ~3 μm when operated with ±150 V.

This STM has been operating very reliably for over 4 years on many samples, including semiconductors, metals, and molecules. In these studies, the instrument has demonstrated that it is capable of measuring surface corrugations below 0.1 Å. Many other STM designs also exist and have been described in the literature.

Our recent work has been very successful in demonstrating the application of the STM to surface structural problems on many different types of samples, including submonolayer coverages of metals on Si(111) (Wilson et al., 1987a,b,c, and 1988), individual atoms of clean metals (Hallmark et al., 1987; Wöll et al., 1989), and individual adsorbed molecules on metals (Ohtani et al., 1988; Chiang et al., 1988b,

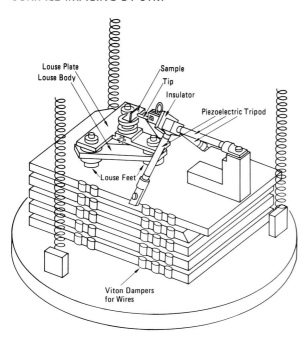

Figure 8.3 Schematic diagram of our STM, showing "pocket" STM hung on double spring stages. The louse piezoelectric walker carrying the sample is on the left, and the piezoelectric tube tripod for scanning the tip is shown on the right. From Chiang et al. (1988a).

1990; Lippel et al., 1989). These samples have all been studied using the UHV STM described above; the details of these studies are discussed in the following sections. Much of the terminology used in this chapter is specialized to surface chemistry and surface imaging. Therefore, the nonspecialist reader may simply wish to study the images rather than to attempt a full understanding.

METALS ON SEMICONDUCTORS

First, we discuss briefly the (7×7) structure of the clean Si(111) surface. (The term (7×7) refers to the surface unit cell dimensions. Atoms on the surface form a reconstructed structure that has two lattice repeat vectors, each seven times greater then the lattice vectors of the unreconstructed silicon below the surface.) The metal-on-semiconductor systems discussed here are as follows: (1) the $\sqrt{19} \times \sqrt{19}R$

23.4° (where R is the rotation angle with respect to the substrate unit cell) overlayer of Ni on Si(111) (Wilson and Chiang, 1987b), (2) Ag on Si(111) in the $\sqrt{3} \times \sqrt{3}R$ 30° structure (Wilson and Chiang, 1987a,c), and (3) the incommensurate Cu/Si(111)(5 × 5) structure (Wilson et al., 1988).

Si(111)(7 × 7)

Plate 8.1 shows a three-dimensional view of the (7 × 7) reconstruction of the clean Si(111) surface across an atomic step, with the characteristic 12 adatoms per unit cell and a corrugation of ~2 Å. (Adatoms are atoms that add onto the substrate to create the reconstructed surface structure.) This particular surface was first observed in the STM by Binnig et al. (1983b); it has now been observed by many groups and is often used as a test sample for new UHV STMs. The accepted model for the surface, Takayanagi's DAS (dimer–adatom stacking fault) model (Takayanagi et al., 1985), was determined by TEM and includes the adatoms observed by the STM. These adatoms are each bonded on top of the surface to three dangling bonds in the top half of the double layer.

Ni/Si(111)($\sqrt{19} \times \sqrt{19}$)$R$23.4°

When a small amount of nickel is added to the Si(111) surface, it is possible to form the $\sqrt{19} \times \sqrt{19}R$23.4° reconstruction. This surface has only one nickel atom per unit cell of 19 silicon atoms (Hansson et al., 1981). The STM image appears to have six large protrusions at the corners of the unit cell, plus one isolated bump in each cell, as shown in the image of unoccupied electronic states in Figure 8.4. By registering the observed STM images with a Si(111) unit mesh, we have successfully made a model, shown in Figure 8.5, that explains this structure and all of the features in our images (Wilson and Chiang, 1987b). This model has six Si atoms bonded to each Ni atom in the corners of the unit cell, as known from surface extended x-ray absorption fine structure (SEXAFS) measurements (Comin et al., 1983). Each of six Si adatoms are then bonded to the dangling bonds of the atoms near the corners of the unit cell. Three more dangling bonds in the structure surround a three-fold hollow site and can bond to an extra adatom (A) shown in the model, while the remaining three dangling bonds do not surround such a site and should appear as a Y-shaped minimum (M) in the observed corrugation. The features (A) and (M) expected from the model do indeed appear in the image in Figure 8.4, although the six adatoms around the corners are not resolved. Figure 8.4 also shows a surface dislocation in the arrangement of the Si adatoms. In this image, when the Ni sites are more closely spaced than normal, an adatom is eliminated (E); where the corner sites are excessively separated, an additional adatom can bind to the surface where a three-fold hollow site (H) is available.

Figure 8.6 shows a larger area STM scan of this structure. Now two domains (R1) and (R2) are observed, together with a surface dislocation (D) and a disordered area

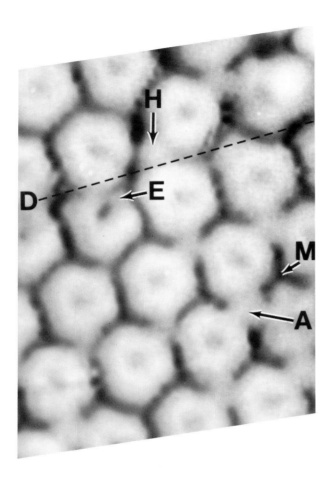

Figure 8.4 STM image of the Ni/Si(111)($\sqrt{19}$ × $\sqrt{19}$)$R23.4°$ surface over a ~60 × 80 Å² region, showing 3 Å corrugation. $I_T = 0.2$ nA and tunneling tip bias $V_T = -1.9$ V for tunneling into empty states of the sample. The ($\sqrt{19}$ × $\sqrt{19}$) cell edges are known to be 16.65 Å long. This STM image shows both ordered and defect regions. In most of the perfect cells an adatom (A) over a three-fold hollow site and the Y-shaped minimum (M) of the corrugation are evident. A surface dislocation (D), an eliminated adatom (E), and an extra adatom (H) are indicated. From Wilson and Chiang (1987b).

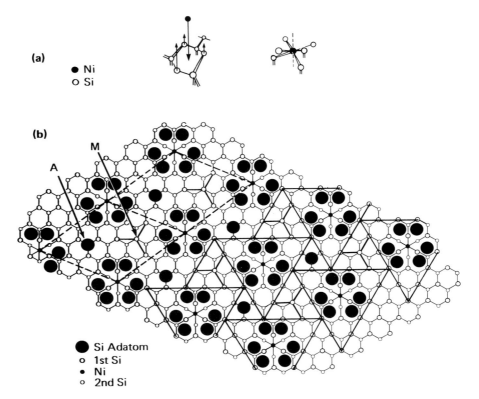

(a)

● Ni
○ Si

(b)

● Si Adatom
○ 1st Si
• Ni
○ 2nd Si

Figure 8.5 (a) Two views of the Si(111) surface before and after the introduction of an intralayer Ni atom. (b) Top view of the Ni site showing surrounding top double layer Si atoms and adatom positions inferred from Figure 8.4. From Wilson and Chiang (1987b).

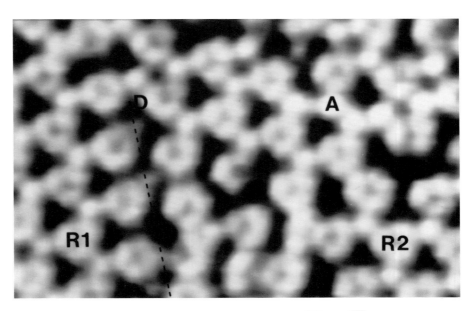

Figure 8.6 Image of a larger region of the Ni/Si(111)($\sqrt{19} \times \sqrt{19}$)$R23.4°$ sample, recorded at 0.2 nA and +1.20 V, showing two domains (R1) and (R2), together with surface dislocation (D) and disordered area (A). From Wilson and Chiang (1987b).

Figure 8.7 STM image of a 90×90 Å^2 region of the $(\sqrt{3} \times \sqrt{3})R \; 30°$ Ag/Si(111) interface. The corrugation of the ordered area is about 1 Å at a tip bias of $V_T = -0.39$ V. White represents elevated features in the gray scale representation. From Wilson and Chiang (1987a).

(A). The disordered area appears to consist of three-fold adatom structures around the corner holes, connected by isolated adatoms. The persistence of the three-fold structures into the disordered region is strong evidence for considering them to be independent stable subunits associated with the Ni atoms.

Ag/Si(111) $(\sqrt{3} \times \sqrt{3})R30°$

We have also studied several reconstructions of silver on the Si(111) surface (Wilson and Chiang, 1987a). Figure 8.7 shows an STM image of the $(\sqrt{3} \times \sqrt{3})R30°$ (hereafter called $\sqrt{3}$) Ag/Si(111) structure, showing a honeycomb array of protrusions on the surface. Several different models for this structure have been suggested, and three of them are shown in Figure 8.8. The honeycomb (H) model (LeLay, 1983) is based on an array of Ag atoms embedded in three-fold hollow sites of the Si surface. Van Loenen et al. (1987) have presented current-imaging tunneling spectroscopy data showing a honeycomb arrangement of protrusions that were interpreted as top-layer Si atoms of an embedded trimer (ET) structure (Horio and Ichimiya, 1985). The third model shown is that of Kono et al. (1987), which consists of a honeycomb arrangement of Ag atoms embedded in a missing top-layer (MTL) structure, where only the lower half of the Si(111) double layer is present.

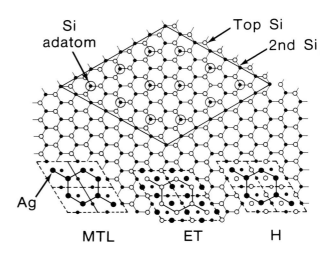

Figure 8.8 Adatom positions for the Si(111)(7 × 7) DAS structure (top), and three models for the $(\sqrt{3} \times \sqrt{3})R \sim 30°$ Ag/Si(111) surface (bottom). From Wilson and Chiang (1987c).

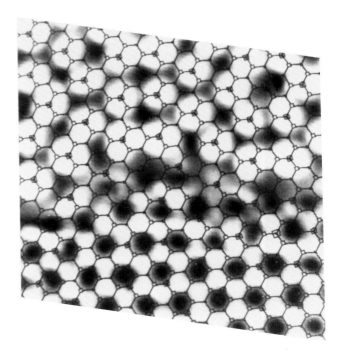

Figure 8.9 Top view of same data shown in Plate 8.2 with dark Si(111) mesh superimposed to show registration. White represents elevated features and small squares on the mesh mark adatom sites of the Si(111) DAS model. From Wilson and Chiang (1987c).

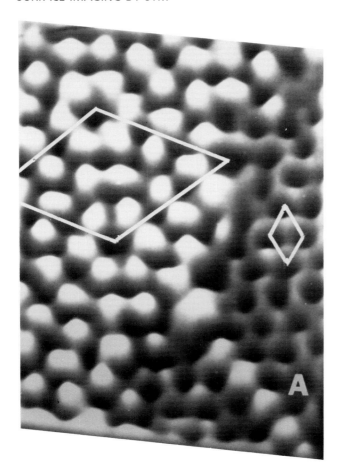

Figure 8.10 Contrast-enhanced STM image, recorded at $V_T = -1.88$ V and $i_T = 2.0$ nA, of a line of protrusions (A) in the lower right-hand corner, which extends from the clean Si(111)(7 × 7) region onto the $\sqrt{3}$ Ag/Si domain. The unit cells are indicated on the three-dimensional view. From Wilson and Chiang (1987c).

By depositing silver on a substrate held at 500°C, it is possible to make domains of the $\sqrt{3}$ structure that coalesce along step edges next to regions of the clean Si(111)(7 × 7) surface (Wilson and Chiang, 1987c). Plate 8.2 shows a three-dimensional view of such domains, after image-processing contrast enhancement (Wilson and Chiang, 1988), which facilitates the display of the 2 Å high Si adatoms in the (7 × 7) region next to the 0.2 Å corrugation of the $\sqrt{3}$ structure. By registering the adatoms of the Si(7 × 7) region with their known binding sites from the Takayanagi DAS model (Takayanagi et al., 1985), we have been able to use a computerized least-squares fit to determine that the protrusions in the honeycomb array observed by the STM in the $\sqrt{3}$ region have the same binding sites as silver atoms in three-fold hollow sites in the H or MTL models (Wilson and Chiang, 1987c). This registration of the STM data with the Si(111) mesh is shown in Figure 8.9.

Figure 8.10 shows a contrast-enhanced three-dimensional view of another domain boundary, where a line of protrusions (A) extends from the (7 × 7) region onto the $\sqrt{3}$ region. The large height, 2 Å, and diameter, 5 Å, of these protrusions strongly resembles the dimensions of Si adatoms in the (7 × 7) region. The registration of this image with a Si(111) mesh shows that the binding site of these atoms is also consistent with those of Si adatoms, making the MTL model unlikely because it would lack dangling bonds to which the Si adatoms could bind in the $\sqrt{3}$ region. Many different groups, however, are still working on the problem of the structure of this Ag/Si system, and it now appears that the correct model may be more complicated than any of those discussed here.

Cu/Si(111) (5 × 5)

The Cu on Si(111)(5 × 5) layer has been described as an incommensurate structure on the basis of LEED observations of nonintegral order spots that fall between those expected for (5 × 5) and (6 × 6) structures (Daugy et al., 1985; Kemmann et al., 1987). An incommensurate structure is formed when a mismatch occurs in the lattice constant of the adsorbate layer, a_a, relative to that of the substrate, a_s, such that regular, repeating subunits are not formed (Villain, 1980). The (5 × 5) structure is complete at coverages of about 1 monolayer, defined as 1 Cu atom per top layer Si atom.

The STM images of this structure typically show regions of hexagonally packed (HP) protrusions, with the same spacing as the (1 × 1) lattice of Si(111), together with some larger triangular depressions, as shown in Figure 8.11. A Si(111) mesh is shown superimposed on the STM image in Figure 8.11(a), with the HP protrusions mapped onto three-fold hollow sites of the mesh. The triangular depressions map onto equivalent positions in the irregularly spaced (5 × 5) subunits shown. These (5 × 5) subunits have lattice spacings varying from 5 to 7 a_s, where a_s = 3.84 Å for the Si(111) surface. The observation of these irregularly spaced subunits in the STM images indicates that substrate–adsorbate interactions dominate lateral interactions within the Cu layer. A three-dimensional view of this same image is shown in Figure 8.11(b).

Figure 8.12 illustrates the difficulty of observing STM images on the Cu/Si(111)(5 × 5) surface as compared with the (7 × 7) surface in that the lateral resolution must be significantly better, and the vertical sensitivity is increased by a factor of 10.

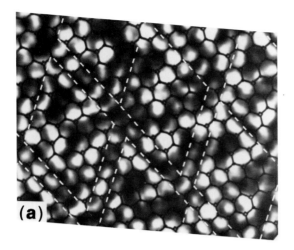

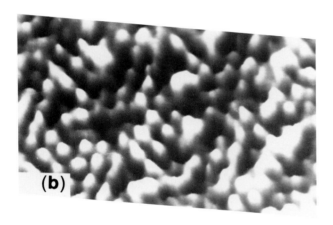

Figure 8.11 (a) STM image of Cu/Si(111)(5 × 5) sample, recorded with $V_T = +0.5$ V and $i_T = 2.0$ nA, overlaid with the black Si(111) mesh. Open (closed) circles represent first (second) layer Si atoms. White dashed lines show a decomposition of the image into (5 × 5) cells and domain walls. (b) A three-dimensional perspective view of a small (5 × 5) region emphasizes the internal structure of the cell. From Wilson et al. (1988).

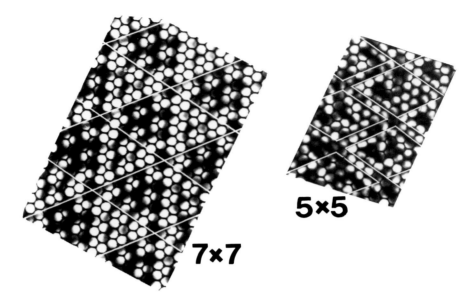

Figure 8.12 STM images of clean Si(111)(7 × 7) and Cu/Si(111)(5 × 5) portions of the surface, after mapping onto the Si(111) mesh. The scan widths are about 90 and 60 Å, respectively. From Wilson et al. (1988).

CLEAN METAL SURFACES

Individual close-packed metal atoms were first observed in our laboratory on Au(111) thin films (Hallmark et al., 1987). This work has recently been extended to determine the atomic positions in the $(23 × \sqrt{3})$ stacking fault reconstruction on a Au(111) single crystal (Wöll et al., 1989).

Au(111) Atoms

Our group made the first STM observations resolving individual close-packed metal atoms on the surface of an epitaxial Au(111) thin film on mica (Hallmark et al., 1987). These measurements were performed both in the constant current mode in ultrahigh vacuum and also in the constant height mode in air. Plate 8.3 shows a three-dimensional view of a constant height image of Au in a thin film measured in air, with an atomic spacing of ~2.8 Å. Previous to these measurements, it had been thought that the STM could not resolve close-packed metal atoms because the electronic wave functions were not sufficiently spatially localized. The observation of an unexpectedly large corrugation, ~ 0.3 Å, presumably results from a significant electronic enhancement whose source is not yet understood. Nonetheless, this work opened the door for high-resolution investigations of metal surfaces by STM. Recently, individual metal atoms have also been observed on Al(111) (Wintterlin et al., 1989), Cu(100) (Lippel et al., 1989), Pt(111) (Eigler, private communication), and Au(100) (Kuk, private communication).

Au(111) (23 × $\sqrt{3}$) Reconstruction

The earlier work on imaging individual Au atoms has now been extended to studies of the (23 × $\sqrt{3}$) stacking fault reconstruction on a single-crystal Au(111) sample (Wöll et al., 1989). Previous TEM images led to a model for the surface atom rearrangement involving an ordered array of boundaries between surface regions with face-centered-cubic (fcc) type stacking (*ABC*) and hexagonal-close-packed (hcp) stacking (*ABA*) (Takayanagi and Yagi, 1983). The atomic-resolution STM measurements showed directly for the first time the atomic occupation of hcp sites on an fcc metal. The good agreement of the STM images with previous models of the reconstruction (Takayanagi and Yagi, 1983; Harten et al., 1985) suggests that the STM measures spatially localized electronic states near the atomic cores. The soliton model from helium atom scattering experiments (Harten et al., 1985) is shown in Figure 8.13. The STM measurements can then be used directly to obtain the precise relative positions of the atoms in this large reconstructed unit cell. Figure 8.14 is a high-resolution STM image showing the lateral displacement of the atoms from the straight line in the [110] direction, analogous to the line in the model of Figure 8.13. Even in images without atomic resolution, the domain boundaries between regions with hcp and fcc type atomic stacking could be clearly observed. These domain boundaries were observed to be ~ 22 Å apart in a unit cell ~ 66 Å wide, and their corrugation height was only 0.15 ± 0.04 Å, in agreement with the helium scattering data of Harten et al. (1985).

ATOMIC-RESOLUTION IMAGES OF ADSORBED MOLECULES ON METALS

Recently, we have made the first high-resolution images of molecules on surfaces by STM that show individual molecules and their internal structure. Two systems have been studied in detail: (1) benzene and carbon monoxide (CO) coadsorbed on Rh(111) in two different ordered overlayers, the (3 × 3) (Ohtani et al., 1988) and the $c(2\sqrt{3} \times 4)$rect (Chiang et al., 1988b, 1990) and (2) copper phthalocyanine on Cu(100) (Lippel et al., 1989).

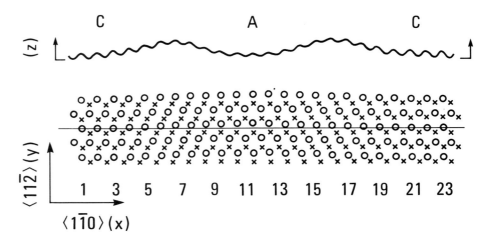

Figure 8.13 Model for the reconstruction of the Au(111) surface (after Harten et al., 1985). The crosses denote the positions of atoms in the second layer, whereas open circles denote the positions of atoms in the reconstructed top layer. C and A mark the regions of *ABC* (fcc) and *ABA* (hcp) stacking, respectively. The lattice defect at the boundary between these two regions corresponds to a bulk Shockley partial dislocation with Burgers vector $\frac{1}{6}$ (11$\bar{2}$). The displacement of the atoms from the straight line that has been drawn in the [1$\bar{1}$0] direction is clear. From Wöll et al. (1989).

Figure 8.14 STM image, taken in UHV at $V_T = 0.611$ V and $i_T = 0.3$ nA, for an epitaxially grown Au(111) thin film on mica, showing a region of 100×40 Å². The atomic resolution of the image allows the determination of the positions of single atoms, and the displacement of the atoms from the straight line in the [1$\bar{1}$0] direction can be directly compared with the model in Figure 8.13 (From Wöll et al. (1989).

Coadsorbed Benzene and CO on Rh(111)

The system of benzene and CO coadsorbed on Rh(111) was selected because the overlayer was known to be extremely stable by previous LEED measurements (Mate and Somorjai, 1985). Because the molecules are strongly chemisorbed in a saturated layer, their diffusion rate was limited, permitting the imaging of the molecules with our room-temperature UHV STM (Chiang et al., 1988a). Ringlike features associated with individual benzene molecules were observed in STM images for two different arrangements of the molecules. In one arrangement, features associated with individual CO molecules were also clearly identified. Translational and rotational domain boundaries, molecules adsorbed near step edges, and evidence for surface diffusion were also observed.

(3 × 3) Structure

The (3 × 3) ordered superlattice of coadsorbed benzene and CO on Rh(111) was studied first (Ohtani et al., 1988). The unit cell shown in Figure 8.15, as determined by dynamical low-energy electron diffraction analysis (Lin et al., 1987) contains one flat-lying benzene molecule and two upright CO molecules, all chemisorbed over hcp-type three-fold hollow sites directly over second layer Rh atoms. Figure 8.16 shows a wide-area STM image of this surface with a variety of steps and defects. Here each bright spot in the ordered array is the image of a single benzene molecule. The sharp, monatomic step in the middle of the image lies along a [011] direction so that the step has the relatively open face of a (100)-type square lattice, as does the double step in this image. The rougher step at the left of the image has a (111)-type close-packed face. The observation of benzene molecules near these steps opens the door for further study of chemistry and diffusion on stepped surfaces.

At tip voltage $V_T = -0.010$ V and tunneling current $i_T = 2$ nA, 2 Å high, three-fold symmetric ringlike structures were observed that could be associated with the benzene molecules, although the CO molecules were not clearly resolved, as shown in Plate 8.4. The observed three-fold symmetry of the molecular images is presumably due to the interaction of the π bonds of the benzene molecules with the rhodium substrate atoms below. For the high-resolution images, the STM was imaging empty states near the Fermi Energy, E_F, which must be mostly metallic, with some contribution from hybridization with molecular states (Garfunkel et al., 1986). Finally, comparison of STM images near a step edge taken about 10 minutes apart showed evidence of benzene molecular diffusion between images (Ohtani et al., 1988). In some cases, molecules that appeared diffuse in the image, probably due to their diffusion during the measurement of the STM image, seemed to shift into their favored (3 × 3) lattice positions.

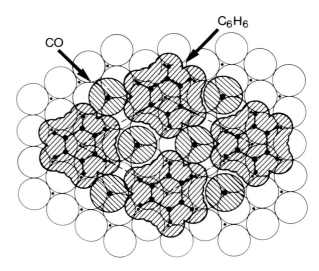

Figure 8.15 The structure of Rh(111)(3 × 3)(C₆H₆ + 2CO) determined from dynamical LEED analysis (Lin et al., 1987). Large circles and small dots represent the first- and second-layer metal atoms respectively. From Ohtani et al. (1988).

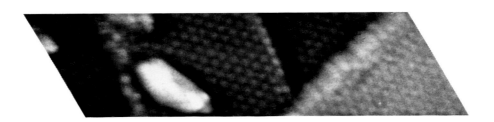

Figure 8.16 Three steps and the (3 × 3) superlattice of benzene molecules on each Rh(111) terrace are visible in this 300 × 90 Å² image. $V_T = -1.25$ V and $i_T = 4$ nA. Statistical differencing (Wilson and Chiang, 1988) has been used to reduce the apparent height of the atomic steps and defects in order to render the (3 × 3) corrugation visible. From Ohtani et al. (1988).

258

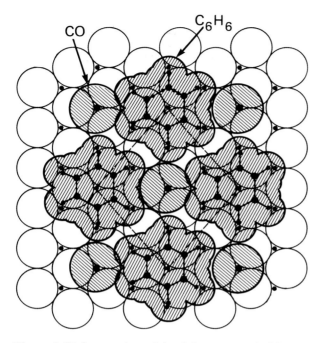

Figure 8.17 Structural model of benzene and CO on Rh(111) in $c(2\sqrt{3} \times 4)$rect array, as determined by dynamical LEED analysis (Van Hove et al., 1986). Large circles and small dots represent first and second layer Rh atoms, respectively. The dashed lines show the primitive unit cell, with benzene molecules at the corners and one CO molecule in the center. From Chiang et al. (1990).

$c(2\sqrt{3} \times 4)$rect Structure

STM studies of the coadsorbed benzene and CO molecular system on Rh(111) were later extended to the $c(2\sqrt{3} \times 4)$rect overlayer (Chiang et al., 1988b, 1990). In those studies, CO molecules were resolved spatially for the first time, with the small protrusions associated with the CO appearing approximately three times smaller than the three-fold ring-like structures resolved for the benzene molecules. These results gave the first demonstration of the ability of the STM to image two types of chemisorbed molecules and distinguish between them, an important step towards real-space imaging of chemical reactions at surfaces.

The primitive unit cell in Figure 8.17 for this structure, as determined by dynamical LEED analysis (Van Hove et al., 1986), contains one flat-lying benzene molecule and only one CO molecule, each chemisorbed over an hcp-type three-fold

(a)   **(b)**

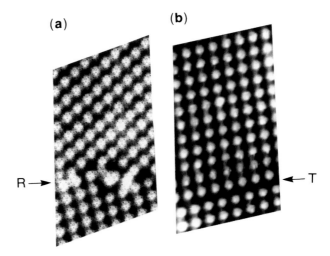

R→   ←T

Figure 8.18 (a) Rotational domain boundary (R) between two domains of $c(2\sqrt{3} \times 4)$rect array of benzene and CO on Rh(111), with $V_T = -0.21$ V and $i_T = 2$ nA. The mesh, with large (small) diamonds indicating top (second) layer Rh atoms, has been overlaid on the data according to the dynamical LEED model. (b) Translational domain boundary (T) between two domains of $c(2\sqrt{3} \times 4)$rect array of benzene and CO on Rh(111), with $V_T = -0.08$ V and $i_T = 2$ nA. From Chiang et al. (1990).

hollow site. Three rotational domains of this overlayer have been observed in STM images, together with translational domain boundaries, atomic steps, and isolated defects (Chiang et al., 1990). Figure 8.18 shows examples of both rotational and translational domain boundaries between arrays of benzene molecules. Figure 8.19, obtained for $i_T = 2$ nA and $V_T = -0.010$ V, shows two views of our best STM image, which clearly resolved both the CO molecules and the hole in the benzene molecules (Chiang et al., 1988b). Figure 8.19(a) shows a gray-scale top view of the $\sim 30 \times 60$ Å^2 area overlaid by a mesh with the large and small diamonds representing the top and second layer Rh atoms, respectively. When the large bright rings, identified as benzene molecules from their spacing, were placed on hcp-type threefold hollow sites according to the LEED model (Van Hove et al., 1986), the small protrusions in the image were identified with CO molecules because they lie exactly at the positions expected for CO in the unit cell. The three-dimensional view shown in Figure 8.19(b) displays more clearly the relative heights of the benzene molecules, ~ 0.6 Å, compared with the 0.2 Å CO protrusions. Again, as in the (3×3) overlayer, the benzene molecules appear to be three-fold symmetric, presumably due to the interaction of the benzene molecular states with those of the rhodium substrate. The observation of features that can be associated with both benzene and

(a)

(b) CO C$_6$H$_6$

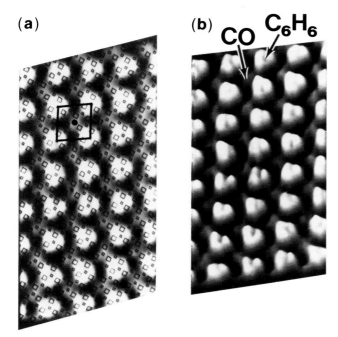

Figure 8.19 (a) Top view of STM image, measured at V_T = -0.010 V and i_T = 2 nA, of $c(2\sqrt{3} \times 4)$rect array of benzene and CO molecules on Rh(111), with brightness proportional to height. The mesh, with large (small) diamonds indicating top (second) layer Rh atoms, has been overlaid on the data according to the dynamical LEED model (Van Hove et al., 1986). The primitive unit cell is shown by solid lines, with benzene molecules at the corners and one CO molecule at the filled circle. (b) Three-dimensional view of the same data shown in (a), with three-fold benzene features ~0.6 Å high and smaller CO protrusions ~0.2 Å high. From Chiang et al. (1988b).

CO molecules indicates that the STM is sensitive to localized states of these molecules which are hybridized with the metallic states to yield states near E_F.

Copper Phthalocyanine on Cu(100)

Recently, we have obtained the first STM images showing the internal structure of isolated molecules. These measurements were performed on copper phthalocyanine (Cu-phth) molecules, which adsorb in a flat orientation on Cu(100) with two different rotational orientations (Lippel et al., 1989). A schematic diagram of the Cu-phth

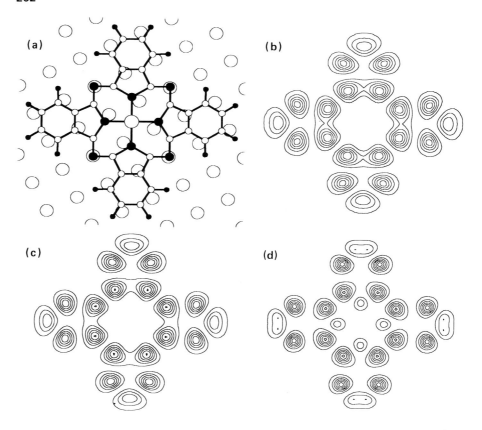

Figure 8.20 (a) Model of the copper-phthalocyanine molecule above a Cu(100) surface. Small (large) circles are C (Cu) atoms and small (large) filled circles are H (N) atoms. The Cu(100) lattice is shown rotated by 26.5°. (b) and (c) Contour plots of the highest occupied molecular orbital (HOMO) and lowest unoccupied molecular orbital (LUMO) 2 Å above the molecular plane. (d) Charge density of the HOMO 1 Å above the plane. From Lippel et al. (1988).

molecule on a Cu(100) surface is shown in Figure 8.20, together with the highest occupied molecular orbital (HOMO) and lowest unoccupied molecular orbital (LUMO) from simple Hückel calculations of the free molecular states. Atomic-scale features within the molecules agree well with the Hückel calculations. Tip-induced motion of isolated molecules has sometimes been observed, as have isolated molecules above an atomically resolved metal surface. Near 1 monolayer coverage, the two domains evident in LEED (Buchholz and Somorjai, 1977) are observed, each corresponding to one of the molecular orientations present at low coverages. Simultaneous observations of molecules and the metal–substrate corrugation indicate that insight into the molecular binding sites can be obtained. Unusual molecular binding sites at step edges have also been observed for the first time, with important implications for the role of surface roughness in molecular catalysis.

Plate 8.5(a), measured at low bias, shows the Cu-phth molecules as delocalized pedestals with weak internal structure, with the fine structure emphasized by baseline subtraction. A gray-scale representation of the calculated HOMO has been embedded in the image. The inequivalent appearance of the two molecular orientations implies a strong asymmetry of the tip. Plate 8.5(b) shows a stepped region with gray levels chosen to highlight the fine structure within the individual molecules. This fine structure appears as four protrusions on each of the four lobes of the molecule, agreeing surprisingly well with the calculations, in which the four predominant protrusions per lobe arise from highly populated carbon atom p states.

Good STM images of Cu-phth were not obtainable for coverages higher than 1 ML on Cu(100), or at submonolayer coverages on Au(111) or Si(111). Instabilities in high coverage images may have been associated with a buildup of molecules on the tip, resulting in poorly conducting layers that reduce the tip–molecule gap. For Au(111), low activation energy barriers allowing rotations or translations of molecules during the time required for imaging may contribute to smearing of the images. The ability to choose a substrate on which the molecules are bound by a highly corrugated molecule–surface interaction potential appears to be essential for high-resolution imaging of isolated molecules. Simultaneous observation of molecules and of the metal–substrate corrugation appears to be difficult because of strong tip–surface interactions accompanying small tunneling gaps.

SUMMARY

The images we have shown demonstrate that the STM is an excellent tool for understanding surface structure on an atomic scale on well-characterized surfaces. A large number of scientists are now working in the field, and developments are occurring rapidly in applying the technique to many types of samples, including organic and biological ones. STM has already been used extensively to study semiconductor systems. Further developments should permit the technique to be used to study metal overlayers on metals at an atomic level. Applications of the instrument to metal–adsorbate surfaces also appear promising for observations of surface chemical processes, such as molecular diffusion, nucleation phenomena, and step- or defect-related reactivity.

ACKNOWLEDGMENTS

We thank our collaborators for their work on the following experiments discussed here: F. Salvan for the Cu/Si(111) work; V. M. Hallmark, J. Rabolt, and J. D. Swalen for the first Au atom observations; Ch. Wöll and P. H. Lippel for the work on the reconstruction of Au(111);

H. Ohtani and C. M. Mate for the observations of benzene and CO on Rh(111); and P. H. Lippel, M. D. Miller, and Ch. Wöll for the work on copper-phthalocyanine molecules. We would also like to thank Ch. Gerber for his aid in building the UHV STM, J. Shyu and B. Hoenig for computer programming assistance, and M. Flickner and W. Niblack for assistance in image processing.

REFERENCES

Baro, A., Binnig, G., Rohrer, H., Gerber, Ch., Stoll, E., Baratoff, A., and Salvan, R. (1984). *Phys. Rev. Lett.* **52**, 1304.

Binnig, G., and Rohrer, H. (1985a). *IBM J. Res. Develop.* **30**, 355.

———, and Rohrer, H. (1985b). *Sci. Am.* **253**, 50.

———, and Rohrer, H. (1987). *Rev. Mod. Phys.* **59**, 615.

———, Rohrer, H., Gerber, Ch., and Stoll, E. (1984). *Surf. Sci.* **144**, 321.

———, Rohrer, H., Gerber, Ch., and Weibel, E. (1982a). *Helv. Phys. Acta* **55**, 726.

———, Rohrer, H., Gerber, Ch., and Weibel, E. (1982b). *Phys. Rev. Lett.* **49**, 57.

———, Rohrer, H., Gerber, Ch., and Weibel, E. (1983a). *Phys. Rev. Lett.* **50**, 120.

———, Rohrer, H., Gerber, Ch., and Weibel, E. (1983b). *Surf. Sci.* **131**, L379.

Buchholz, J. C., and Somorjai, G. A. (1977). *J. Chem. Phys.* **66**, 573.

Chiang, S., and Wilson, R. J. (1987). *Anal. Chem.* **59**, 1267A.

———, Wilson, R. J., Gerber, Ch., and Hallmark, V. M. (1988a). *J. Vac. Sci. Techol.* **A6**, 386.

———, Wilson, R. J., Mate, C. M., and Ohtani, H. (1988b) *J. Microsc.* **152**, 567.

———, Wilson, R. J., Mate, C. M., and Ohtani, H. (1990). *Vacuum* **41**, 118.

Comin, F., Rowe, J. E., and Citrin, P. H. (1983). *Phys. Rev. Lett.* **51**, 2402.

Daugy, E., Mathiez, P., Salvan, F., and Layet, J. M. (1985). *Surf. Sci.* **154**, 267.

Eigler, D., private communication.

Fowler, R. H., and Nordheim, L. (1928). *Proc. R. Soc. London* **A119**, 173.

Garfunkel, E. L., Minot, C., Gavezzotti, A., and Simonetta, M. (1986). *Surf. Sci.* **167**, 177, and E. L. Garfunkel, private communication.

Gerber, Ch., Binnig, G., Fuchs, H., Marti, O., and Rohrer, H. (1986). *Rev. Sci. Inst.* **57**, 221.

Hallmark, V. M., Chiang, S., Rabolt, J. F., Swalen, J. D., and Wilson, R. J. (1987). *Phys. Rev. Lett.* **59**, 2879.

Hamers, R. J., Tromp, R. M., and Demuth, J. E. (1986). *Phys. Rev. Lett.* **56**, 1972.

Hansma, P. K., and Tersoff, J. (1987). *J. Appl. Phys.* **61**, R1.

Hansson, G. V., Bachrach, R. Z., Bauer, R. S., and Chiaradia, P. (1981). *Phys. Rev. Lett.* **46**, 1033.

Harten, U., Lahee, A. M., Toennies, J. Peter, and Wöll, Ch. (1985). *Phys. Rev. Lett.* **54**, 2619.

Horio, Y., and Ichimiya, A. (1985). *Surf. Sci.* **164**, 589.

Kemmann, H., Muller, F., and Neddermeyer, H. (1987). *Surf. Sci.* **192**, 11.

Kono, S., Higashiyama, K., Kinoshita, T., Miyahara, T., Kato, H., Ohsawa, H., Enta, Y., Maeda, F., and Yaegashi, Y. (1987). *Phys. Rev. Lett.* **58**, 1555.

Kuk, Y., private communication.

LeLay, G. (1983). *Surf. Sci.* **132**, 169, and references therein.

Lin, R. F., Blackmann, G. S., Van Hove, M. A., and Somorjai, G. A. (1987). *Acta Crystallogr.* **B43**, 368.

Lippel, P. H., Wilson, R. J., Miller, M. D., Wöll, Ch., and Chiang, S. (1989). *Phys. Rev. Lett.* **62**, 171.

Mate, C. M., and Somorjai, G. A. (1985). *Surf. Sci.* **160**, 542, and references therein.

Ohtani, H., Wilson, R. J., Chiang, S., and Mate, C. M. (1988). *Phys. Rev. Lett.* **60**, 2398.

Stroscio, J. A., Feenstra, R. M., and Fein, A. P. (1986). *Phys. Rev. Lett.* **57**, 2579.

Takayanagi, K., and Yagi, K. (1983). *Trans. Jpn. Inst. Mat.* **24**, 337.

———, Tanishiro, Y., Takahashi, S., and Takahashi, M. (1985). *Surf. Sci.* **164**, 367.

Van Hove, M. A., Lin, R. F., and Somorjai, G. A. (1986). *J. Am. Chem. Soc.* **108**, 2532.

Van Loenen, E. J., Demuth, J. E., Tromp, R. M., and Hamers, R. J. (1987). *Phys. Rev. Lett.* **58**, 373.

Villain, J. (1980). In *Ordering in Strongly Fluctuating Condensed Matter Systems*. T. Riste (ed.), 221, Plenum, N.Y.

Wilson, R. J., and Chiang, S. (1987a). *Phys. Rev. Lett.* **58**, 369.

———, and Chiang, S. (1987b). *Phys. Rev. Lett.* **58**, 2575.

———, and Chiang, S. (1987c). *Phys. Rev. Lett.* **59**, 2329.

———, and Chiang, S. (1988). *J. Vac. Sci. Technol.* **A6**, 398.

———, and Chiang, S., and Salvan, F. (1988). *Phys. Rev. B* **38**, 12696.

Wintterlin, J., Wiechers, J., Brune, H., Gritsch, T., Hofer, H., and Behm, R. J. (1989). *Phys. Rev. Lett.* **62**, 59.

Wöll, Ch., Chiang, S., Wilson, R. J., and Lippel, P. H. (1989). *Phys. Rev. B.* **39**, 7988.

9

Atom Probe Field-Ion Microscopy: Imaging at the Atomic Level

M. K. MILLER AND M. G. BURKE

Sam Weller, in Dickens' *Posthumous Papers of the Pickwick Club,* responded to a question from Serjeant Buzfuz as to why he saw nothing by saying: "If they wos a pair o' patent double million magnifyin' gas microscopes of hextra power, p'raps I might be able to see . . . ." Over a century later his idea was to become a reality. In 1951, Professor Erwin W. Müller unveiled a powerful new type of microscope that used a gas to produce images with over a million times magnification. This instrument, known as a field-ion microscope, enabled individual atoms to be seen for the first time. An example of the type of image produced by this instrument is shown in Figure 9.1. In this micrograph, each spot is a highly magnified image of a single atom. A few years later, a mass spectrometer was added to the instrument so that the elemental identity of the atoms could also be determined. The combined instrument is known as an atom probe field-ion microscope (APFIM). A modern instrument combines several variants of the atom probe into a single vacuum system, as shown schematically in Figure 9.2. The APFIM has the highest spatial resolution in chemical analysis of any microanalytical tool yet devised.

PRINCIPLE OF FIELD-ION MICROSCOPY

The field-ion microscope is remarkable for its simplicity. A schematic diagram of a field-ion microscope is shown in Figure 9.3. The specimen, in the form of a very sharp needle, is mounted on a cryostat (liquid-nitrogen cooled support) in the center

266

Figure 9.1 A true atomic-resolution micrograph of tungsten. Each spot in this field-ion micrograph is the image of a single atom.

of a vacuum chamber and points towards a fluorescent screen. The microscope is filled to a pressure of about 10^{-3} Pa with an inert gas such as helium or neon. A high positive voltage, in the kilovolt range, is then gradually applied to the specimen. The gas atoms close to the specimen are polarized because of the electric field and then are drawn toward the apex of the needle. Eventually, if the electric field is high enough, gas atoms just above the specimen surface become ionized and are repelled from the specimen toward the fluorescent screen, where they produce a magnified image of the specimen surface. Since the ionized image gas atoms are repelled radially from each point on the specimen surface, the resulting image is a simple projection of the curved specimen surface onto the flat screen. A channel-plate image intensifier is usually incorporated into the screen assembly to increase the brightness of the image. A normal specimen-to-screen distance R is 50 mm, and the radius of the specimen r is typically 50 nm. The magnification of the image is defined simply by the ratio of the distance that the specimen is from the screen divided by the end radius of the specimen (i.e., $R/r = 10^6$ times).

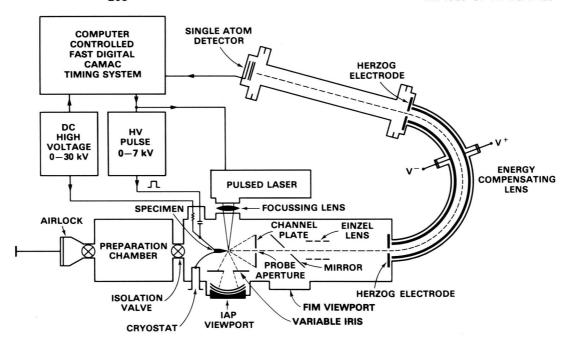

Figure 9.2 Schematic diagram of the main components of a modern atom probe with a field-ion microscope, a time-of-flight energy-compensated mass spectrometer, an imaging atom probe, and a pulsed laser atom probe. The operation of the instrument is controlled by a microcomputer.

A field-ion micrograph of an Ni_4Mo specimen is shown in Figure 9.4. In addition to resolving individual atoms, many sets of concentric rings are evident in field-ion micrographs. These rings are a characteristic feature of field-ion micrographs and may be understood by examination of the atomic details of the surface. Schematic diagrams of a small portion of the curved surface of a field-ion specimen, where each atom is represented by a circle, are shown in Figure 9.5. The upper portion in Figure 9.5 is a view from above the surface, and the lower portion is a cross-sectional view through the specimen. The rings in the field-ion image arise because of preferential ionization of the image gas above sites that protrude from the surface. The most protruding sites are the edges of the atomic planes where they intersect the curved surface. These positions are indicated in the cross-sectional view by the filled circles. By considering only these edge sites, that is, the filled circles, in the top view, the rings become apparent. Computer simulations may also be performed to demonstrate this imaging behavior, as shown in Figure 9.6. In this simulation, only the edge atoms within a very thin surface region are permitted to contribute to the image. These simulations permit the full image to be modeled for different crystal structures and specimen orientations. The many sets of concentric ring patterns are evident in the simulation. Each set of rings corresponds to a different set

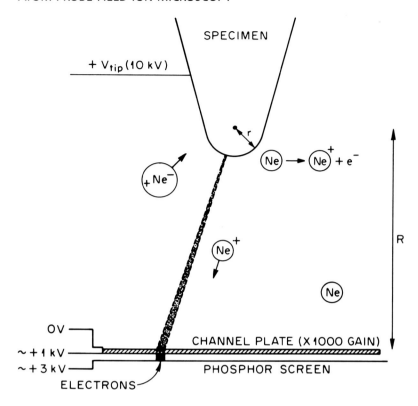

Figure 9.3 Principle of field-ion microscopy. Image gas atoms, usually neon or helium, are ionized near the apex of the needle-shaped specimen owing to the high electric field. These ions are then radially projected from the specimen to a channel plate and phosphor screen, where they produce a spot of light. The magnification is given by the specimen-to-screen distance R divided by the end radius of the specimen r. Typical magnifications are in the millions.

of atomic planes in the crystal. The position, symmetry, and prominence of these planes provide information concerning the crystal structure of the specimen. In practice, however, the identity of the region under investigation and therefore its crystal structure are usually inferred from the composition as measured in the atom probe or by other means. The crystallographic orientation of the specimen may be determined from the position of the planes in the image since the angles between planes are established for a given crystal structure.

One of the most significant contributions that field-ion microscopy has made to the understanding of metals is the ability to image the structure of boundaries and interfaces at the atomic level. Field-ion micrographs revealed that perfect crystallinity is maintained right up to the boundary, and there is no disordered or amorphous region at the boundary. This extremely narrow region of atomic disturbance is demonstrated in Figure 9.7 for a grain boundary in tungsten. Similar results

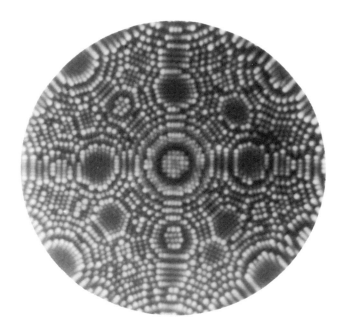

Figure 9.4 Field-ion micrograph of long-range-ordered Ni$_4$Mo specimen showing individual atoms. Also evident are the rings characteristic of field-ion micrographs.

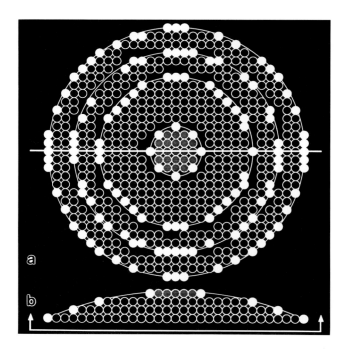

Figure 9.5 Schematic diagrams of a small section of a field-ion specimen in which each atom is represented by a circle. The upper portion (a) is a view from above the specimen, and (b) is a cross-sectional view through the specimen. The filled circles represent atoms at the edge of the plane that protrude most from the surface. Preferential ionization of the image gas occurs at these sites. These sites give rise to the characteristic rings in the field-ion image.

Figure 9.6 Computer simulation of a (110) centered field-ion image of a body-centered-cubic crystal. These simulations reveal the ring structure over the entire specimen.

are observed at interfaces between dissimilar crystal structures, as illustrated in Figure 9.8. In this example, the interphase interface is between an ordered hexagonal BeFe phase and a body-centered-cubic iron-rich phase containing approximately 2.7 at. % beryllium. The orientation relationship between the two grains or phases may be determined from the position of the planes in the micrograph.

The orientation relationship may be deduced directly from the image if there is a continuation or coherency of the planes from one phase to another, as illustrated in Figure 9.9. In this micrograph, a brightly imaging $B32$-ordered iron-beryllium (FeBe) phase is surrounded by a ferrite matrix. The atoms in both phases lie on a body-centered-cubic lattice. In the $B32$ case, the iron and beryllium atoms are ordered, whereas they are randomly arranged in the ferrite. The coherent nature of

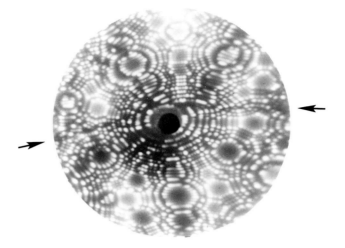

Figure 9.7 Helium field-ion micrograph of a grain boundary in tungsten, showing the region of atomic disorder at the boundary is extremely narrow.

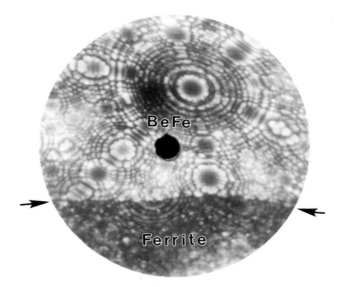

Figure 9.8 Atomically sharp interphase interface between ordered hexagonal BeFe phase and body-centered-cubic ferrite in an Fe–25 at. % Be alloy.

272

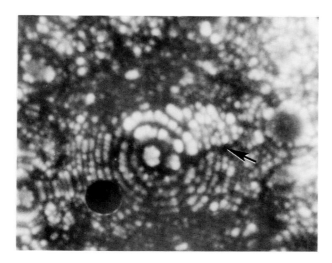

Figure 9.9 Coherency between a brightly imaging $B32$-ordered FeBe phase and body-centered-cubic ferrite is demonstrated by the continuation of the rings between the two phases.

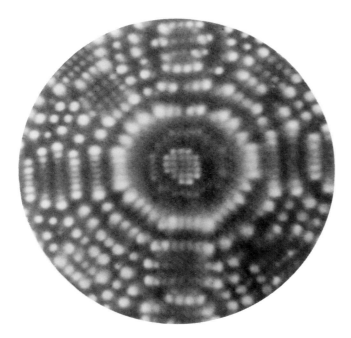

Figure 9.10 Field-ion micrograph illustrating that different atoms may exhibit distinctive contrast. The nickel atoms on the top two planes image less brightly than the molybdenum atoms on the third plane in long-range-ordered Ni_4Mo.

these two phases is evident from the continuation of the rings or planes from one phase to the other.

The imaging conditions may be adjusted in some cases to distinguish the different types of atoms, as shown in Figure 9.10. In this micrograph of the ordered Ni_4Mo intermetallic compound, the conditions were selected so that the atoms in the top two pure nickel layers image less brightly than the atoms in the third pure molybdenum layer.

FIELD EVAPORATION

The process of field evaporation is the key phenomenon that makes three-dimensional structural and chemical analysis possible with the atom probe. During field evaporation, the most prominent surface atoms are removed by applying a slightly higher field than that required for producing the field-ion image. The surface atoms are ionized and projected away from the specimen by the same processes as the image gas atoms described above. Unfortunately, removing atoms from the specimen in this manner makes atom probe field-ion microscopy a destructive technique. An example of the fine control that is possible is shown in Figure 9.11. In the first image, nine atoms are visible on the topmost plane of the surface. Between each subsequent micrograph, a single atom was removed from this plane by field evaporation. In each case, an atom on the edge of the plane was removed. It should be noted that atoms will also be removed from other regions of the specimen. The rate of field evaporation can be precisely controlled so that only a few atoms or many layers are removed from the specimen surface.

The size and morphology of a particular feature, such as a second phase, may be accurately determined with field evaporation. As atoms are field evaporated from the specimen, the rings shrink in diameter because atoms at the edge of the plane are removed first. Once all the atoms on that layer have been evaporated a fixed amount of material has been removed from the specimen. The interplanar distance or thickness of that plane may be calculated by simple geometry, since the unit cell dimensions are known for most materials. If micrographs are taken after a known number of layers have been removed, a three-dimensional view of any features in the volume imaged may be reconstructed. In this manner, the two brightly imaging precipitates shown in Figure 9.12 were determined to have spherical and disc-shaped morphologies, respectively.

The morphology of complex interconnected microstructures may also be determined. A field-ion micrograph of an iron–beryllium specimen is shown in Figure 9.13. In this micrograph, the brightly imaging regions are an iron-rich α phase and the darkly imaging regions are a beryllium-enriched $B2$-ordered phase. These darkly imaging regions are in the form of horizontal, vertical, and circular sets of bands. These bands are aligned with the $<100>$ directions of the crystal because of mismatches in atomic spacings between the two phases. This orthogonal arrangement is apparent from the horizontal and vertical bands. The image of the third set

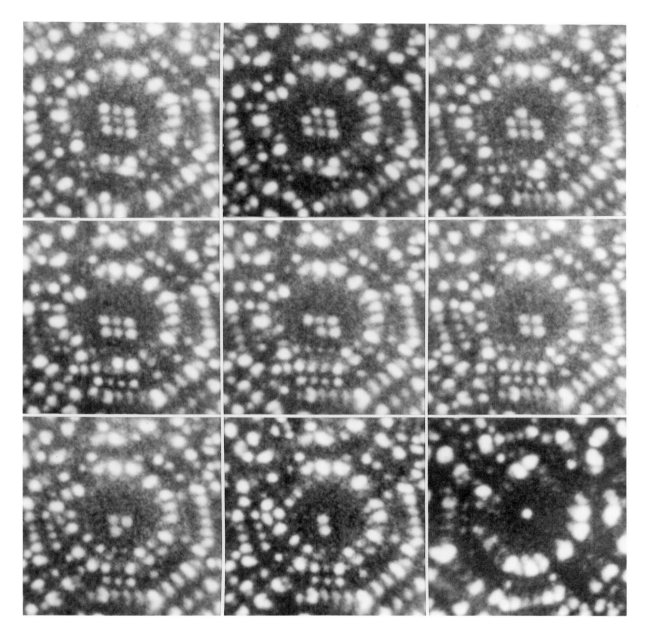

Figure 9.11 Field evaporation sequence of the top layer of a nickel–zirconium specimen. A single atom was field-evaporated between each micrograph.

275

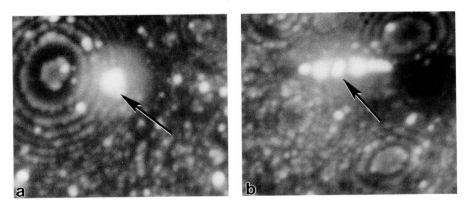

Figure 9.12 Brightly imaging spherical and disc-shaped precipitates in irradiated A533B pressure vessel steel welds. Atom probe analysis revealed that the spherical precipitate was a Mo_2C carbide and the disc-shaped precipitate was a MoN nitride.

Figure 9.13 Crystallographically aligned microstructure in an Fe–25 at. % Be alloy aged 20 min at 400°C. The darkly imaging regions correspond to a $B2$-ordered beryllium-enriched phase, and the brightly imaging regions to an iron-rich phase. The microstructure shows clear crystallographic alignment with the orthogonal <100> directions from the horizontal, vertical, and circular sets of dark bands. Field evaporation revealed that the iron-rich phase forms a fairly regular array or macrolattice of particles in the continuous beryllium-enriched matrix.

276

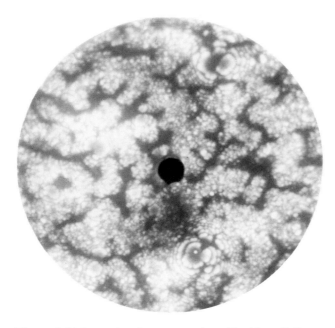

Figure 9.14 Isotropic microstructure in an Fe–45 at. % Cr alloy aged 485 h at 525°C. The darkly imaging regions are a chromium-enriched α′ phase, and the brightly imaging regions are an iron-rich α phase. Field evaporation revealed that both phases were fully continuous.

of bands is circular because of the curved surface of the field-ion specimen, as discussed above for the rings. The average spacing of these bands was determined to be 4.3 nm. Field evaporation revealed that the three darkly imaging bands form a continuous matrix and the brightly imaging iron-rich phase forms a fairly regular array of roughly cubic islands. In the iron–chromium alloy shown in Figure 9.14, no crystallographic alignment was apparent, since the lattice parameters and crystal structures of the two phases are similar. In this material, the darkly imaging regions are a chromium-enriched α′ phase, and the brightly imaging regions are an iron-rich α phase. Field evaporation revealed that the darkly imaging regions are completely continuous, as are the brightly imaging regions. The morphology of this two-phase mixture may best be described as an interconnected sponge. In these complex or spatially extended cases, serial sectioning is generally performed by video recording the evaporation sequence over prolonged distances.

Without field evaporation, the atom probe technique would be restricted to surface studies with limited applications. Field evaporation removes fine-scale surface irregularities from the specimen preparation stage and also any thin surface films that may have formed prior to examination. Field evaporation may also be used to produce an atomically clean surface so that surface studies may be performed.

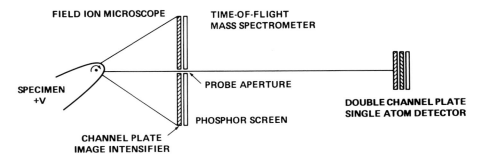

Figure 9.15 Principle of operation of the atom probe. The image is formed in the field-ion microscope on the left, and a chemical analysis is performed in the time-of-flight mass spectrometer on the right. A small entrance aperture defines the area on the surface of the specimen that is to be analyzed. The detector in the mass spectrometer has single-atom sensitivity.

PRINCIPLE OF ATOM PROBE FIELD-ION MICROSCOPY

Field evaporation also enables the identity of atoms to be determined. In order to perform this type of analysis, a mass spectrometer is added to the instrument. A small aperture is also incorporated in the fluorescent screen to function as the entrance aperture to the mass spectrometer. A schematic diagram of this configuration is shown in Figure 9.15. A single atom is selected for analysis by positioning the specimen so that the image of that atom falls over the aperture. The specimen is then field evaporated to remove the atom. Atoms are also removed from the entire specimen, but only those trajectories that enter the mass spectrometer are analyzed. The atom is field evaporated by applying a nanosecond-duration high-voltage pulse to the specimen so that the atom is removed at a precisely defined time. In most instruments, a time-of-flight type of mass spectrometer is used. Therefore, the time t that the atom takes to travel a known distance d from the specimen to a single-atom detector is measured by a high-speed timing system. Just before the atom is field evaporated it has a potential energy neE, because of the applied voltage E. Immediately after field evaporation, this potential energy is converted to kinetic energy md^2/t^2. The mass-to-charge ratio m/n may be determined by equating these two energies and rearranging the variables (i.e., $m/n = cEt^2$, where the constant c incorporates the electronic charge e and the flight distance). Since each atom has a mass that is characteristic of one of the isotopes of that element, the identity of the atom is known. Most modern instruments have sufficient mass resolution to resolve the isotopes of each element completely. The composition of a small region is determined by simply collecting and identifying the atoms in that volume of material. This method is therefore the most fundamental way to determine composition. The atom probe has no mass limitations and is equally sensitive to all elements.

SINGLE-ATOM IDENTIFICATION

The true uniqueness of the atom probe is demonstrated by its ability to identify a single atom. In the micrograph of the ordered Ni_4Mo specimen shown in Figure 9.10, all the atoms on the top nickel plane exhibited approximately the same brightness. However, in the micrograph shown in Figure 9.16(a), a single bright atom is observed on a predominantly dimly imaging plane. Although the nature of the bright atom may be inferred from the image, its identity may be absolutely determined in the atom probe. This identification is performed by aligning the field-ion image of that atom over the entrance aperture to the mass spectrometer and carefully field evaporating the specimen until the atom is removed and characterized. This brightly imaging atom was found to be molybdenum. The opposite case, where a dimly imaging site is observed on a brightly imaging plane, is shown in Figure 9.16(b). In this case, it is not possible to ascertain from the field-ion image alone that the darkly imaging site is a nickel atom or a vacant site. The two possible cases may be distinguished by whether or not a nickel atom is collected in the mass spectrometer.

The ability to identify a single atom may also be used to investigate segregation to defects, interfaces, boundaries, and other microstructural features. These features often play a very important role in determining the mechanical properties of a material. Small additions of boron have been found to improve the ductility and prevent the nickel aluminide, Ni_3Al, from fracturing at the grain boundaries when a stress is applied. Field-ion images of grain boundaries in boron-doped Ni_3Al are frequently decorated with bright spots, as shown in Figure 9.17. This micrograph reveals that the bright spots are confined to a very narrow region precisely at the boundary. Bright spots have also been observed to decorate dislocations in this material, as shown in Figure 9.18. In both cases the bright spots have been identified as individual boron atoms with the atom probe. Similar behavior has also been observed in this material at other types of defects, including low-angle boundaries, stacking faults, and antiphase boundaries.

In boron-doped nickel aluminides containing 25 and 26 at. % aluminum, brightly imaging features have also been observed in the matrix not associated with any defects, as shown in Figure 9.19. Atom probe analysis has revealed that these features were small clusters containing between two and ten boron atoms.

The observation that bright spots are not necessarily individual atoms may also be demonstrated from a decorated grain boundary observed in an irradiated pressure vessel steel used in a nuclear reactor. In the evaporation sequence shown in Figure 9.20, a grain boundary almost perpendicular to the specimen axis exhibits a high density of brightly imaging features. Atom probe analysis revealed that these brightly imaging features were ultra-fine molybdenum carbide and nitride precipitates that formed a semicontinuous film on the boundary. It should be noted that not all decorated boundaries show bright spots and more than one element may be segregated. In this example, the nickel, manganese, and phosphorus levels at the boundary were significantly higher than in the matrix.

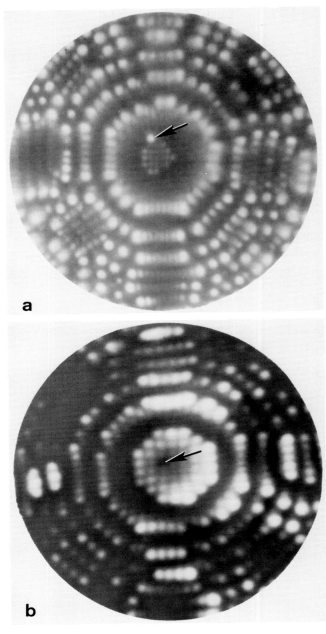

Figure 9.16 Field-ion micrographs of antisite defects in Ni$_4$Mo. (a) A brightly imaging molybdenum atom on a nickel plane. (b) A dimly imaging nickel atom or a vacant site on a brightly imaging molybdenum plane. Atom probe analysis is required to distinguish between the two possibilities.

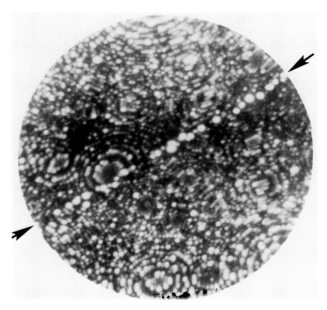

Figure 9.17 Decorated grain boundary in boron-doped $Ni_{76}Al_{24}$ alloy. Atom probe analysis revealed that each brightly imaging spot of the boundary was a boron atom.

Figure 9.18 Decorated dislocation in boron-doped $Ni_{76}Al_{24}$ alloy. The bright spot was determined to be a single boron atom.

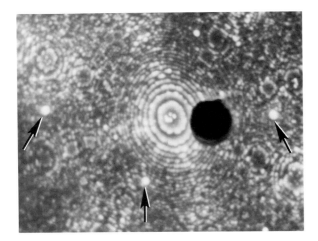

Figure 9.19 Brightly imaging boron clusters in a boron-doped Ni$_3$Al alloy. Atom probe analysis revealed that these clusters each contained between two and ten boron atoms.

An example of segregation to a boundary in a 4130-type steel is shown in Figure 9.21. This experimental steel was designed to minimize hydrogen sulfide gas from cracking boundaries in the steel casing used to line oil wells. This was achieved by the addition of approximately 1 at. % palladium to the steel. The bright spots along the lath boundary were identified as individual palladium atoms. A small manganese–palladium precipitate was also seen in this section of the boundary. The contrast exhibited by this precipitate was very distinct in that the interior of the precipitate imaged darkly whereas the rim imaged very brightly. This behavior and atom probe chemical analysis indicated that there was palladium decoration at the precipitate–matrix interface.

SURFACE ANALYSES

Since many different crystallographic facets are exposed on its curved surface, a field-ion specimen is suitable for surface reaction studies. Therefore, the atom probe may be used to investigate the formation of thin films and the surface chemistry of catalysts. In these types of experiments, the specimen is first field evaporated in the field-ion microscope to produce an atomically clean surface, and then the specimen is exposed to a controlled environment, usually at elevated temperatures. This stage is often performed in a separate chamber to avoid contaminating the vacuum system. The specimen is reimaged in the field-ion microscope, and a composition profile is obtained from the surface into the interior of the metal. Field-ion micro-

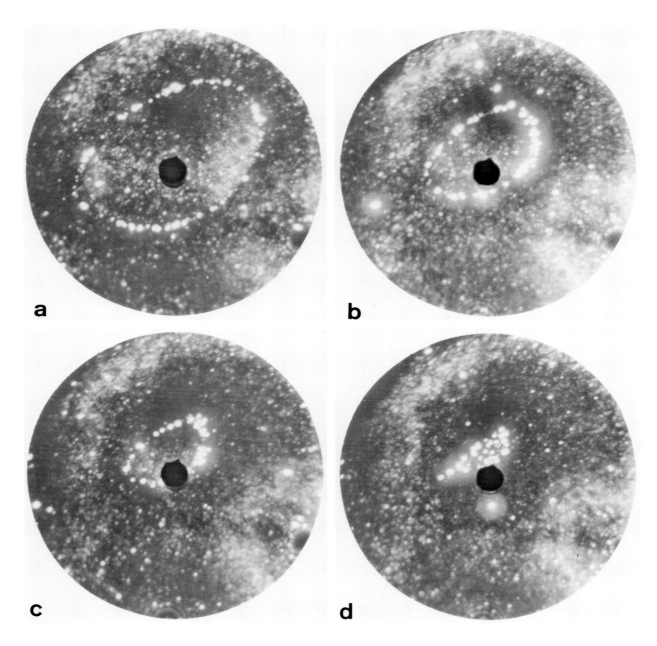

Figure 9.20 Field evaporation sequence of a specimen containing a grain boundary that exhibited bright spot decoration in an irradiated pressure vessel steel weld. As material is removed from the specimen, the roughly circular image of the boundary shrinks. The brightly imaging spots are produced by ultra-fine molybdenum carbide and nitride precipitates.

283

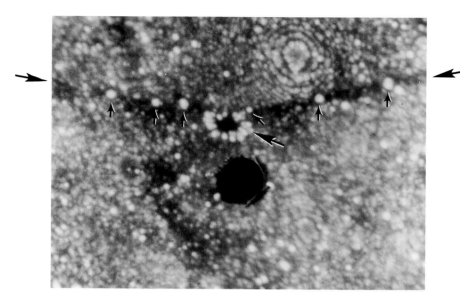

Figure 9.21 Decorated lath boundary in an experimental 4130 steel containing 1 at. % palladium. The brightly imaging spots on the boundary were found to be palladium atoms. A small manganese–palladium precipitate on the boundary exhibited a darkly imaging interior and a brightly imaging rim. This behavior indicated that there was palladium decoration at the precipitate–matrix interface.

graphs of an oxidized nickel–zirconium specimen before and after field evaporation through to the metal are shown in Figure 9.22. The nonuniform imaging behavior reveals that the surface oxide film is very rough on a near-atomic scale. The approximate thickness of the film is determined from the number of atoms collected during atom probe analysis through the film. This number is converted into a distance with the use of calibration experiments on well-characterized materials taken under similar conditions. This oxide film was determined to be 4 nm thick. The atom probe composition profile revealed several distinct regions in this thin film, including a zirconium-enriched surface, an oxygen-depleted layer, and a broad diffuse metal–oxide interface.

OTHER MATERIALS

Most of the examples presented so far have depicted features in metallic systems. The atom probe field-ion microscope is not limited solely to this class of materials. The limiting constraint is that the sample has sufficient electrical conductivity for the field ionization and field evaporation processes to take place. In practice, the

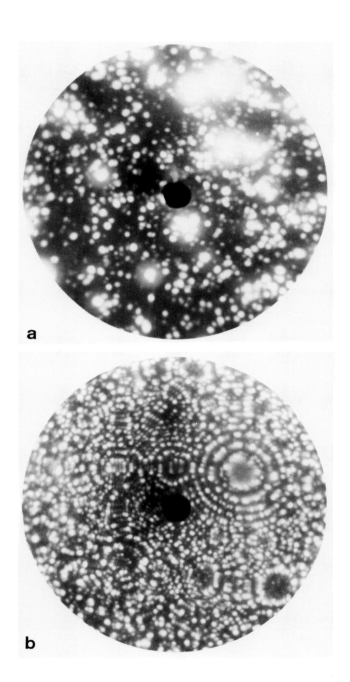

Figure 9.22 Field-ion micrographs of an ∼ 4-nm-thick oxide film and the field-evaporated metal surface of a nickel–zirconium specimen. The nonuniform imaging behavior of the oxide indicated a rough surface.

285

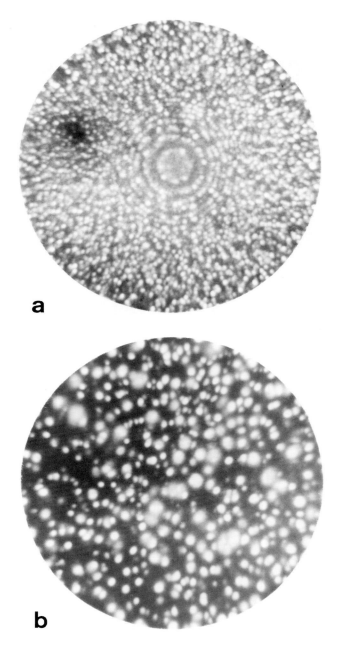

Figure 9.23 Field-ion micrographs of (a) silicon and (b) silicon carbide semiconducting specimens.

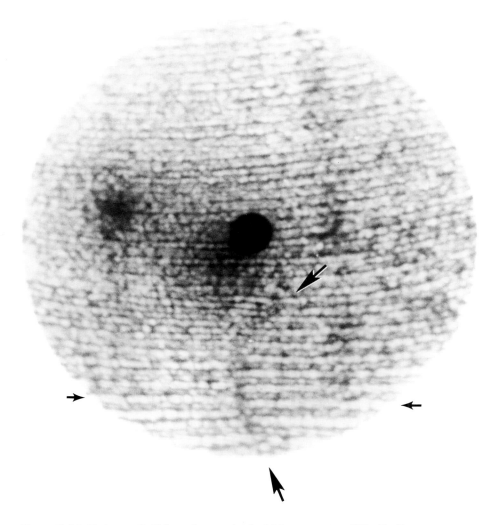

Figure 9.24 Hydrogen field-ion micrograph of a high-temperature $YBa_2Cu_3O_{7-x}$ supercon-ducting oxide specimen taken during field evaporation. A twin boundary is indicated by the large arrows. The small relative displacement of the planes is indicated by the small arrows.

specimen may be a poor semiconductor. Field-ion micrographs of two semiconductors, silicon and silicon carbide, are shown in Figure 9.23. The image quality in this class of materials is generally inferior to most metallic specimens. Some ring structure is evident in the silicon specimen, whereas little atomic detail is discernible in the silicon carbide specimen. The quality of the image does not affect the ability of the atom probe to perform an elemental analysis. Since a high-voltage field evaporation pulse would be seriously attenuated as it propagates through the specimen, it is replaced by a short-duration laser pulse in these high-resistance materials. This variant of the atom probe is known as a pulsed laser atom probe.

Conductive ceramic oxides, such as the new class of high-temperature superconductors, have also been successfully imaged. An example is shown in Figure 9.24. This micrograph was taken while the specimen was being field evaporated. Because of selective imaging and the relatively large number of atoms per unit cell, the normal ring structure is not strong in these materials. The horizontal stripes are the manifestation of the edges of the *c* planes. This evaporation technique, although blurring out some atomic detail, enabled the boundary to be observed more clearly and revealed a small relative displacement of the planes either side of a twin boundary. Defects such as these will impair the electrical properties of these materials.

SUMMARY

The atom probe field-ion microscope is a very powerful instrument, capable of detecting and analyzing atomic-scale features that are not accessible by any other technique. The unique ability to analyze individual atoms, boundaries, or fine second phases makes atom probe field-ion microscopy a most valuable tool in microstructural characterization. The reader is referred to the bibliography for additional references on atom probe field-ion microscopy.

ACKNOWLEDGMENTS

This research was sponsored by the Division of Materials Sciences, U.S. Department of Energy, under contract DE-AC05-84OR21400 with Martin Marietta Energy Systems, Inc., and through the SHaRE program under contract DE-AC05-76ORO0033. The authors would like to acknowledge the contributions of P. Angelini, J. A. Horton, E. A. Kenik, A. J. Melmed, K. L. More, E. M. Perry, K. F. Russell, and T. A. Zagula to the research described.

REFERENCES

Bowkett, K. M., and Smith, D. A. (1970). *Field Ion Microscopy*. North-Holland, Amsterdam.

Miller, M. K. (1987). *Materials Review* **32**, 221.

————, and Smith, G. D. W. (1989). *Atom Probe Microanalysis: Principles and Applications to Materials Problems*. Mater. Res. Soc., Pittsburgh, Penn.

Müller, E. W., and Tsong, T. T. (1969). *Field Ion Microscopy: Principles and Applications*. Elsevier, N.Y.

Smith, G. D. W. (1986). *Metals Handbook* **10**, R. E. Whan (ed.), ASM, Metals Park, Ohio, p. 583.

Wagner, R. (1982). *Field Ion Microscopy in Materials Science*, Vol. **6**. *Crystals: Growth, Properties and Applications*, Springer-Verlag, Berlin.

10

Compositional Mapping of the Microstructure of Materials

D. E. NEWBURY, R. B. MARINENKO,
R. L. MYKLEBUST, AND D. S. BRIGHT

Recent developments in electron probe microanalysis and secondary ion mass spectrometry techniques enable the analyst to prepare compositional maps of the microstructure of materials. The technique of compositional mapping involves performing a complete quantitative analysis at each beam location in the imaged field of view on the specimen. A digital image processor is used to view the resultant compositional maps in the form of images in which the gray or color scale is related directly to the concentrations of the elemental constituents that are present in the specimen. The images directly convey the sense of the spatial distribution and compositional interrelationships of the elemental constituents in the microstructure. The images are supported at every picture element by numerical values of the concentrations, as well as data on the statistical precision of the measurement. The combination of images and conventional numerical concentration data provides a powerful tool for microstructural analysis.

MAPPING ELEMENTAL DISTRIBUTIONS BY ELECTRON PROBE MICROANALYSIS

Microbeam analysis by electron microprobe provides a major resource for the characterization of the elemental composition of materials on a scale of micrometers (Goldstein et al., 1981; Newbury et al., 1986b). Electron probe microanalysis

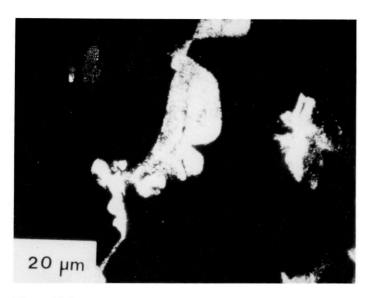

Figure 10.1 X-ray area scan (dot map) showing the distribution of zinc at the grain boundaries of polycrystalline copper. Sample courtesy of Daniel Butrymowicz, National Institute of Standards and Technology.

(EPMA) is based upon the physics of the interaction of energetic (ca. 20 keV) electrons with a solid specimen target. The interaction of the electrons can ionize an inner-shell electron of an atom, and the subsequent de-excitation can produce a characteristic x ray, whose energy identifies the elemental species present in the beam interaction volume. In most quantitative applications of electron probe micro-analysis, the beam is scanned to a selected position on the specimen, and this position is maintained constant during the spectrometric measurement of the analytical signal. The measured wavelength or energy of the characteristic radiation identifies which elemental species are present (qualitative analysis), and the intensity of the radiation is related to the concentrations of the elemental constituents within the volume of the specimen that is excited by the beam. Because of the nature of the physics of electron interaction and x-ray generation and propagation, corrections are needed to convert the measured x-ray intensities into elemental concentrations. These point beam analyses are often complemented by images that either directly or indirectly map the spatial distribution of the elemental constituents. For the technique of electron probe microanalysis, indirect compositional imaging is possible by means of backscattered electron imaging, which is sensitive to the atomic number of the target. Direct compositional imaging is possible by means of x-ray area scanning or "dot mapping," in which an image is created that is related to the detection of characteristic x rays (Cosslett and Duncumb, 1956). Figure 10.1 shows an example of an x-ray area scan of the zinc distribution at the grain boundaries of polycrystalline copper, which results from the phenomenon of diffusion-induced grain boundary migration (DIGM) (Piccone et al., 1982).

While the dot mapping method has proven extremely useful and is widely applied in practical analytical applications in materials science, it must be recognized that such maps are really qualitative in nature and are subject to significant limitations. The maps only depict the presence of an elemental constituent at a particular location in a microstructure and reveal nothing about the amount of that constituent. In the process of recording the map, the count rate information critical to quantitative analysis is lost. In the case of x-ray area scans, the limitations are particularly severe.

The technique of x-ray area scanning or dot mapping has changed relatively little from the procedure introduced by Cosslett and Duncumb (1956); the salient characteristics and limitations can be summarized as follows:

1. The generation of characteristic x rays is a relatively inefficient process, because of a small cross section for inner-shell ionization and low detection efficiency since the collection angle of the x-ray spectrometer is less than 0.1 steradian. Only about 1 in 10,000 electrons incident on the target generates a detectable x ray. As a result, the x-ray signal detected from a minor or trace element is discontinuous in nature, even with high incident beam currents. In order to make an x-ray area scan map, the amplifier pulse associated with the detection of each x ray, either by wavelength dispersive spectrometry (WDS) or energy dispersive spectrometry (EDS) must be recorded. A full brightness (white) dot is written on the display screen whenever an x ray is detected. The recorded map therefore has only two gray levels, black and white; a continuous gray scale is not possible.

2. An important consequence of the low x-ray signal rate is that x-ray area scans have poor detection sensitivity. A statistically valid map requires between 100,000 and 1,000,000 counts, depending on the spatial distribution of the element within the field of view. Major elements (10% or higher in concentration) require approximately 10 minutes to map, minor elements (1–10%) approximately 1–5 hr, and trace elements (<1%) may not be accessible at all in a practical measurement time. The x-ray area scan shown in Figure 10.1 required more than 6 hr of scanning and depicts Zn concentrations as low as 1%, as confirmed by individual point analyses.

3. Because it is not possible to produce a range of gray levels except in the most extraordinary situations, the contrast sensitivity of the technique is also poor. Thus, while it is possible to observe 1% Zn regions against a background with 0% Zn, as shown in Figure 10.1, it is extremely difficult to observe a modulation of 5% against a general distribution of 50%.

4. The map is recorded directly onto film, which is a relatively inflexible medium for subsequent processing, such as overlaying maps for several different elements from the same area.

5. Because most analog imaging systems usually permit only one channel of recording, it is difficult to combine x-ray information from multiple spectrometers or from several different detectors. Although the x rays for several different elements of interest may be detected in parallel with an energy spectrometer or with multiple wavelength spectrometers, single-channel film recording makes the process extremely inefficient with respect to time utilization if maps of

several elements are needed. This problem is somewhat alleviated for an EDS system coupled with a computer-based multichannel analyzer where digital recording rather than film may be employed. However, the relatively poor resolution of EDS compared with WDS leads to a poor peak-to-background (at least a factor of 10 lower for EDS than WDS), which directly translates into much poorer detection sensitivity for EDS mapping.

These limitations on the conventional mapping technique often result in the analyst being forced to accept qualitative maps with a minimum of x-ray counts, which may be so poor as to preclude a statistically valid interpretation of the elemental distributions. The development of quantitative compositional mapping provides a powerful new tool that overcomes most of the limitations of dot mapping (Fiori et al., 1984; Marinenko et al., 1985, 1987).

QUANTITATIVE ELECTRON MICROPROBE COMPOSITIONAL MAPPING

The basis of quantitative compositional mapping with the electron microprobe is to perform, under computer control, a complete, fully quantitative analysis at each beam location in the area scan (Marinenko et al., 1985, 1987). The first step in the mapping procedure is to record at each picture element in the image, or "pixel," the count rate for each element as measured with wavelength spectrometers, an energy spectrometer, or a combination of both. The measured count rates are corrected for dead time and background. Standardization is performed by the classic technique of forming an intensity ratio, the so-called k ratio, between the intensity measured on the unknown to the intensity measured on a known standard, which may be a pure element, a simple compound such as GaAs, or a multielement standard such as a glass. Complete matrix corrections are then applied to the k values at each pixel by one of the established procedures: ZAF, α factors, or $\phi(\rho z)$ (Goldstein et al., 1981) to yield the concentrations at that pixel.

The resulting matrices of numerical concentration values, with one matrix array for each element measured, can be assembled into images with a digital image-processing computer that is coupled to an image display monitor capable of 8-bit intensity range for each of the primary colors (red, green, and blue). Gray or color scales can be generated in these images, with the gray level or color value assigned to each pixel according to the measured numerical concentration values, thus creating a "compositional map." The best choice of gray scales, color scales, or image-processing operations depends on the nature of the problem to be studied. This topic has been addressed by several authors and remains an area of fruitful research (Heinrich and Fiori, 1984; Marinenko et al., 1987; Bright et al., 1988).

An example of a quantitative compositional map is shown in Plate 10.1, which depicts the same region as Figure 10.1. The concentration data for the concentration range from 0 to 10 wt % have been encoded with a special color scale, the "thermal scale," which is the sequence of colors emitted from a blackbody when it is heated.

The thermal scale provides an intuitive sequence of colors that the observer auto-matically recognizes in the proper order (see Chapter 11 for more detail). This depiction of the concentration data for Figure 10.1 as a compositional map shows distinct concentration levels visible and recognizable from 0.1% (note narrow red boundary at bottom of field, which was not visible in the conventional dot map) to 10% (white regions), a range of two orders of magnitude.

Quantitative compositional mapping is highly dependent on the efficient use of computer control of the microprobe for beam positioning and x-ray data collection (Fiori et al., 1985). The x-ray count rate depends only on the beam energy, electron dose, and spectrometer geometric and quantum efficiencies. The compositional mapping procedure cannot improve the rate of x-ray counting, so that image ac-cumulation times are no shorter than in the conventional dot mapping procedure to record a desired number of x-ray pulses. To map major constituents (>10%) re-quires approximately 1–2 hr of accumulation time. Mapping minor constituents (1–10%) requires 2–5 hr, while trace constituents (<1%) can require 5–20 hr, depend-ing on the level of concentration that is sought. However, by efficiently recording the x-ray peaks in parallel from a combination of an EDS and several WDS detec-tors, maps for 10 or more elements can be obtained simultaneously, which makes the time penalty for compositional mapping more acceptable. Parallel recording of the elements is actually a necessity, since accurate calculation of the matrix effects by a technique such as ZAF requires that all major elements, at least, be measured and included in the calculations.

Special correction techniques have been developed to compensate for the major instrumental artifacts, wavelength spectrometer defocusing and energy spectrometer decollimation, which are manifest as severe intensity decreases under scanning beam conditions when the x-ray intensity measurement is made off the optic axis of the microprobe (Marinenko et al., 1985, 1987; Myklebust et al., 1986; Swyt and Fiori, 1986). Proper correction of spectrometer defocusing or decollimation is vital if the analyst is to obtain valid measurements even on major constituents. An example of the intensity distribution in a three-element map before and after correc-tion for spectrometer defocusing is shown in Plate 10.2, which shows a color overlay of compositional maps for Mg, V, and Co in a ceramic prepared with wavelength-dispersive spectrometers. In Plate 10.2(a) a significant part of the con-trast in the image arises from the effect of spectrometer defocusing and not from the actual compositional microstructure. In Plate 10.2(b) the defocusing has been cor-rected by the defocus modeling procedure described by Myklebust et al. (1986), so that all of the contrast observed in the image is due to the true compositional contrast of the specimen and not to instrumental artifacts.

For minor and trace elements, a second important factor must be considered, that of local, compositionally dependent variations in the background. Several methods are available for background correction in images (Myklebust et al., 1987). An example of the misleading artifact that uncorrected or improperly corrected back-ground can introduce is shown in Figure 10.2. The specimen is a directionally solidified aluminum–copper eutectic, as shown in Figure 10.2(a) for Al and Figure 10.2(b) for Cu. The map taken for trace scandium, Figure 10.2(c), appears to show an inhomogeneous distribution of this element at the level of 0.2% (maximum) with

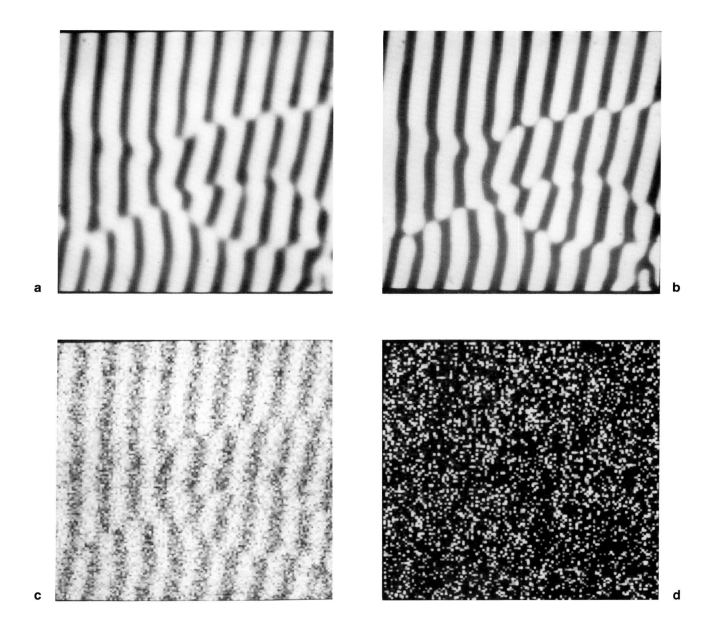

Figure 10.2 Correction of background effects in quantitative electron probe compositional mapping: (a) Al concentration map. (b) Cu concentration map. (c) Uncorrected map at the position of Sc showing apparent concentrations of 0.1–0.2 wt %. (d) Corrected map for Sc, with contrast expansion at 500 ppm (white). Image field width, 50 μm.

a distribution that corresponds to that of the copper-rich phase, Figure 10.2(*b*). This apparent scandium constituent is merely an artifact of the atomic number dependence of the bremsstrahlung background radiation. With proper correction of the background, Figure 10.2(*d*), the apparent scandium distribution vanishes, even with contrast expansion applied at the level of 500 ppm (white dots).

Quantitative compositional mapping provides the powerful combination of images that are supported at every pixel by the numerical values of the concentration, and the statistical precision of the measurement at that pixel. The presentation of numerical concentration information in the form of images provides the major utility of the mapping technique. The interpretation of information in the form of images is a major intuitive strength of the human mind. Although computer-aided imaging and interpretation are beginning to provide automatic tools for image analysis, the field of digital imaging still depends very heavily on the observer as interpreter. It is often possible to recognize patterns readily in data presented as an image that would be difficult to derive from those same data in tabulated numerical form, or indeed, to develop a computer program that could determine the pattern automatically. The following examples will illustrate the application of quantitative compositional mapping to several problems in materials science and will demonstrate a number of special aspects of the new technique, including the use of various image-processing and enhancement algorithms to aid the analyst.

APPLICATIONS OF QUANTITATIVE COMPOSITIONAL MAPPING WITH THE ELECTRON MICROPROBE

Detection of a Small Concentration Change against a High Average Level

The problem of detecting small concentration changes against a zero background is well illustrated by the mapping of DIGM in Cu-Zn, as shown in Figure 10.1 and Plate 10.1. When carefully applied, even conventional dot mapping can reveal low concentrations, approximately 1%, against a zero background. Plate 10.1 shows that concentration levels of 0.1% (1000 ppm) can be readily detected against a zero background in the compositional mapping mode. In favorable circumstances, such as low x-ray absorption situations in light element matrices that have low bremsstrahlung background, the detection limit can be reduced below 100 ppm (Garruto et al., 1985). However, when the concentration of the element of interest forms a high background level, it becomes impossible to detect, by dot mapping, the modulation of the image caused by the compositional contrast. Concentration matrices obtained by quantitative compositional mapping can be manipulated for display purposes by a simple processing algorithm that involves assigning the gray scale to the data such that the zero is offset, and the threshold for black is set to a high value, while the remaining values span the complete gray-scale intensity range of the display. This digital contrast expansion is equivalent to analog differential amplification of "black level" used in scanning electron microscopy. An example of

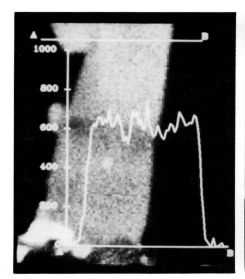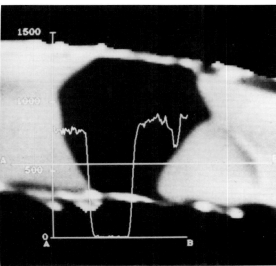

a   b

Figure 10.3 Detection of image effects near background. Sample: Fe-Zn (a) Compositional map at the surface of an Fe-Zn specimen showing a region of continuous DIGM. Image field width, 200 μm. (b) Compositional map of a cross section of this specimen showing an apparent hexagonal void. A compositional trace is plotted for the vector **AB** which crosses the "void," showing statistical scatter in the background in the void region, which is actually pure Fe and which has not taken up Zn during the DIGM process. Image field width, 50 μm. Sample courtesy of Carol Handwerker, National Institute of Standards and Technology.

the results of this operation is shown in Plate 10.3(a), which shows DIGM in an alloy of silver–gold, in which the enrichment of the silver in the boundary amounts to a 7% increase above a general background of 65% Ag (Butrymowicz et al., 1984). The corresponding decrease in the gold concentration in the DIGM zone is shown in Plate 10.3(b).

Imaging Concentrations near Background

The use of careful background correction permits the study of compositional structures at low concentration. Figure 10.3(a) shows a compositional map of Zn that reveals the zone of DIGM in the Fe-Zn system. After this specimen was prepared in cross section, a compositional map revealed an apparent hexagonally shaped void, Figure 10.3(b). The question could be posed as to whether the background correction procedure overcorrected this grain by subtracting off too large a value and incorrectly suppressing the gray scale. When the numerical concentration data were examined across this region, as selected by the vector AB, the statistical scatter in the region across the "void" reveals in fact that the background correction was performed correctly and that the iron grain did not take up any Zn during the DIGM process, despite the presence of an overlayer of Zn deposited on both sides of the grain. This example also illustrates the utility of having numerical data that can be interrogated at any pixel, vector of pixels, or array of contiguous pixels.

Primary Color Superposition for Recognition of Constituent Interrelationships

Since the earliest applications of the conventional dot mapping procedure, primary color superposition has been utilized to examine spatial relationships among two or three images of different constituents from the same field of view (Duncumb, 1957). By the rules of primary color addition, the appearance of secondary colors reveals the locations where two constituents coexist (red plus blue equals magenta; red plus green equals yellow; green plus blue equals cyan), and the appearance of white indicates all three constituents are present. Digital maps of constituents from the same field of view can be readily superimposed since the pixel correspondence is ensured by the method of simultaneous data collection. An example of the primary color overlay technique is shown in Plate 10.4. The Ba, Cu, and Y components of a high-T_c ceramic superconductor are depicted in Plates 10.4(a)–(c) individually as gray-scale images, and the color overlay of these three components is shown in Plate 10.4(d). The appearance of small regions of primary color reveals locations where the unreacted pure oxide components remain.

Compositional maps can be enhanced as gray-scale images prior to the application of the color overlay technique. This procedure is illustrated in Plate 10.5, which depicts an early stage of DIGM in the Cu-Zn system. The Zn composition map, Plate 10.5(a), has been enhanced so that the maximum concentration of 2 wt % corresponds to full white. The corresponding Cu composition map, Plate 10.5(b), in which the concentration values are near 100%, has also been enhanced, and reveals distinct regions of decreased copper concentration. By superimposing these images in primary colors, Plate 10.5(c), with Zn in red and Cu in green, the region of Zn overlap with the Cu can immediately be recognized by the appearance of the secondary color yellow. However, adjacent to the yellow DIGM zone, there is clearly a region of Cu deficiency that is not due to the presence of Zn. This decrease was subsequently shown to be due to the presence of oxygen, as will be described in the section on secondary ion mass spectrometry (Newbury et al., 1986a).

Recognition of Constituent Interrelationships: Concentration–Concentration Histograms

While gray-scale or color-scale images are useful for displaying compositional maps, the images generally suffer from a lack of quantitative information. It is difficult to recognize specific shades of gray in a gray-scale depiction. The thermal color scale significantly improves this situation, but it is only useful for one image at a time. If concentration data for several different elements recorded from the same region are to be viewed simultaneously, there is, at present, no ready way to superimpose the data and maintain a quantitative color or gray scale for each image in the overlay.

An alternative approach is the "concentration–concentration histogram (CCH)," a multidimensional type of scatter plot (Bright et al., 1988) that permits a viewer to recognize quantitative relationships in images. This technique is also discussed in Chapter 11. A binary CCH is prepared by taking two compositional maps from the same area and replotting the concentration data on a pixel-by-pixel basis into a new, two-dimensional concentration space. When all of the pixels in the original images have been interrogated, the CCH consists of an array in which the number at any location in the array represents the number of pixels in the original images that had that particular set of concentration values. To view the CCH, the array is plotted as an image in which the values are encoded with the thermal scale to permit the viewer to recognize high-frequency and low-frequency events. The CCH procedure is illustrated in Plate 10.6, which uses the Ba and Cu compositional maps shown in Plate 10.4 as the starting point. The resulting CCH, shown in Plate 10.6(a), has one spot of high occupation, and three bands of low frequency leading away from this spot. The relationship of a region in the CCH to the original images can be found by using a function known as TRACEBACK, which locates the pixels in the original images that are selected from a specified concentration range in the CCH. When the bright spot in Plate 10.6(a) is selected, TRACEBACK locates the pixels shown in the color superposition shown in Plate 10.6(b), which depicts the majority superconducting phase in this material. When the band at constant Cu concentration is chosen, Plate 10.6(c), the minority phase shown in Plate 10.6(d) is highlighted. Selection of the band with Cu and Ba both varying, Plate 10.6(e), highlights the edges of the voids in the original images, Plate 10.6(f), probably as a result of penetration of the electron beam through thin material.

Other Applications

Several additional examples of applications of quantitative electron microprobe compositional mapping will serve to illustrate the wide range of possibilities for this new technique.

Plate 10.7 shows an application of compositional mapping to the study of the interaction zone between a steel screw and an aluminum wire in a high-temperature failure of a residential electric power outlet (Newbury, 1982). The critical step in the failure process has been found to be the reaction of the iron and aluminum to form a zone of Fe-Al intermetallic compounds at the current-carrying interface, which significantly increases the resistance of the junction, leading to further heating. The existence of the intermetallic compound zone as the secondary color yellow is readily apparent in the color overlay image (Fe, red; Al, green) of Plate 10.7.

Plate 10.8 shows a compositional map of the cross section of a fracture surface from a failed pressure vessel that operated in a high sulfur environment. The color overlay (Fe, red; S, green) shows the existence of a zone of penetration of sulfur below the original fracture surface. Close examination of the zone of penetration reveals the presence of sulfur at two distinctly different concentrations, as evidenced by the different colors.

Plate 10.9 shows the microstructure of an explosively prepared GaAs compound that was fabricated by shock compression of the elemental Ga and As constituents in a sealed container. After reaction, inclusions were observed in the microstructure. The compositional mapping procedure revealed that these inclusions originated from a Cr contaminant. The color overlay map of Plate 10.9 clearly shows the presence of inclusions of an As-Cr phase (yellow) in a general matrix of GaAs (magenta).

Plate 10.10 contains a series of individual element compositional maps of Si, Al, and C for a system consisting of silicon carbide cemented with aluminum. The complex microstructure can be seen to contain a precipitate of Si-Al (yellow) in the aluminum-rich phase (red), while adjacent to the SiC (cyan), there exists an Al-C containing phase (magenta). While these phases could be located and analyzed by the conventional single-point approach, the compositional mapping technique provides in a single image a comprehensive view of the various components of the microstructure.

Plate 10.11 contains three different presentations of quantitative compositional maps that depict the microstructure of the ceramic superconductor $YBa_2Cu_3O_{7-x}$ (Blendell et al., 1987). Plates 10.11(a)–(d) show individual elemental maps for Ba, Cu, Y, and O (calculated by stoichiometry) depicted with a thermal scale. A quantitative trace through the Ba map is shown in Plate 10.11(e). A color superposition of the three single elemental images is shown in Plate 10.11(f). Examination of this set of images reveals that the superconducting phase forms a continuous matrix that is interrupted by second-phase particles and voids. Particular compositions of the second-phase particles that are observed include a Ba-Cu phase with no Y component, and a Y-rich phase that is partially depleted in Ba. Zones of Y enrichment are also observed. The composition around the numerous voids is seen to vary in a complex fashion.

COMPOSITIONAL MAPPING BY SECONDARY ION MASS SPECTROMETRY

While the electron probe compositional mapping procedure has proved to be a powerful technique with wide application to materials science, the technique is nevertheless subject to limitations. The chief limitation is the significant time period that is necessary to accumulate a map, particularly for minor and trace constituents, which necessarily places a limit upon the practical application of electron probe compositional mapping. Moreover, because of the nature of x-ray generation, the sensitivity of the technique is limited to concentration levels of several hundred parts per million in all but the most favorable cases, and then only at the expenditure of accumulation times in excess of 10 hr per image field. There are many important applications for which sensitivity at the ppm or ppb level is important.

Secondary ion mass spectrometry provides a powerful compositional mapping technique that complements the electron microprobe technique. In secondary ion mass spectrometry (SIMS) (see Chapter 3 by Levi-Setti et al., and Benninghoven et al., 1987) an energetic (ca. 10 keV) ion beam, which is typically O, Ar, or Cs, bombards the target and causes sputtering, which is the direct ejection of atoms lying at the surface. A small fraction of the sputtered atoms is ionized, and these so-called secondary ions can be collected with an electrostatic field and analyzed with a mass spectrometer to determine the elements present in the specimen. Ion detection can be in the digital form of single-ion counting when an electron multiplier is used as the detector, or in the analog form of a current if a Faraday cup is used as the detector.

SIMS can operate in a microanalytical mode by either of two approaches. In the microprobe method, the primary ion beam can be focused to form a small probe in the micrometer or even nanometer size range to limit the size of the region selected for analysis (Liebl, 1975). Images can be prepared with the same type of scanning technique used in the electron microprobe and SEM (see Chapter 2 by Joy and Goldstein).

Alternatively, because the secondary ions are charged and can be focused, a direct imaging microscope approach can be used (Slodzian, 1975). In the ion microscope, an area several hundred micrometers in diameter on the sample is illuminated with a primary ion beam, and secondary ions emitted from all points in the illuminated area are simultaneously passed through the mass spectrometer and brought to a focal plane where a true image of the surface is created. The limiting lateral resolution in the direct imaging microscope mode is of the order of 500 nm.

SIMS offers several attractive features that make it a powerful method of compositional mapping. (1) While it varies significantly from element to element in the periodic table, the limiting sensitivity of SIMS can be quite extraordinary and is virtually always better than that for the electron microprobe by two orders of magnitude or more. Sensitivity in the ppm or ppb region can be achieved for many elements of interest because the background is negligible in most cases. (2) The signal rate of SIMS typically exceeds 10^6 counts per second for a major constituent. (3) The high signal rate coupled with the high sensitivity translates into rapid imaging compared with the electron microprobe. In the ion microscope, which is optimized for the imaging mode of analysis, images for major elements can typically be monitored at television rates. At minor and trace levels, images can generally be obtained in 1–100 sec. This rapid imaging is a great advantage when an extensive area of the unknown must be covered to locate an area of interest. (4) Because the sputtering process effectively "peels" the specimen a few atom layers at a time, SIMS can be used to study the distribution in depth of a constituent in the so-called depth profiling mode. By combining lateral imaging with depth profiling, a three-dimensional compositional map can be developed through the use of computer-aided imaging to reconstruct desired cross-sectional views of the specimen from a series of lateral images taken as a function of depth.

SIMS suffers a number of drawbacks as an analytical and compositional mapping tool:

1. The variable elemental sensitivity and lack of an accurate model for the physics of secondary ion formation makes quantitative analysis difficult. The most successful approach to quantitative SIMS has been by means of the empirical technique of relative sensitivity factors determined on multielement standards similar in nature to the unknown. Relative errors of the order of 10% have been achieved by this method. Quantitative analyses based upon physical models of ion emission are generally subject to uncertainties of the order of a factor of 2 or more. When images are to be quantified, additional aspects of instrumental artifacts must be considered in order to achieve quantitative results (Newbury and Bright, 1988a,b).

2. Most existing SIMS instruments are only capable of measuring a single mass-to-charge value at a time. Although rapid switching among mass peaks is possible under computer control, ion count or imaging data for several elements cannot be obtained exactly in parallel. Because the specimen must be constantly eroded to generate secondary ions, the data for several elements collected sequentially actually come from slightly different locations on the specimen with respect to depth. If the relative distribution of the elements changes rapidly, for example as an interface between two different phases is approached, then it may be difficult to achieve proper registration among a series of SIMS images.

APPLICATIONS OF COMPOSITIONAL MAPPING BY SECONDARY ION MASS SPECTROMETRY

The following series of examples will illustrate several of the useful features of compositional mapping by SIMS. One aspect of these applications that cannot be truly appreciated in a series of static images is the sense of the rapid and dynamic flow of image information that an ion microscope is capable of producing. In most cases, the images shown in these applications were obtained at a television imaging rate, which permitted ready surveying of complex specimens to locate areas of interest, much as electron imaging in the scanning electron microscope and electron microprobe is used to search an unknown specimen rapidly for the area to receive detailed and slow x-ray compositional mapping.

Trace Constituent Compositional Mapping

An example of trace-level compositional mapping by SIMS is shown in Plate 10.12 (Newbury et al., 1986a). Reaction-bonded silicon carbide is a material that is formed by heating grains of SiC in Si. As the temperature is increased, the Si melts, and the molten Si begins to dissolve the grains of SiC. If the temperature is lowered before dissolution is complete, the dissolved SiC will precipitate out on the partially melted SiC grains, which act as nucleation sites for the new growth of SiC from the melt. The final microstructure consists of the original grains of SiC that form the core upon which the redeposited SiC has grown, with the composite grains cemented together by the intergranular Si. When this material is examined in the scanning electron microscope, anomalous contrast is observed in images prepared with an Everhart–Thornley detector, which is sensitive to both secondary electrons and backscattered electrons. Strong contrast is observed among grains, and even within grains as banding. This contrast is not observed in an image prepared with a backscattered electron detector, and is evidently not due to concentration changes involving differences in composition on the percent scale, since such differences should give rise to contrast in the BSE image. SIMS compositional mapping was used to examine the distribution of trace elements in this material. SIMS images of the trace aluminum constituent were found to have a similar form to the SE/BSE SEM image. Plate 10.12(a) shows an SE/BSE SEM image, and Plate 10.12(b) the ion microscope image from the same area. In order to test the degree of the match between the two images, it was desirable to superimpose one upon the other. However, the relative distortions between the two images were quite severe, and it was necessary to apply an image warping function to one image, which was arbitrarily taken to be the SIMS image. The manually chosen correspondence vectors between the two images are superimposed on the SEM image in Plate 10.12(a). The warped SIMS image was found to match well with the SEM image, as shown in the color overlay in Plate 10.12(c). The unusual contrast in the SE/BSE image is believed to arise from electronic effects due to impurity modification of the band structure of the semiconductor SiC. The impurity distribution in the unmelted and the redeposited SiC is different, leading to differences in secondary electron emission. The maximum Al concentration is approximately 100 ppm.

Light Element Compositional Mapping

Aluminum–Lithium Alloys

Alloys of aluminum and lithium are promising structural materials for the next generation of aerospace applications, offering an improvement in strength-to-weight ratio over existing aluminum alloys. Microanalysis of these alloys, which contain less than 10 at. % Li, is virtually impossible by any technique other than SIMS (Williams et al., 1988). The characteristic x ray from Li is so low in energy as to be effectively undetectable, and Auger electron spectroscopy and electron energy loss spectrometry do not possess sufficient sensitivity. However, SIMS performed with a primary ion beam of oxygen ions provides extraordinary sensitivity to Li, with the detectability limits in the ppm or ppb range in almost all analytical situations.

Plate 10.13 shows a series of Al and Li images, superimposed by primary color overlay with Al assigned to red and Li to green, as a function of depth into the specimen. A striking change in the lateral distribution of Li is observed as the depth into the specimen is increased. Li is apparently leached to the surface of the specimen during mechanical polishing in water. In order to observe a microstructure that is truly representative of the specimen's condition with this preparation, it is necessary to remove a layer approximately 0.1 μm in thickness. The final image in the series, Plate 10.13(c), reveals a grain boundary enriched in Li and a distribution of Li-rich precipitates throughout the matrix of the grains.

Plate 10.14 shows an example of a fully quantified SIMS compositional map for the Al-Li system (Newbury and Bright, 1988a). The image data were obtained with a digital television camera and were converted into the equivalent ion count rates by means of a calibration curve that related the digitized camera value to the ion count measured by the pulse-counting electron multiplier ion detector. The relative ion count rates for Li and Al measured at a particular pixel were converted into a concentration ratio by means of a relative elemental sensitivity factor measured on a series of solutionized, homogeneous Al-Li alloys. The compositional maps, Plate 10.14(a) for Li and Plate 10.14(b) for Al, are encoded with a thermal scale that spans 0–100 at. % of each component. In Plate 10.14(a) the second-phase particles can be seen to be approximately 50 at. % Li and correspond to the Al-Li intermetallic compound in this system. When a quantitative line trace is taken along the vector AB, which intersects several of these particles, the particles are indeed seen to have a concentration of 50 at. % Li.

Distribution of Oxygen and Carbon in a High-T_c Ceramic Superconductor

While the electron probe can measure light elements for atomic number 4 and greater, the sensitivity is poor for atomic numbers of 9 and less because the x rays of these light elements are low in energy and are consequently strongly absorbed by the matrix. Mapping oxygen and carbon in highly absorbing matrices such as $YBa_2Cu_3O_{7-x}$ requires a long accumulation time, and the sensitivity, especially to small changes in concentration, is poor. When SIMS is performed with a primary beam of Cs ions, and negative secondary ions are measured, oxygen and carbon are found to be two of the most sensitive elements in the Periodic Table. Oxygen is believed to play a critical role in establishing the correct superconducting phase in the Y-Ba-Cu-O system, and carbon is believed to be present as a critical contaminant in $YBa_2Cu_3O_{7-x}$ synthesized from starting reagents containing carbonate.

Ion microscope images that reveal the oxygen distribution in a $YBa_2Cu_3O_{7-x}$ bulk superconductor are shown in Plate 10.15 with a primary color superposition (Ba, red: O, green; Y, blue). A complex microstructure is observed, with the continuous superconducting phase interrupted by numerous small particles of several different phases, as evidenced by the different colors in the composite image. The particles that appear white have high signals for all three elements, O, Cu, and Y. The principal problem with SIMS intensity maps is the difficulty in interpreting the images intuitively. An observed increase in the ion intensity may not actually correspond to an increase in concentration. Quantitative corrections, however, require a standard similar in composition to the unknown in order to determine appropriate elemental sensitivity factors. Thus, an intensity image such as that shown in Plate 10.15 provides intriguing information, but concentration relationships must not be inferred unless corrections are applied. An example of a SIMS

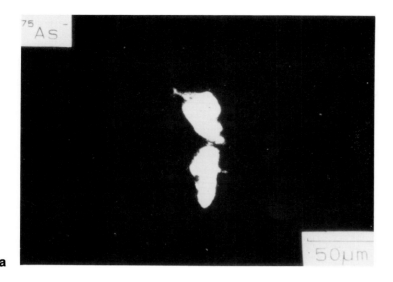

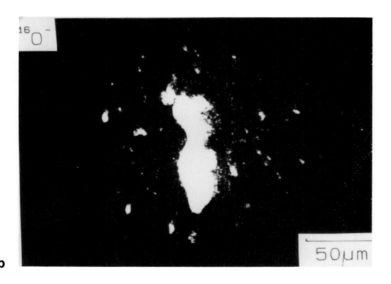

Figure 10.4 Application of secondary ion mass spectrometry compositional mapping to the study of the microstructure of diffusion-induced grain boundary migration in the Cu-As system. (a) As image. (b) O image from the same region. Image field width, 150 μm.

intensity map of the distribution of carbon is shown in Plate 10.16, is subject to the same cautions on interpretation.

The Role of Oxygen in Diffusion-Induced Grain Boundary Migration
As illustrated in the example of Plate 10.5, electron microprobe compositional mapping of the DIGM phenomenon in the Cu-Zn system revealed an anomalous compositional contrast that suggested that a light element not easily measured by an x-ray technique was present in the DIGM zone. Similar observations were made for the Cu-As binary system. SIMS compositional mapping provided direct evidence that oxygen was incorporated into the alloy during the thermal treatment, which led to the DIGM process (Butrymowicz et al., 1985). By using a cesium primary ion beam and detecting negative secondary ions, good sensitivity was obtained for both arsenic and oxygen. Figure 10.4 shows ion images for the arsenic and oxygen distributions in a Cu-As alloy that reveal a strong localization of oxygen in the DIGM region; the correspondence is not exact, in part because of the sequential collection of the image data for the two species and the consequent difference in the exact region of the specimen that was sampled. These SIMS maps support the electron microprobe observations, providing an example of the synergistic use of the two compositional mapping techniques.

SUMMARY

Quantitative compositional mapping by electron probe microanalysis and secondary ion mass spectrometry provides an important new tool for the materials scientist to employ in the characterization of microstructures. The tools are highly complementary, with electron probe compositional mapping offering quantitative analysis with a high degree of accuracy and reasonable sensitivity for most of the elements in the Periodic Table, but suffering from long image accumulation times and a lack of sensitivity for the light elements. SIMS compositional mapping offers an extraordinary advantage in the speed of image collection and profound sensitivity for most elements. SIMS suffers from such strong matrix effects that raw intensity images cannot be interpreted intuitively, and simple relative intensity relationships may be misleading. Quantitative SIMS analysis does not offer the flexibility in treating arbitrarily selected unknowns that quantitative electron probe analysis provides. Quantitative SIMS currently relies on an empirical sensitivity factor approach that demands multielement standards similar in composition to the unknown.

The chief advantage of compositional mapping comes in the analysis of unknowns about which little prior information is available. Compositional mapping uniformly samples every point in an image matrix without bias, which unfortunately is not usually the case when an analyst is selecting the sample locations for point analyses. As a result of this unbiased sampling, compositional mapping is often able to reveal unusual and unexpected events that might be overlooked in conventional analysis. The combination of images supported by quantitative concentration values at every picture element in the image gives a comprehensive view of the chemical microstructure.

REFERENCES

Benninghoven, A., Rudenauer, F. G., and Werner, H. W. (1987). *Secondary Ion Mass Spectrometry*. Wiley, N.Y.

Blendell, J. E., Chiang, C. K., Cranmer, D. C., Freiman, S. W., Fuller, E. R., Drescher-Krasicka, E., Johnson, W. L., Ledbetter, H. M., Bennett, L. H., Swartzendruber, L. J., Marinenko, R. B., Myklebust, R. L., Bright, D. S., and Newbury, D. E. (1987). *Ad. Ceramic Materials* **2**, 512.

Bright, D. S., Newbury, D. E., and Marinenko, R. B. (1988). *Microbeam Analysis–1988*. San Francisco Press, San Francisco, p. 18.

Butrymowicz, D. B., Cahn, J. W., Manning, J. R., Newbury, D. E., and Piccone, T. J. (1984). Diffusion-Induced Grain Boundary Migration. In *Character of Grain Boundaries*. M. F. Yan and A. H. Heuer (eds.), Advances in Ceramics 6, 202.

————, Newbury, D. E., Turnbull, D., and Cahn, J. W. (1985). *Scripta Met.* **18**, 1005.

Cosslett, V. E., and Duncumb, P. (1956). *Nature* **177**, 1172.

Duncumb, P. (1957). In *X-ray Microscopy and Microradiography*. V. E. Cosslett, A. Engstrom, and H. H. Pattee (eds.), Academic Press, N.Y., p. 435.

Fiori, C. E., Leapman, R. D., and Gorlen, K. E. (1985). *Microbeam Analysis–1985*. San Francisco Press, San Francisco, p. 219.

————, Swyt, C. R., and Gorlen, K. E. (1984). *Microbeam Analysis–1984*. San Francisco Press, San Francisco, p. 179.

Garruto, R. M., Swyt, C., Fiori, C. E., Yanagihara, R., and Gajdusek, D. C. (1985). *Lancet* ii, 1353.

Goldstein, J. I., Newbury, D. E., Echlin, P., Joy, D. C., Fiori, C. E., and Lifshin, E. (1981). *Scanning Electron Microscopy and X-ray Microanalysis*. Plenum, N.Y.

Heinrich, K. F. J., and Fiori, C. E. (1984). *Microbeam Analysis–1984*. San Francisco Press, San Francisco, p. 175.

Liebl, H. (1975). *The Ion Microprobe—Instrumentation and Techniques*. In *Secondary Ion Mass Spectrometry*, K. F. J. Heinrich and D. E. Newbury (eds.), National Bureau of Standards (U.S.) Special Publication 427 (Washington), p. 1.

Marinenko, R. B., Myklebust, R. L., Bright, D. S., and Newbury, D. E. (1985). *Microbeam Analysis–1985*. San Francisco Press, San Francisco, p. 159.

————, Myklebust, R. L., Bright, D. S., and Newbury, D. E. (1987). *J. Microscopy* **145**, 207.

Myklebust, R. L., Newbury, D. E., Marinenko, R. B., and Bright, D. S. (1986). *Microbeam Analysis–1986*. San Francisco Press, San Francisco, p. 495.

————, Newbury, D. E., Marinenko, R. B., and Bright, D. S. (1987). *Microbeam Analysis–1987*. San Francisco Press, San Francisco, p. 25.

Newbury, D. E. (1982). *Anal. Chem.* **54**, 1059A.

————, and Bright, D. S. (1988a). *Secondary Ion Mass Spectrometry—SIMS VI*, A. Benninghoven, A. H. Huber, and H. W. Werner (eds.), Wiley, New York, p. 389.

————, and Bright, D. S. (1988b). *Microbeam Analysis–1988*. San Francisco Press, San Francisco, p. 105.

————, Bright, D. S., Williams, D. B., Sung, C. M., Page, T., and Ness, J. (1986a).

Secondary Ion Mass Spectrometry—SIMS V, A. Benninghoven, R. J. Colton, D. S. Simons, and H. W. Werner (eds.), Springer-Verlag, Berlin, p. 261.

———, Joy, D. C., Echlin, P., Fiori, C. E., and Goldstein, J. I. (1986b). *Advanced Scanning Electron Microscopy and X-ray Microanalysis.* Plenum, N.Y.

Piccone, T. J., Butrymowicz, D. B., Newbury, D. E., Manning, J. R., and Cahn, J. W. (1982). *Scripta Met* **16**, 839.

Slodzian, G. (1975). *Secondary Ion Mass Spectrometry,* K. F. J. Heinrich and D. E. Newbury (eds.), National Bureau of Standards (U.S.) Special Publication 427 (Washington), p. 33.

Swyt, C. R., and Fiori, C. E. (1986). *Microbeam Analysis–1986.* San Francisco Press, San Francisco, p. 482.

Williams, D. B., Levi-Setti, R., Chabala, J. M., and Newbury, D. E. (1988). *J. Microscopy* **148**, 241.

11

Processing Images and Selecting Regions of Interest

D. S. BRIGHT, D. E. NEWBURY, R. B. MARINENKO,
E. B. STEEL, AND R. L. MYKLEBUST

Over the past decade, the development of laboratory computers has made possible the application of digital imaging techniques to a wide variety of photon, electron, and ion microscopes. Characterization of the microstructure of materials by these microscopy techniques can provide information on morphology, crystal structure, and elemental or molecular chemistry on a spatial scale ranging from micrometers to nanometers. Traditionally, image information from these microscopes was recorded exclusively by analog means such as film. The added dimension of digital imaging, where the image is recorded directly as digitized arrays of signals derived from detectors or is digitized from previously recorded analog images, gives the microscopist/analyst access to a powerful and wide-ranging suite of tools for the acquisition, enhancement, and interpretation of information derived from the microscope. Many of these tools were impractical or impossible in the realm of analog image manipulation. A second burst of development in the digital imaging field is now taking place with the advent of "computer-aided microscopy." Computer-aided microscopy augments conventional digital imaging techniques with algorithms adapted from various fields of computer science. The aim of this approach is to develop tools that can function automatically to carry out image-analysis tasks that are too complex or time consuming to permit conventional interpretation by direct viewing of the image on a screen. Digital images lend themselves to a variety of techniques for enhanced display and automated analysis. Images produced by the Microanalysis Group of the Center for Analytical Chemistry at the National Institute of Standards and Technology will be used to illustrate several techniques for enhancing images or extracting the chemical or morphological information contained in them. The images are micrographs from electron and ion microprobes and electron and light microscopes. Some electron diffraction patterns from analytical electron microscopes are also included. Techniques are illustrated by examples where they have proved useful and fall into three groups: enhancement and processing of single images, multiple-image relationships, and feature analysis using transforms.

ENHANCEMENT AND PROCESSING OF SINGLE IMAGES

The typical digital image from an analytical instrument often has a far greater dynamic range than can be displayed or rendered in a photographic print. Techniques to alter the contrast or display selected gray levels in carefully chosen colors can help to visualize more of the image information. The thermal scale (see also Chapter 10) has proven to be a useful color selection scheme, and histogram equalization is useful for visualizing certain types of detail. Individual scan lines or intensity profiles along a selected line in the image can always be obtained after the fact, as the information is inherent in the digital representation of the images. Also, both the magnitude and the direction of the gradient of the image (discussed later) can bring out image detail that otherwise might be missed.

Contrast Enhancement

The dynamic range of many images is such that a simple contrast enhancement can render visible much detail that could otherwise not be seen. For example, Plate 11.1 is a transmission light micrograph of asbestos fibers. The image contrast is low. Plate 11.2 results from enhancing the contrast by assigning minimum brightness (light gray) in the micrograph to black and maximum brightness (slightly lighter gray) to white; other intensities are interpreted linearly. This image shows the fibers more clearly, and also shows that the illumination is not even. The contrast-enhanced image also shows a small contouring effect—the steps between adjacent brightness levels are now visible as concentric rings around the periphery of the image. This effect, due to the digital or quantized nature of the stored image, shows that the useful limit for contrast enhancement has been reached.

Color Scales

Pseudocoloring is similar to contrast enhancement, except that intensity levels are translated, more or less arbitrarily, into colors rather than gray levels. If done with care, this increases the amount of information that can be displayed. An example of a useful pseudocolor scale is the thermal scale, which looks like the sequence of

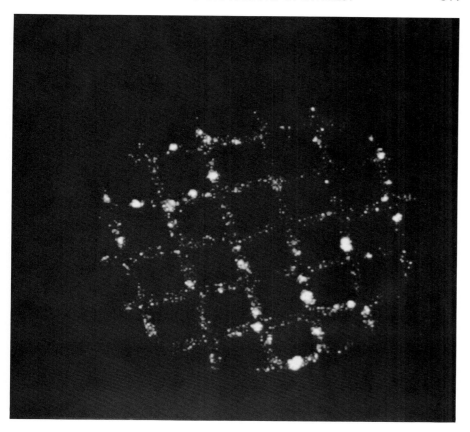

Figure 11.1 Benzalkonium chloride image (molecular weight, 304) taken with an ion microscope. Field width, 150 μm.

colors produced when a black body is heated: deep red, red, cherry red, orange, yellow, white. This sequence is perceived by a viewer in a "logical" fashion such that the "attention value" of the colors is in order of increasing intensity.

Most display devices map intensity values into brightness or color using a lookup table (LUT). The bar along the bottom of some of the images shows what color or brightness is used to display each intensity level, from zero (dark) on the left to 255 (bright) on the right. Many images have a dynamic range greater than 256 levels and have been scaled or otherwise compressed to 256 levels for display on the particular device used here. Even 256 brightness levels are more than what is absolutely necessary, because the difference between one brightness level and the next becomes visible only when the total number of brightness levels is as few as 32 (see the following discussion on dynamic range). Figure 11.1 is a molecular secondary

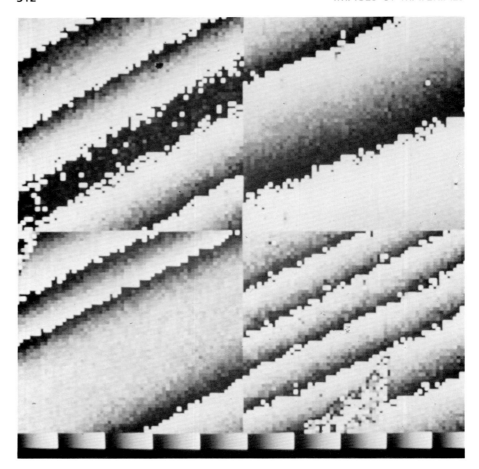

Figure 11.2 Four electron probe maps (400 ×) of homogeneous metal samples (upper left, Cr; upper right, Fe; lower left, Ge; and lower right, Sc). Banding due to defocusing effect is shown with contrast enhancement—ten gray ramps rather than the usual single gray ramp. Field width for each map, 250 μm.

ion microscope image of the benzalkonium cation (molecular weight 304), sputtered from the chloride salt, which was deposited from a saturated ethanol solution onto a copper transmission electron microscope grid (25 μm between bars). The deposition was clumpy. The image has a small dynamic range—about 20 gray levels—but the bright spots (the clumps) need to be shown clearly, along with the grid, which has very low intensity. The grid is discernible in Figure 11.1, but is clearly seen in the thermal scale rendition in Plate 11.3. The clumps of benzalkonium chloride are just as clearly seen in the thermal scale image. (See Chapter 3 by Levi-Setti et al. for a discussion of the ion microscope.)

Occasionally, contrasting colors help to visualize needed information. Figure

11.2 shows electron probe x-ray intensity maps of four homogeneous metal samples. The maps are shown with contrast enhancement, to exaggerate the contrast to better locate the position of the intensity ridge, which is seen as the broad white band in the first map. The intensity ridge is an instrumental artifact caused by the electron beam being slightly misaligned with respect to the crystal spectrometer which selects the x rays used to create the map (see Chapter 10). The location of the ridge is used in correcting for the instrumental artifact. The gray scales are adequate for the Cr and Ge maps on the left, but, for the asymmetric Fe and Sc maps on the right, gray scales can be confusing. Plate 11.4 shows the same maps but rendered with contrasting color bands to show that the intensity ridge due to defocusing is in the field of view in the maps on the left, and out of the field of view in the maps on the right.

Histogram Equalization

Sometimes the LUT is calculated from the image. One such scheme assigns gray levels such that each level has the same number of pixels, or the same area as each other gray level. This scheme, called *histogram equalization,* tends to bring out detail in the background or homogeneous areas, and to lose detail in the smaller features. Figure 11.3 is a TEM image of particles on a 0.4 μm polycarbonate filter. The equalized display, Plate 11.5, shows the pores of the filter, some dimmer particles, and the drop off in illumination intensity as the image was scanned top to bottom. Details of some of the smaller dark particles are lost. Figure 11.4 is a bright-field light micrograph of particles mounted on a TEM grid. The equalized image, Plate 11.6, shows some structure of the film supporting the particles, as well as some low-level reflection of light off the grid itself.

Dynamic Range

64 gray levels are enough to display images without visible intensity steps or any contouring effect. Figures 11.5 (*a*)–(*d*) show four versions (with four numbers of gray levels) of an x-ray map with smoothly varying intensity. The map [rendered with all 256 gray levels, Figure 11.5(*a*)] has just enough noise so that 32 gray levels [Figure 11.5(*d*)] are sufficient to display the map with no noticeable contouring effect. 8 [Figure 11.5(*b*)] and 16 [Figure 11.5(*c*)] gray levels do show the contouring effect.

Figures 11.6 and 11.7 and Plates 11.7 and 11.8 show different versions of an x-ray map of the distribution of zinc near a grain boundary. This map (Figure 11.6) has enough noise to mask the contouring effect, even when only 16 gray levels (4 bits) are used (Figure 11.7). However, comparison of the pseudo-colored images shows slight differences between the original 256-bit image (Plate 11.7) and the compressed 16-bit image (Plate 11.8). Some information is lost in using only 16 gray levels.

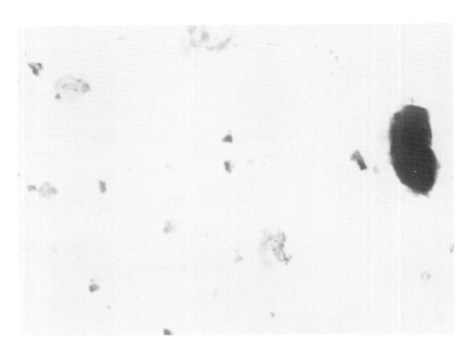

Figure 11.3 TEM image of particles on a polycarbonate filter. Field width, 20 μm.

Figure 11.4 Bright-field light micrograph of particles on a TEM grid. Field width, 150 μm.

314

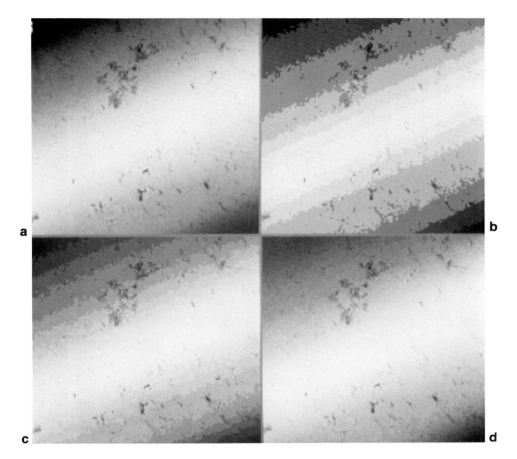

Figure 11.5 Electron probe map of a metal alloy shown with the same contrast, but four numbers of colors: 256, 8, 16, and 32. Contouring or banding is seen when too few gray levels are used. Field width for each map, 500 μm.

315

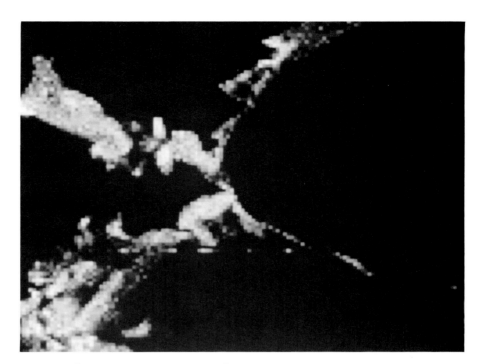

Figure 11.6 Electron probe map (Zn) of grain boundary, 256 gray levels. Field width, 250 μm.

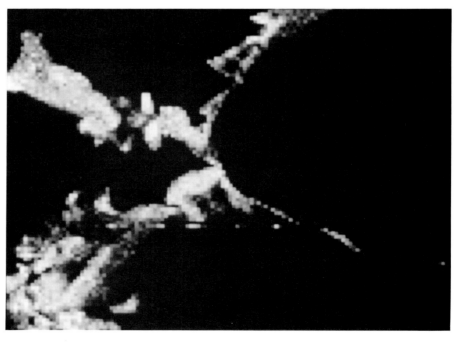

Figure 11.7 Figure 11.6 rendered with only 16 gray levels. Discrepancies from Figure 11.16 are small. Field width, 250 μm.

Scan Lines

Having images in digital form facilitates the display of intensity profiles. Plate 11.9 is a ^{138}Ba isotope image of a particle field taken on an ion microscope. The intensity profiles for the line A-B are shown for five isotope images of Ba (magenta, ^{n}Ba, $n = 138, 137, 136, 135, 134$), as well as for two "background" images (green).

Gradient

The gradient of an image is related to the change in brightness between a given pixel and its neighbors. Analogously to the gradient of a two-dimensional mathematical function, the gradient of an image has a magnitude and a direction for each pixel. Unlike the precisely defined gradient of a continuous mathematical function, the image gradient is usually a quickly calculated approximation based on the differences in the brightness value of adjacent pixels (Rosenfeld and Kak, 1982). The magnitude of the gradient is used to enhance edges, or act as a "high-pass filter." Figure 11.8 is a reflected light micrograph of a polished binary alloy, with the one phase appearing dark. The grain boundaries are shown in Figure 11.9 in the gradient magnitude image. Figure 11.8 is ideally suited for use of the gradient magnitude. Figure 11.10 is an example of a different type of image where the edges are not continuous and are less distinct. It is a bright-field light micrograph of glass shards. The gradient magnitude image in Figure 11.11 looks like a dark-field image, but the edges appear as double lines, which are not continuous around some of the shards.

The gradient also has a direction associated with it, denoting the direction of the maximum intensity change from a pixel to its neighbors. Images of the gradient direction have been used for distinguishing fuzzy edges of objects from the adjacent background, and for visualizing details in the background (usually due to instrumental effects). This is largely because changes in brightness of the gradient direction image are not related at all to brightness or sharpness of edges in the original image, but only to the direction of the maximum intensity change, no matter how large or small that change may be. Figure 11.12 (Bright, 1987a) shows a way to display the gradient direction (Smith et al., 1977). A polar plot of the display intensity versus gradient direction is superimposed on the gradient direction of a computer-generated intensity cone. This display makes objects appear to be illuminated from the upper left. Figure 11.13 is an ion microscope lithium image of a particle field taken with a digitizer from a television camera signal. The camera is viewing the output of a dual channel plate ion detector. Characterization of blooming due to the camera and channel plates versus the surface diffusion of the lithium and the noise due to the digitizer were of interest. Figure 11.14, the gradient direction image, shows the shape of the field of view, the texture of the background, and the extent of blooming of the images of the particles. Light microscope images are rarely noise free. Figure 11.15 is a bright-field light micrograph of particles. Pseudo-coloring the brighter intensity levels (Plate 11.10) to show small changes in background level visualizes the texture in the background (presumably due to

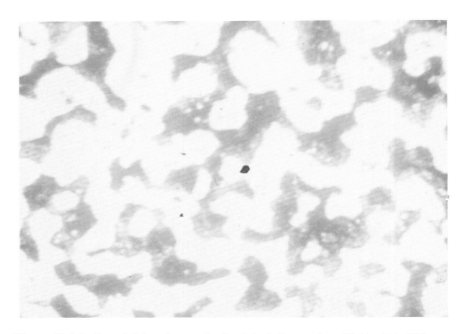

Figure 11.8 Reflected light micrograph of polished alloy surface. Field width, 200 μm.

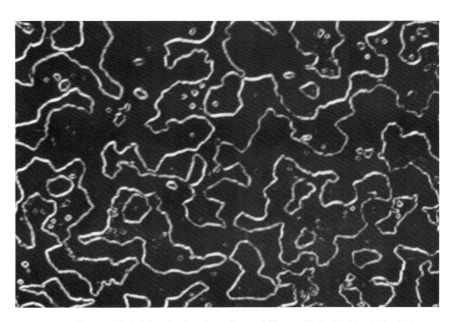

Figure 11.9 Magnitude of gradient of Figure 11.8. Field width, 200 μm.

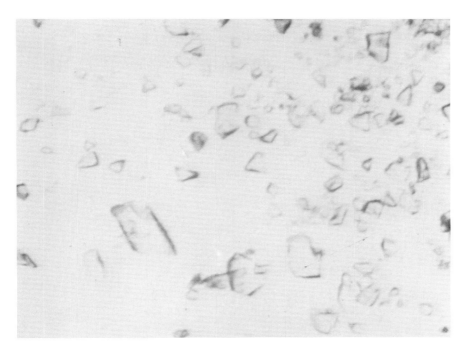

Figure 11.10 Bright-field light micrograph of glass shards. Field width, 100 μm.

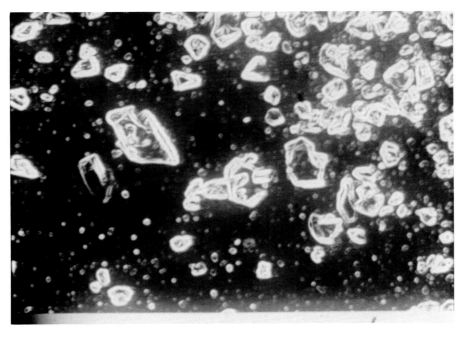

Figure 11.11 Magnitude of gradient of Figure 11.10. Field width, 100 μm.

319

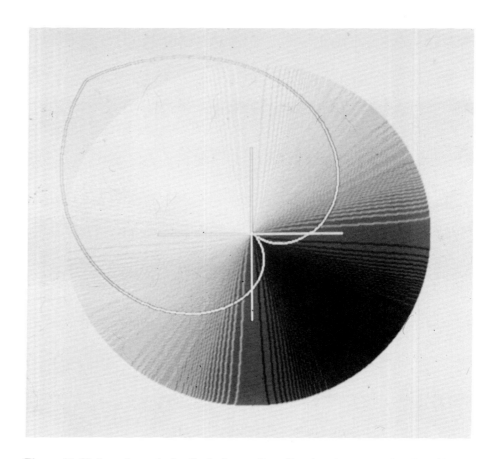

Figure 11.12 Intensity scale for displaying gradient direction shown as polar plot of intensity (radius) versus angle of gradient direction. Gradient direction of computer-generated intensity cone also shown displayed with this scale. Field width, 512 pixels.

320

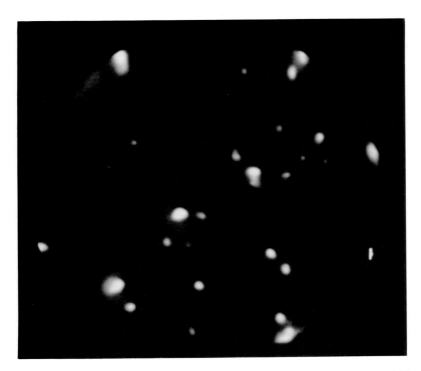

Figure 11.13 Lithium image of particles taken with an ion microscope. Field width,150 μm.

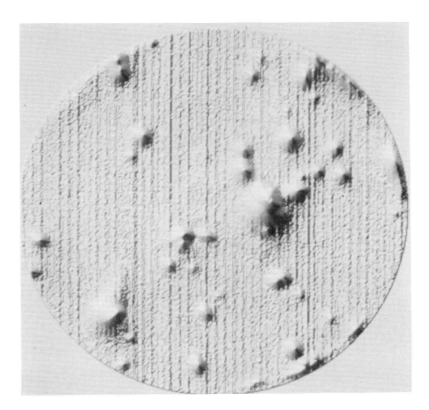

Figure 11.14 Gradient direction image of Figure 11.13. Field width, 150 μm.

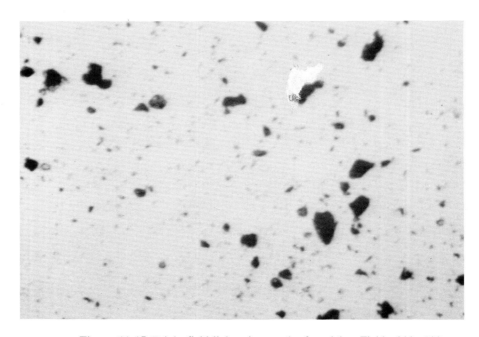

Figure 11.15 Bright-field light micrograph of particles. Field width, 100 μm.

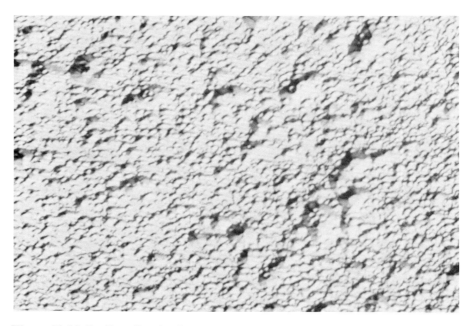

Figure 11.16 Gradient direction image of Figure 11.15 showing small-scale inhomogeneities in the background (imperfections in optics or slide mount). Field width, 100 μm.

imperfections in the sample mounting and the microscope lenses, as well as the circular banding due to the slightly inhomogenous illumination level). The gradient direction shown in Figure 1.. shows the background texture in a more intuitive way, but not the variations in illumination level. The latter changes only over longer distances across the image and thus has low spatial frequency, to which this gradient measurement is not sensitive.

MULTIPLE-IMAGE RELATIONSHIPS

Images taken with different modalities (signals) can be used to relate morphological and chemical information. The electron probe takes several x-ray maps simultaneously, ensuring registration of each pixel. These images can be examined by using color overlays and concentration–concentration histograms.

Color Overlays

When multiple-element x-ray maps or isotope ion images are obtained from complex samples, spatial relationships are often more clearly seen in a color overlay of the maps that are not obvious when looking sequentially at the individual maps. To illustrate this, Plates 11.11 through 11.13 show somewhat astigmatic Cu, Ba, and Y maps (upper left, upper right, and lower left, respectively) and the color overlays (lower right) of the maps from three superconducting thin film ceramics. Qualitative differences between the samples can be seen at a glance from the color overlays. The differences can also be seen from the individual maps after some study.

Figures 11.17 and 11.18 show Au and Pt maps, respectively, for a Au-Pt-Cu alloy. Copper is a minor constituent. The color overlay in Plate 11.14 shows yellow borders on the Au grains, which indicate an overlap due to the size of the beam interaction volume and perhaps also due to diffusion of elements across the grain boundaries. The product of the two maps in Figures 11.17 and 11.18, shown in Figure 11.19, is very sensitive to the boundaries, because only the boundaries have large pixel counts on both maps. The gradient of an image (calculated from the difference in counts between neighboring pixels) also highlights the edges, as seen by the gradient of the Pt image in Figure 11.20, but the product image shows some edges and small cracks that are not seen in the gradient image. These details appear blue (product image only) rather than bluish green (both images) in the color overlay (Plate 11.15). Occasionally color overlays are useful for superimposing images. Figures 11.21 and 11.22 are scanning electron micrographs of asbestos fibers, taken in the scanning (SEM) and transmission (STEM) modes, respectively. Thus the SEM image is formed by electrons scattered *back* from the sample whereas the STEM image is formed from electrons transmitted *through* the sample. Plate 11.16 shows the superposition of the two, with the SEM image displayed in green and the inverted STEM image in red. Yellow shows the overlap of the two. The end

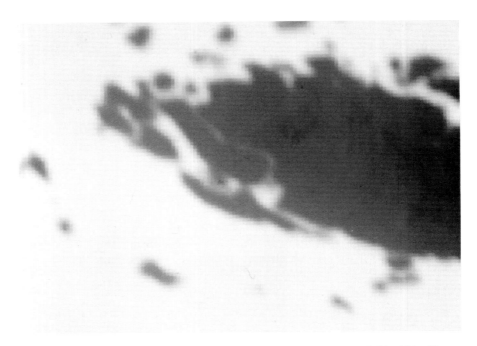

Figure 11.17 Gold map of Au-Pt-Cu alloy. Field width, 50 μm.

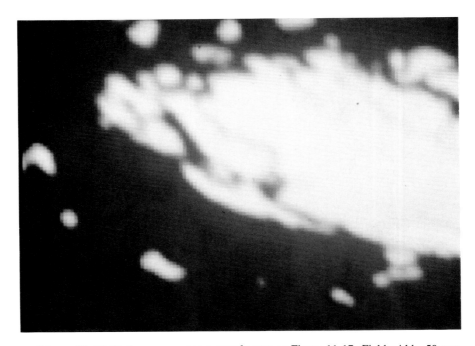

Figure 11.18 Platinum map, same sample area as Figure 11.17. Field width, 50 μm.

324

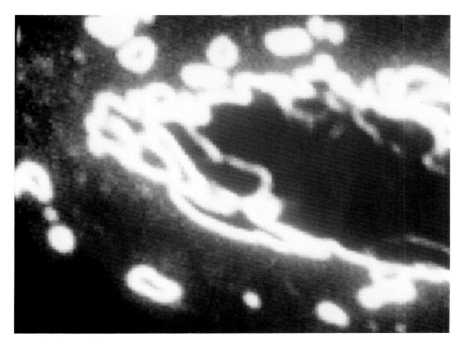

Figure 11.19 Pixel-by-pixel product image of Figures 11.17 and 11.18. Field width, 50 μm.

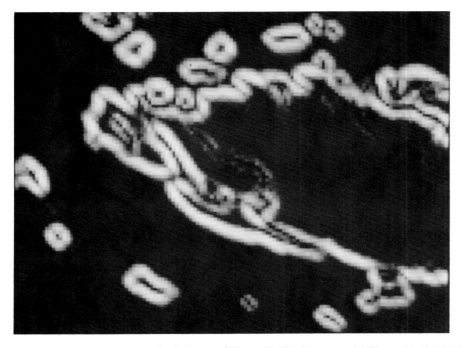

Figure 11.20 Gradient magnitude image of Figure 11.18. Compare with Figure 11.19. Field width, 50 μm.

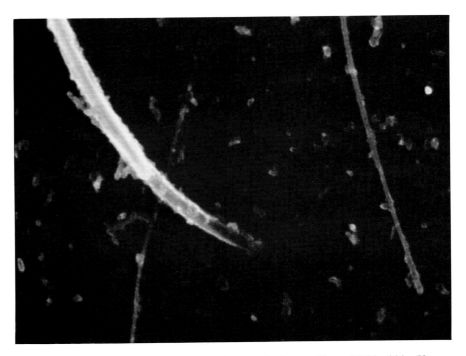

Figure 11.21 SEM image of asbestos fibers. Field width, 50 μm.

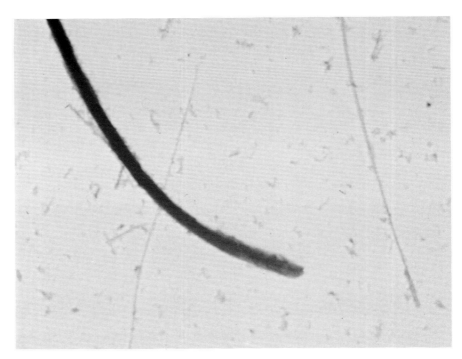

Figure 11.22 STEM image of same fibers as in Figure 11.21. The end of the large fiber is beneath the substrate surface, and is therefore missing from the SEM image, Figure 11.21. Field width, 50 μm.

of the large fiber is missing from the SEM image because it is beneath the surface of the substrate, but it is shown clearly in the STEM image and thus appears bright red in this overlay.

Concentration–Concentration Histograms

The color overlay of two or three quantified (Marinenko et al., 1987) x-ray maps can be augmented with a two (CCH) or three (CCCH) -dimensional histogram of the corresponding concentration pixel pairs or triplets. The concentration–concentration histogram (CCH) is a two-dimensional analog of the one-dimensional histogram and is a type of scatter diagram (Browning, 1987; Prutton, 1987). The CCH has two concentration axes defined so that a point in the plane is determined by the concentration values for any corresponding pair of pixels in a pair of x-ray maps. The accumulated number of counts for each of these points corresponds to the number of pixel pairs in the maps with those particular concentrations.

A computer simulation of electron probe x-ray maps is shown in Plate 11.17. The "maps" are in the upper half of the figure, and the corresponding CCHs are in the lower half. The left map CCH pair is free from noise, while the right has Gaussian noise added to the maps. Except for the boundary between the "phases," the noise-free CCH has counts in only two bins, which correspond to the two "phases." These points are shown as white squares for emphasis, as each has many counts. The yellow line between the white points is due to a random part of the interaction volume of the beam (simulated) for any edge pixel representing one phase and the rest representing the other phase. The added noise (to simulate counting statistics) makes the CCH appear like those of real samples. The two white points are now oval clusters, and the line between them has expanded to form a band. Plate 11.18 is a 128 × 128 pixel electron probe concentration map of a sample of a Cu-Ti alloy, analogous to the computer simulation.

Figure 11.23 shows the CCH as a graph, with increasing Cu concentration back and to the right, increasing Ti concentration back and to the left, and counts per bin upward. This display is analogous to a one-dimensional histogram. The CCH is shown in image form in Plate 11.19, using a thermal scale similar to Plate 11.12, so that the bins with only one count are visible (bright red) as well as the bins with many counts (white). Such a visual display of the CCH allows the analyst to recognize quickly and quantitatively those compositional value pairs that occur commonly in the original single-band images.

Figure 11.24 and Figure 11.25 show x-ray maps (Ba, Cu, Y, and O by difference—upper left, upper right, lower left, and lower right, respectively) of two areas of a superconducting ceramic sample. The areas differ somewhat in morphology, as seen in the color overlays in Plates 11.20 and 11.21, but the Cu-Y CCHs (Plates 11.22 and 11.23) more clearly show that the bulk of the map areas are clustered near the same concentration ratio and that Plates 11.21 and 11.23 lack pixels that register high in Y.

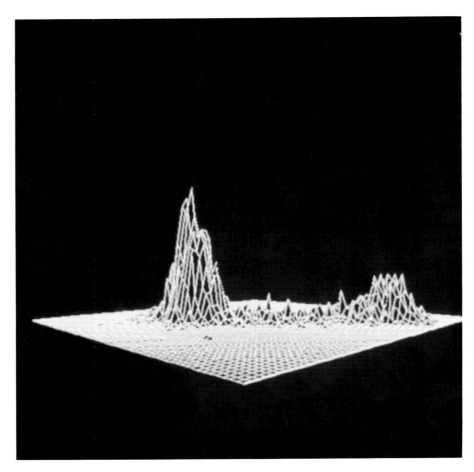

Figure 11.23 A mesh plot of CCH for Cu-Ti maps in Plate 11.18 (see text). Right front edge of square—Cu, 25–59 wt. %. Left front edge of square—Ti, 39–75 wt. %.

328

The CCCH can be used to select areas of interest. Plate 11.24 shows Ba, Cu, and Y maps (elements shown as in Plate 11.11) for a bulk ceramic superconductor. Regions that are outlined (explanation below) are high in Ba. The region that is high in Ba compared with the matrix can be distinguished by its reddish hue in the magnified superposition (Plate 11.25). Three elements require a three-dimensional analogy to the CCH, that is, the CCCH. A perspective projection of the CCCH is shown in Plate 11.26, and a duplicate of this figure, with the histogram also projected to the left, rear, and bottom face of the cube, is shown in Plate 11.27. Instead of stereo pairs (which we also use), the projections on the faces of the cube help discern the three-dimensional shape of the CCCH. In these figures, the Ba axis is toward the viewer and to the lower left, the Cu axis is in the plane of the page and to the right, and the Y axis is upward. Plate 11.27 shows that most of the structure of the histogram is preserved in the projection on the bottom face, the Ba-Cu projection, which is shown again in the lower left corner of Plate 11.28. The right horizontal arm, representing pixels high in Ba (with constant concentration in Cu) is outlined in white, and the areas in the original maps that correspond to the outlined arm are also outlined in white. Note that there is no visible difference in Cu concentration in crossing the white outline (Cu map, upper right, Plate 11.24), as expected. The left projection of the CCCH (Plate 11.27) shows a bright portion at the high Ba end on the Ba axis and no points above (off the axis) this bright portion. The outlined region can be expected to be almost zero in Y concentration, which is consistent with the white outline of the Y map (lower left) in Plate 11.24. The outline does not enclose all of the low Y areas because some low Y areas are outside the outlined area as they are also low in Ba or Cu, or both. For more discussion of this system, see Figure 10.9(a)–(f) in Chapter 10 in this book.

FEATURE ANALYSIS USING TRANSFORMS

Automated analysis of images requires the identification of features and places in the image on which to perform the measurement. The blob-splitting algorithm is used to select simple features, while the linear Hough transform is used as a preprocessing step to simplify locating lines.

Blob-Splitting Algorithm

A particularly useful transform of an image is one that selects out blobs (contiguous groups of pixels) that denote regions of interest. This can sometimes be done with a simple setting of the intensity threshold. When one threshold setting is inadequate, an algorithm that uses multiple thresholds (Bright, 1987b) often works. This algorithm selects objects or areas of interest by the lack of internal structure: The algorithm starts at the lowest threshold of the image and keeps track of the blobs

Figure 11.24 Electron probe maps of superconducting ceramic—Ba, Cu, Y, and O. The oxygen map was calculated by stoichiometry. Field width for each map, 114 μm.

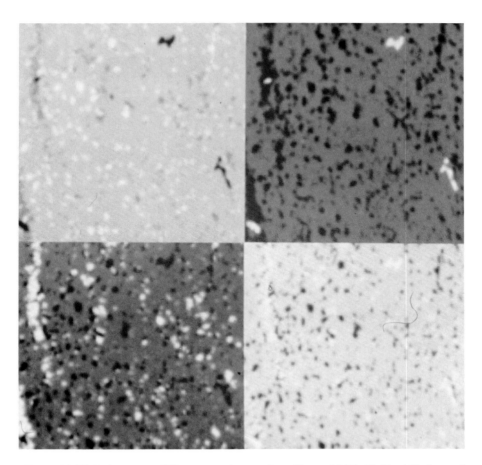

Figure 11.25 Probe maps, different sample area from Figure 11.24. Field width for each map, 114 μm.

(continuous areas or groups of pixels) as the threshold is increased. As the threshold increases, blobs both shrink and split into smaller blobs. The areas of interest are defined as blobs that will not split any more as they continue to shrink down to nothing.

The particles in images such as Figure 11.13 and Plate 11.9 are located by this algorithm, as are diffraction spots, as shown below. The algorithm is also useful for selecting areas of complex particles for further analysis, much as a human operator might. Figure 11.26 is a TEM micrograph of a typical agglomerate particle. The outlines of the larger and more dense regions of interest are shown in green in Plate 11.29, superimposed on the negative or inverse of Figure 11.26. The algorithm looks for bright objects on a dark background, rather than the reverse. The outlines of less dense regions chosen by the algorithm are shown in red and, in this case are mostly fluctuations in the background intensity. Figure 11.27 shows the histogram of intensity of the regions; group A regions are shown in green in Plate 11.29, and group B in red.

Linear Hough Transform

Preprocessing an image with certain transforms can often ease the job of extracting features. Figure 11.28 is a transmission electron micrograph of a metallic shaded replica of an optical diffraction grating. The spacing of the lines was needed for calibrating the microscope magnification factor. Because the dot spacing in the fast Fourier transform (FFT) (Figure 11.29) is inversely proportional to the line spacing in the micrograph, only the FFT spot spacings need be measured, but some precision is lost due to close FFT spot spacings for large image line spacings. The linear Hough transform (LHT) (Pratt, 1978; Dyer, 1983) also maps lines into spots, as shown in Figure 11.30. This image was calculated from the LHT of the brightest five percent of the pixels of the gradient magnitude of Figure 11.28 (Bright, 1987a). The abscissa represents the angle of the perpendicular to the line, and the ordinate represents the length of the perpendicular from the center of the original image. The vertical row of spots of Figure 11.30 represents the grating lines, which are all parallel. The vertical spot spacing in the LHT is equal to the line spacing of the grating (Bright and Steel, 1985). The LHT can also aid in characterizing electron diffraction patterns. The upper left portion of Figure 11.31 is a pattern from an asbestos fiber, where the diffraction spots fall along lines, and the line spacing is the measurement of interest. The blob-splitting algorithm was used to locate the spots, as they varied greatly both in size and in intensity. Figure 11.31, upper right, shows the outlines of the spots found by the algorithm. Figure 11.31, lower left, shows the LHT of the image with spot centers marked with a small disc. The centers of the bright spots making up the vertical row in the LHT correspond to the lines plotted on the diffraction pattern in Figure 11.31, lower right. Thus, the LHT finds the interplanar spacing between the rows of points. Figure 11.32 illustrates the use of the LHT to find fibers. The upper left image is a TEM micrograph of some asbestos

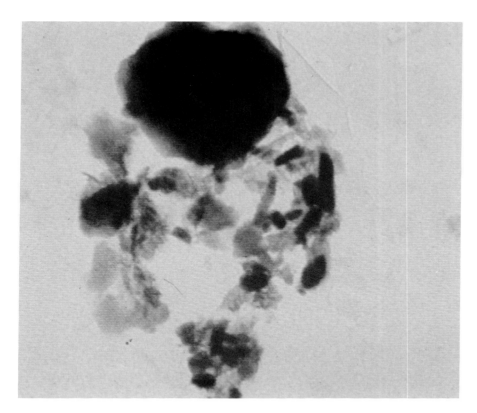

Figure 11.26 TEM image of a complex agglomerate. Field width, 20 μm.

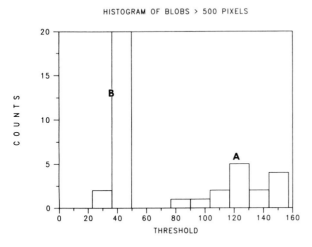

Figure 11.27 A histogram of intensities of areas of Plate 11.29: A. green, B. red. Full scale intensity, 255. Intensities for outlined areas are threshold intensities, that is, intensities at the border of each area.

333

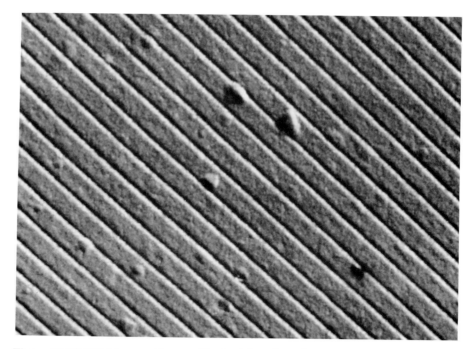

Figure 11.28 TEM image of a metallic shaded replica of a diffraction grating. The line spacing is 2160 per μm. Image is 512 × 512 pixels.

Figure 11.29 FFT of Figure 11.28, 512 × 512 pixels, displayed as a logarithm of the power spectrum. Zero spatial frequency corresponds to the center of the transform. A distance of n pixels for spots closest to the center corresponds to a $512/n$ pixel line spacing for the grating in the original image. Resolution, 512 × 512 pixels.

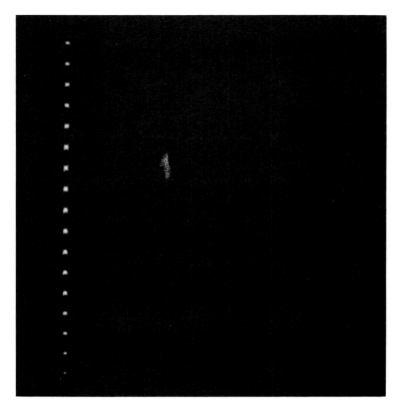

Figure 11.30 LHT (Hough transform) of Figure 11.28. Ordinate—line position in pixels, same scale and resolution as original image. Abscissa—line orientation, 0–179.5°, 1 pixel = 0.5°.

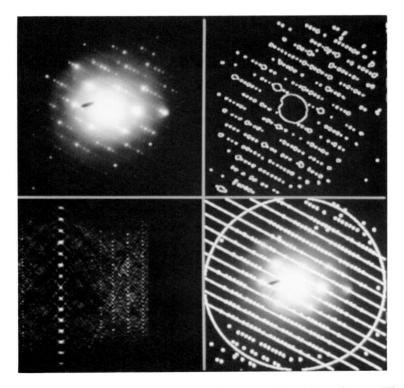

Figure 11.31 Illustration of finding streaks or lines of spots in an electron diffraction pattern: upper left—diffraction pattern, upper right—spot outlines as found by the blob-splitting algorithm (see text), lower left—LHT of spot centers (see text), lower right—lines found from LHT, superimposed on the pattern. Pattern images are 512 × 512 pixels. LHT scale same as Figure 11.30.

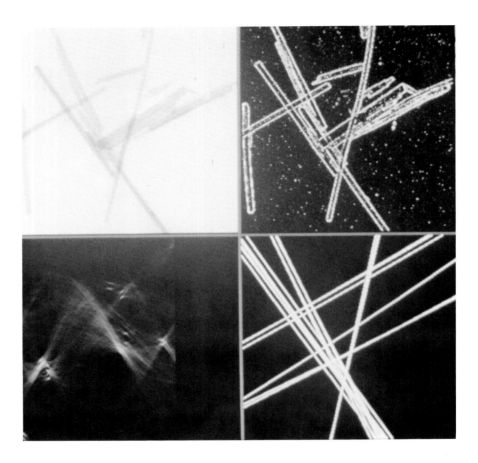

Figure 11.32 Illustration of finding lines: upper left—TEM image of asbestos fibers, upper right—gradient magnitude image of the fiber image, lower left—LHT of the gradient magnitude image, lower right—lines found from LHT (compare with upper left; see text). Field width for all images, 2.1 μm. LHT scale same as Figure 11.30.

fibers. The upper right shows the gradient magnitude image of the upper left. The lower left shows the LHT of the gradient image, and the lower right, the lines corresponding to the bright spots in the LHT. The bright LHT spots were used to calculate the fiber locations, which were then used to plot the lines as a check.

SUMMARY

We have illustrated a variety of techniques that have been found useful in the laboratory for enhancing and analyzing digital micrographs. The techniques include several types of enhancement and pseudo-coloring, which are applicable to most of the images that we use, as well as intensity profiles and histograms, which are more useful on concentration maps. The linear Hough transform and blob-splitting algorithm are useful for finding objects in TEM images with relatively low noise.

REFERENCES

Bright, D. S., and Steel, E. B. (1985). *Microbeam Analysis–1985*. San Francisco Press, San Francisco, p. 155.

———. (1987a). *J. Microscopy* **148**, 51.

———. (1987b). *Microbeam Analysis–1987*. San Francisco Press, San Francisco, p. 290.

Browning, R. (1987). *Analytical electron microscopy–1987,* San Francisco Press, San Francisco, p. 311.

Dyer, C. R. (1983). IEEE Transactions on Pattern Analysis and Machine Intelligence, **PAMI-5,** 621.

Marinenko, R. B., Myklebust, R. L., Bright, D. B., and Newbury, D. E. (1987). *J. Microscopy* **145**, 207.

Pratt, William K. (1978). *Digital Image Processing*. John Wiley, N.Y., p. 523.

Prutton, M., et al. (1987). *Analytical electron microscopy—1987,* San Francisco Press, San Francisco, p. 304.

Rosenfeld, A., and Kak, A. C. (1982). *Digital picture processing,* 2nd ed. Vol. 2. Academic Press, N.Y.

Smith, K. C. A., Unitt, B. M., Holburn, D. M., and Tee, W. J. (1977). *Scanning Electron Microscopy/1977,* Vol. I, Scanning Electron Microscopy, Inc. P.O. Box 6657, AMF O'Hare, Ill. 60666, p. 49.

12

Image Analysis of the Microstructure of Materials

J. C. RUSS

Image analysis is broadly defined as extracting a few (useful) numerical values from an image. The source of the image in the applications of interest here is generally a microscope of some kind discussed already in this book (light, electron, acoustic, etc.), but even unconventional images (e.g., electron diffraction patterns, as in Chapter 6) can be used. Images of this sort contain a great amount of information. One measure of this is the fact that they typically require from 50 kilobytes to several megabytes of computer storage to hold the brightness values of each pixel (picture element). Extracting the desired information from this mass is not easy. Even as sophisticated and powerful a computer as the human visual system cannot always accomplish it without aid. The computers used in most image-analysis systems are far less powerful, but with appropriate programming, and some guidance by a human who is knowledgeable about the material and how the image was obtained, they can often be used to obtain the desired result.

Computerized image analysis has several advantages, including a resistance to being distracted by extraneous information that may be present. More important, humans are not particularly good at measuring things visually. Instead, we rely on comparison. This may be done either directly by bringing two things together (one of which may be a ruler or template of some kind), or indirectly by comparing current images to remembered ones.

A frequent source of confusion in this field comes from the terminology. Image analysis may involve a number of steps to recognize, select, extract, measure, and interpret the desired values. Figure 12.1 indicates a typical flow chart of the steps involved. Image processing, one of the early steps that modifies the contrast in the

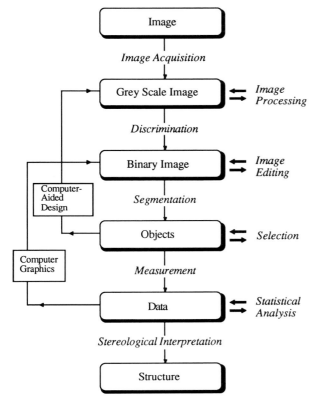

Figure 12.1 The steps involved in image analysis.

original image according to various rules or algorithms that enhance some information at the expense of others, is only a small part (and one that is not always needed, as the examples will show). Like word processing and food processing, image processing manipulates images, rearranges them, and may make them more useful or attractive, but it does not reduce (or increase) the amount of data present (comprehensive references are available in Pratt, 1978; Castleman, 1979; Rosenfeld and Kak, 1982 as well as Chapter 11 in this book).

The first basic step in image analysis is acquisition (usually by digitization from a sequentially scanned image such as a standard video signal or a scanning electron microscope image, Chapter 2). The grey scale (or color) image may be subjected to processing but is ultimately reduced to features, for instance by setting brightness thresholds to separate figure from background. This "binary" image defines the features of interest either as a set of pixels, or their boundaries. It may also require some editing to separate touching features, complete broken ones, or to smooth rough outlines. Whether this is required or not, the next step is generally measurement of the important parameters. These may be either global (pertaining to the image as a whole, and by extension to the entire sample) or feature specific. Examples of each will be shown. Many of these measurements (especially the global ones) can also be performed by hand, using human judgment to perform the recognition and discrimination steps. The computer often performs the measurements in different ways than people do, but the results obtained are the same.

The measured data must then be interpreted in terms of the type of sample and image. The two most important types are polished sections (the typical metallographic specimen) or projected images (either views through thin transparent samples, or particles dispersed on a substrate). These are discussed further in Chapters 1 and 5 respectively. Stereology provides the rules for making the proper interpretation. The results must be subjected to proper statistical interpretation to make valid comparisons between different samples.

Notice in the diagram that two additional topics are shown, computer graphics and computer-aided design. Both of these involve going from a small number of values (e.g., data on dimensions of objects) to images. Some of these graphical techniques are used to communicate the results of image analysis, but they are not central to the process.

GLOBAL PARAMETERS

Historically, the most common stereological parameters determined on materials' microstructures have been global ones. Several important structural parameters that apply to the entire microstructure (Rhines, 1986) can be obtained in a rather straightforward manner. Indeed, they have been measured by hand by many materials scientists for some time. Probably the best known of these global parameters is the volume fraction of a selected phase or structure. Incidentally, the word *phase* is used here in a more generic sense than usual in materials science, and can apply to any recognizable structure, such as a eutectic, which is not a single phase in the metallurgical sense, or to pores, coatings, etc.

It is well known that the volume fraction of a phase is identical to the area fraction that it covers on randomly oriented plane sections. This fraction is also equal to the fraction of the length of randomly drawn lines on those planes, and the fraction of the number of points randomly placed on the planes that lie within the phase boundary. In stereological notation this is stated as $V_V = A_A = L_L = P_P$ (Underwood, 1970), and it illustrates a basic feature of stereological methods: Three-dimensional information can be obtained from measurements made in a lower-dimension space. Manual determination of volume fraction is generally performed using point counting, which is remarkably efficient and has well-characterized precision. All of the lower-dimensionality measurement techniques have been used at one time or another to determine the desired three-dimensional volume fraction, V_V.

Point counting can be accomplished simply by placing a grid of points over the specimen (using an eyepiece reticule directly in the microscope or an overlay on a micrograph). The points can lie in a regular grid provided that the microstructure is itself random. Then the number of points, P_P, within the phase is determined, and from this the volume fraction is known. There are simple modifications of this procedure when the microstructure is oriented or otherwise nonrandom. The great economy and power of manual stereological tools that has led to their widespread use in characterizing materials comes from the fact that for the most part they require counting rather than measurement. Not only is this much easier for users, it is also less prone to systematic errors, and it has statistical errors that are well understood.

In computerized image analysis (Russ, 1986), instead of a sparse array of points in which two grid points rarely fall within the same grain, we have instead an array of pixels that densely sample the image plane. The result is that counting the number of pixels that lie within the phase is essentially a measure of the area of the phase, so that it is A_A that is determined. This is also equal to V_V, of course, but the error analysis is different. Instead of an error proportional to the square root of the number of points counted, we instead find that the error is proportional to the length of the perimeter around the phase, where pixels may be assigned to one phase or another based on their brightness.

Simple brightness discrimination or thresholding is often used, because it is easy to accomplish electronically and because human adjustment using the eye as a control mechanism is familiar. Figure 12.2 shows an example in which it is successful. The sample is a four-phase metal alloy, in which the phases are distinct in brightness in this light microscope image. This is shown by the brightness histogram of the image, which gives the area fraction (number of pixels) having each brightness value. Positioning discriminators on this histogram manually, or using an automatic algorithm (Weszka, 1978; Russ and Russ, 1988a) to optimize their location, allows the selection of any phase.

As shown in the figure, when the light grey phase is selected in this way, the resulting "binary" image (in which pixels are turned "on" when they lie within the phase or features of interest) also contains pixels along the boundary between the lightest and darkest phases. This is because pixels have a finite area, and where they straddle a phase boundary, the brightness is averaged. These false pixels can often be removed using an image-processing or editing step on the binary image. An "erosion" to turn off any pixel whose number of "on" neighbors is less than a preset threshold will remove those isolated pixels. Subsequently, a "dilation" to turn on a pixel whose number of "off" neighbors is less than the value will restore the pixels around the borders of the remaining phase regions. This combined operation [Figure 12.2(c)] also smooths boundaries and removes noise in the image (Serra, 1982; Coster and Chermant, 1985). In this case, it allows straightforward measurement of the area fraction of the phases. Averaging the readings over several fields on this alloy yields the following results for volume fraction: white phase, 2.5%; light grey phase, 39.8%; dark grey phase, 50.1%; black phase, 7.5%.

In many cases, simple brightness thresholding of the original grey-scale image from a light or electron microscope is not adequate to distinguish the phase of interest. As an example, consider the SEM image shown in Figure 12.3. The sample is a cast dental alloy, containing proeutectic grains and eutectic regions. (The

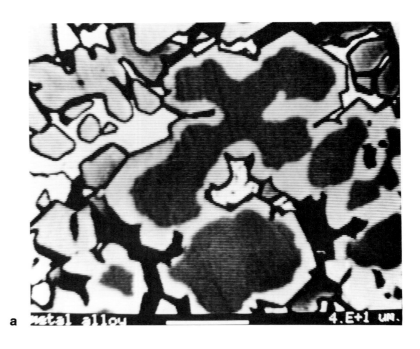

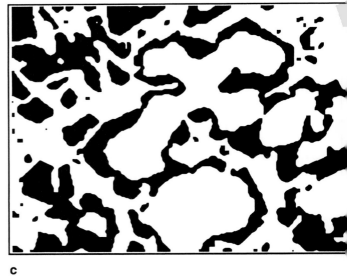

a

b

c

Figure 12.2 (a) A multiphase metal alloy in which the four different phases are distinguishable by their brightness. (b) Binary image obtained by automatically setting thresholds above and below the peak corresponding to the light grey phase in (a). (c) Result of applying an "opening" to (b) to remove pixels that are not part of the light grey phase but had the same brightness resulting from averaging the brightness values of white and dark phases that they happened to straddle in the grey-scale image.

342

proeutectic is the phase that forms first from the parent liquid before the eutectic mixture.) The response of the material to chemical etching is important, because this is used to prepare the surface for bonding. The eutectic etches to produce a rough surface that bonds well, and so it is important to know the volume fraction of the eutectic and proeutectic regions. This would also be important in many other metallurgical studies, since it would enable the direct determination of phase diagrams from metallographic examination.

However, the brightness in the image does not discriminate the two "phase" regions. This is partly due to the fact that the eutectic region includes the same material as is present in the single-phase proeutectic region. It is also compounded by the imaging contrast of the SEM. To accomplish an automatic discrimination of the phases in the computer-based image analyzer, it is helpful to think about how the human eye of the observer distinguishes the regions. It is the textural difference that we perceive as a "roughness" of the surface that is important here. Rigaut (1988) has shown that discrimination based on the fractal dimension of the grey-scale image produces results similar to human perceptions. Russ and Russ (1987) have shown that for both light and SEM images, the local pattern of contrast variation for each pixel in the image is directly correlated with its physical roughness, and that this can be used for discrimination purposes. In this case, a very simple procedure suffices to separate the two regions. To follow it, we need to introduce the idea of local operators for image processing.

A local operation is one in which the pixel and its immediate neighbors (for instance, the 8 touching neighbors in a 3 × 3 pixel region, or the 24 pixels in a 5 × 5 region) are used to produce a new image in which the pixel corresponds spatially to the original, but has a new brightness value that emphasizes some different aspect of the information. A common example (actually little used in this sort of image analysis) is image smoothing to reduce noise, by averaging together the brightness of each pixel with its neighbors.

In this instance a different local operation is used, called a *rank operator*. Each pixel in a 5 × 5 region is ranked according to its brightness, and from that list one brightness value is chosen to replace the original value. When the median value is used, the result is to smooth local variations in the image [but without blurring or shifting true edges and boundaries, as caused by the averaging method (Nagao and Matsuyama, 1979)]. It is also possible to use the minimum or maximum value. If we save the difference between the minimum and maximum brightness values in the region as the new pixel brightness, for each pixel in the original image, the result is that shown in Figure 12.3(b). This is a simple yet effective measure of the local texture in the original, since the smooth proeutectic regions become dark (there is a small difference between the brightest and darkest pixels in the neighborhood), and the eutectic regions have pixels with a greater brightness difference. The resulting image can now be thresholded to separate the two regions. Figure 12.3(c) shows the binary image in which the proeutectic region has been selected. Averaged over several fields, this produces the result that the sample consists of 41% proeutectic.

Another example in which brightness by itself does not suffice for discrimination is shown in Figure 12.4. This cast iron contains both flakes and nodules of graphite. In order to determine the volume fraction of each, it is necessary to first obtain a

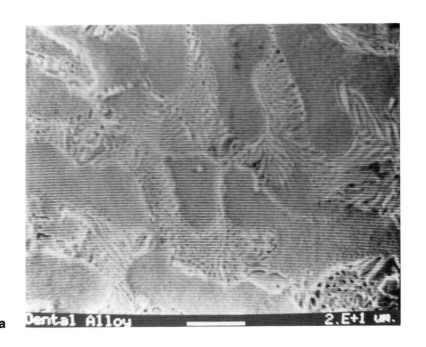

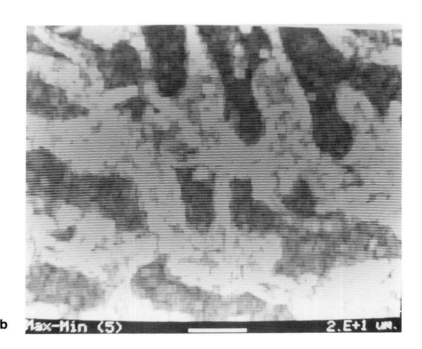

344

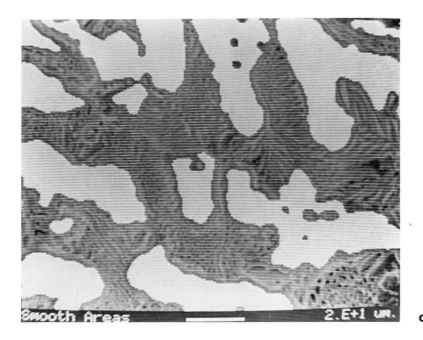

c

Figure 12.3 (a) SEM image of dental alloy showing "smooth" single-phase regions and "rough" eutectic regions. (b) Texture image from (a), in which pixel brightness is the range of values in each local 5 × 5 pixel region in the original. (c) Selection of regions in (a) by discrimination, using brightness thresholds in (b).

345

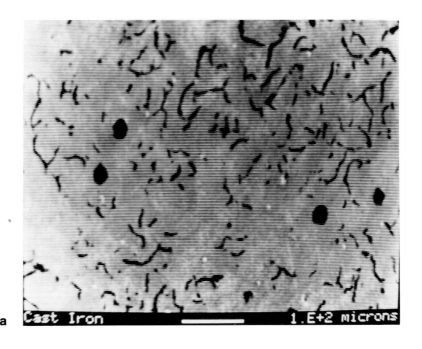

Figure 12.4 (a) Cast iron containing flakes and nodules of graphite. (b) The discriminated binary image from (a) after nodules have been rejected based on their size and shape.

346

binary image showing all of the graphite, by thresholding. This is straightforward as the graphite is darker than the metal. Then, the individual features must be examined in order to select those that are round from those that are elongated and represent sections through flakes. In a later section the measurement of individual features will be discussed. In this application, the individual data are not saved, but those features with a shape factor beyond a preset limit, perhaps combined with a constraint on the permitted size of a nodule or flake, can eliminate one class of features. The result [as shown in Figure 12.4(b)] can then be measured as before to determine the volume fraction of just the selected phase.

Figure 12.5 shows another cast iron, this one containing graphite nodules, some of which are entirely within regions of ferrite (almost pure iron). This illustrates that it may also be interesting to measure the volume fraction of two otherwise identical phase regions on the basis of their neighbor relationships. In this case, this can be accomplished by obtaining two binary images, one for the ferrite [Figure 12.5(b)] and the other for the graphite. This is readily accomplished by brightness thresholding. By filling all of the holes in the ferrite image [Figure 12.5(c)], and using Boolean logic to AND the resulting image with the graphite image, only those regions of graphite that were completely contained within the ferrite remain. The resultant image from these procedures is shown in Figure 12.5(d). In this example, the total volume fraction of the graphite is 17.8%, of which 12.1% lies within the ferrite.

S_V AND GRAIN SIZE

After volume fraction, the quantitative metallographic tool most often encountered is the "grain size." There are several different standard definitions (ASTM standard E112) and methods for determining this, and, for structures in which the grains are not all of the same size, they do not entirely agree. One of the simplest methods, and also the most meaningful for complex microstructures (containing a mixture of grain sizes or nonequiaxed grains), is the linear intercept method (Heyn, 1903). It is carried out by drawing lines of known length on the microstructure and counting the number of intersections they make with grain boundaries. The result is converted to a log scale and becomes the common ASTM grain size number.

Actually, this is a measure not of the "size" of the grains but rather of the amount of surface area of the grain boundaries. This helps to explain the well-known effects of grain size upon mechanical and electrical properties of materials.

Automatic measurement of grain size is carried out in a similar way. Instead of manually counting the number of points per length of test line (P_L) and calculating S_V as $2P_L$ (the constant comes from integrating the probability of the lines intersecting the boundaries at various angles), automatic systems usually measure the total length of grain boundary per unit area of the image (B_A). Then $S_V = (4/\pi)B_A$, where again the numerical factor arises because the boundaries may not be perpendicular to the planar surface examined.

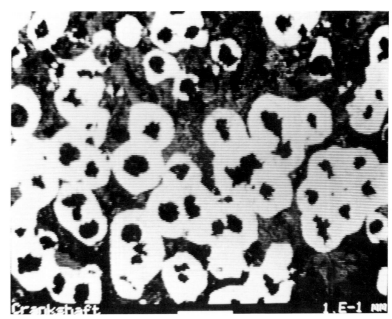

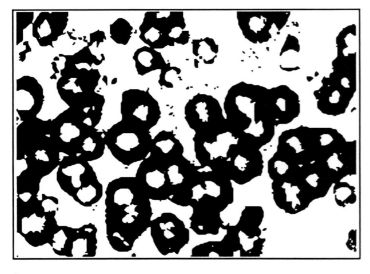

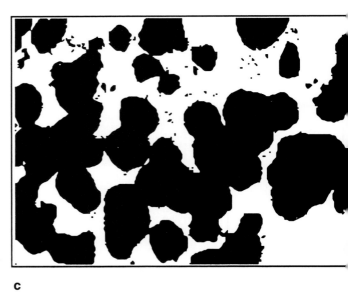

b

c

348

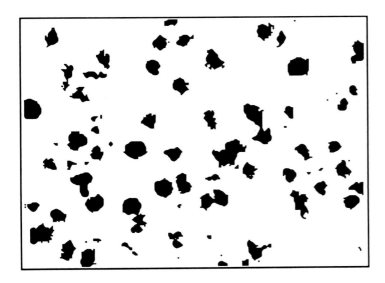

d

Figure 12.5 (a) Light microscope image of cast iron containing ferrite (light grey), pearlite (dark matrix), and graphite (dark round features). (b) Binary image of the ferrite in (a). (c) Result of automatically filling holes in (b). (d) Result of applying a Boolean AND to combine (c) with the binary image obtained by setting brightness thresholds for the graphite in (a).

Figure 12.6 shows a Hastelloy specimen in which the measurement (performed either manually or automatically) is straightforward and gives values for $S_V = 0.1814$ μm^{-1}, corresponding to ASTM grain size number 9.7, and mean intercept length (the reciprocal of P_L) 11.02 μm. However, in many cases the grain boundaries are not so ideally defined by etching as in this example. Figure 12.7 shows a low alloy steel in which the grain boundaries are indistinct, with some portions missing. Furthermore, the contrast varies so that setting brightness thresholds to select the grain boundaries is difficult. In this situation binary image processing can often be used to reconstruct the "missing" boundaries (similar to what the human operator must do when counting manually). As shown in Figure 12.7(b), this is done by starting with the skeleton of the thresholded grain boundary image. Then the assumption is made that no lines should simply end. Hence, all line ends are extended in an iterative process that halts when they touch each other or another boundary line. The result [Figure 12.7(c)] can then be measured to determine that the mean intercept length is 12.18 μm, the ASTM grain size is 9.4, and the specific surface area $S_V = 0.1641$ μm^{-1}.

In other types of materials images, the grain boundaries are not revealed by etching, but by contrast between adjacent grains. Figure 12.8 shows a light micrograph of an aluminum alloy, as an example. The grains have different brightnesses because of their different crystallographic orientation (which affects chemical etching). In other types of images, for instance from the scanning or electron micro-

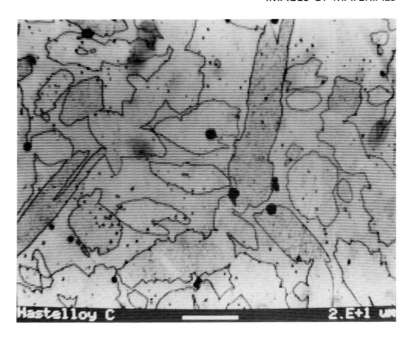

Figure 12.6 Light microscope image of Hastelloy in which the grain boundaries are well delineated after polishing and etching.

scopes, as discussed in Chapters 2 and 5, similar grain-to-grain contrast develops.

In this case, the boundaries can be delineated by using a gradient operator to locate the boundaries by defining them as places where the brightness changes rapidly in any direction on the image. The most common gradient operators (e.g., Sobel and Kirsch operators) are local or neighborhood operators, as we encountered before. They replace the original brightness value of each pixel with a value proportional to the difference between the pixel and its neighbors. As shown in Figure 12.8(*b*), this produces an image in which all of the grains are uniformly dark, and the boundaries have greater brightnesses that depend upon the contrast formerly present across the boundary. This processed image can then be thresholded and the grain boundaries imaged for measurement, as shown in Figure 12.8(*c*).

Notice that there is still some room for decision about which boundaries to include. Some of the relatively low-contrast boundaries have a smaller gradient image brightness, and may be left out. Images in which twin boundaries are present may require manual editing to delete or ignore these lines. This is no different from a manual measurement in which the human decides whether a boundary is present in a particular location, except that the results tend to be more consistent because the decision is made explicitly and the operator can see the results directly on the image.

Closely related to S_V measurements is the idea of contiguity. In a three-phase

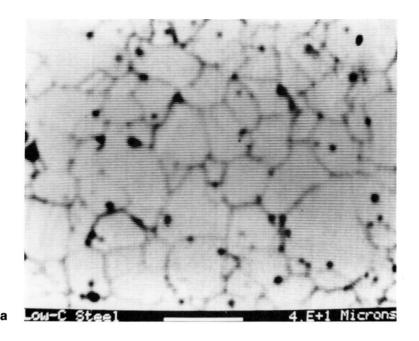

a Low-C Steel   4.E+1 Microns

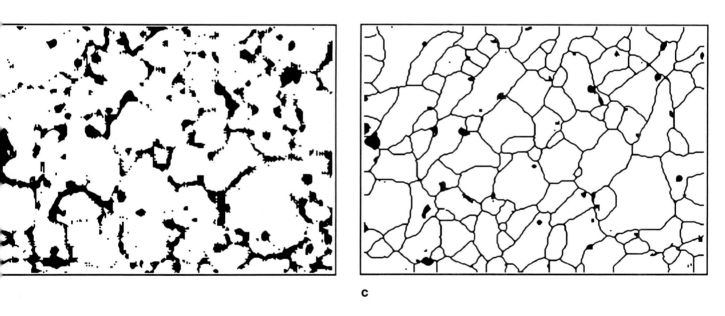

c

Figure 12.7 (a) Video image of low-carbon steel showing an incompletely delineated grain structure. (b) Binary image produced by selecting all dark pixels in (a). (c) Binary image of grain boundaries and carbides suitable for grain size measurement, obtained by skeletonizing (b) and extending all lines.

351

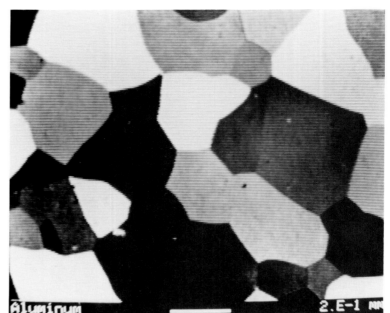

a   Aluminum   2.E-1 NM

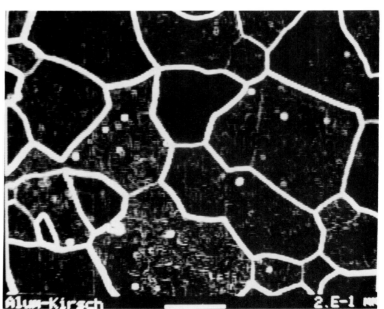

b   Alum-Kirsch   2.E-1 NM

352

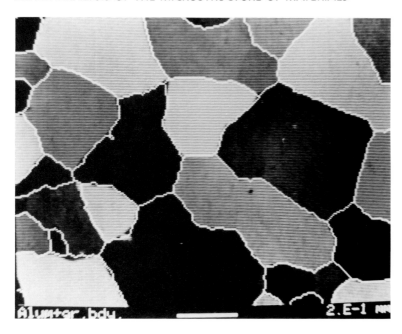

c

Figure 12.8 (a) Light metallographic image of an aluminum alloy. (b) Result of applying a Kirsch edge-finding (gradient) operator to (a). Brightness is proportional to the maximum value of a derivative in each of eight directions. (c) Superposition of the grain boundaries from (b) on the original image of (a).

alloy, for instance, there can be six different kinds of boundaries (α-α, α-β, α-χ, β-β, β-χ, χ-χ) between the different phases. It may be important to measure these separately, and thereby determine which phases are neighbors and which are not. Figure 12.9(a) shows an example of such a three-phase alloy. In manual operation, this would require counting separately the number of intercepts per unit line length for each type of boundary. As shown in Figure 12.9(b), this can be readily accomplished by the automatic methods by first obtaining binary images for each phase and using dilation to enlarge each phase region (left side of the figure). Subsequently, Boolean logic is used to combine the various phase outline regions (right side of the figure) with an AND function to obtain just the boundaries that lie between selected phases [Figure 12.9(c)]. The normalized fraction of these line lengths then gives the fraction of phases that touch. In this example, about 66% of the white phase boundary touches the grey phase and 34% touches the black phase, whereas 52% of the black phase boundary touches the white phase versus 48% touching the grey phase.

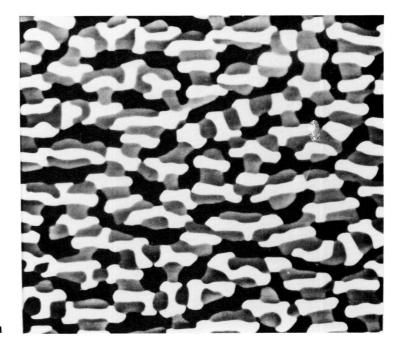

a

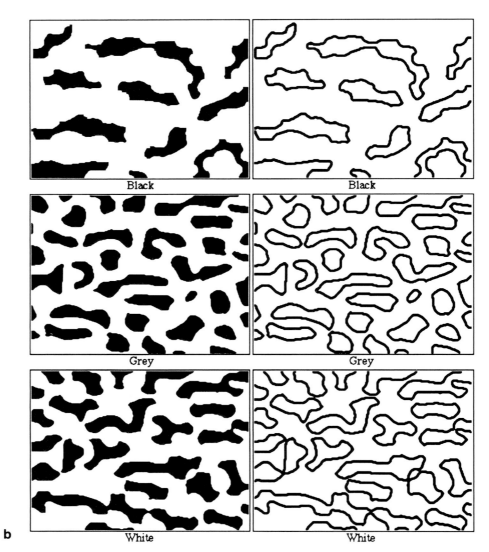

Black   Black

Grey   Grey

White   White

b

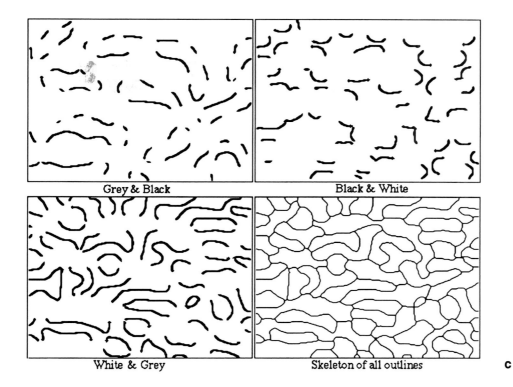

Grey & Black   Black & White

White & Grey   Skeleton of all outlines   **c**

Figure 12.9 (a) SEM backscattered electron image of a three-phase alloy (Ball and Mc-Cartney, 1981). (b) Binary images of individual phases in the alloy in (a) and the outlines around each phase obtained by dilating and Ex-ORing with the original image. (c) AND combinations of pairs of phase outlines to measure contiguity, and the skeleton of the OR of all phase outlines.

FEATURE SPECIFIC MEASUREMENTS

Although global measurements have been shown to describe concisely and power-fully the structural parameters in a variety of materials applications, there is also a need for measurement and analysis of individual objects. The feature viewed in the microscope image (either a section through the object or its projected shadow dimension) cannot be related uniquely to the object dimension for a single object and feature. But by measuring large number of features, a statistically valid rela-tionship between the feature measurements and the object dimensions can be deter-mined based on geometric probability (Kendall and Moran, 1963; Matheron, 1975; Miles and Serra, 1978).

Measurements on features can be grouped into four basic categories. The first and most familiar of these is size, but several different measures can be used. For instance, in a two-dimensional image, the size of a feature can be taken as the area within the feature outline including or excluding any internal holes, the perimeter (length of the outline), the area within a convex outline (the so-called taut-string boundary) or the length of that boundary, the longest chord (maximum Feret's diameter or caliper diameter), the minimum caliper diameter, etc. These can in turn be related to a variety of object size parameters, including volume, surface area, equivalent spherical diameter (the diameter of a sphere with the same volume), fiber length, and so forth. Choosing the parameters that are most relevant to a particular study (e.g., comparing samples, or finding the parameter that correlates best with treatment history or service performance) is an important and highly nontrivial step we will not consider here.

In addition to these measures of size, there are measures of shape, position, and brightness, the remaining three categories of properties. Brightness (including color, for a light microscope image) is often related to phase identification, depend-ing on sample preparation. In the SEM, the backscattered electron image has a brightness that is a monotonic function of mean atomic number (as discussed in Chapter 2). For rough surfaces (fractures, particulates, etc.) the texture (variation of brightness among neighbor pixels in the image) has been shown to correlate with the physical surface roughness and the surface fractal dimension (Pentland, 1983; Russ and Russ, 1987). Of course, as was shown before, image texture can also be used to discriminate features from their surroundings, either for global or local measure-ments (Haralick et al., 1973).

Position parameters include both the location and the orientation of features and are important for studies of gradients and uniformity. Examples will be shown below. Shape is a major subject in its own right (Exner, 1986), which for practical reasons of space will not be examined in detail here. Measures of shape are less familiar to humans because we do not have well-understood words in our common English vocabulary to express these concepts. But the recognition of shape is an important visual capability strongly selected by evolution, and physiological tests indicate that shape recognition begins right in the eye, long before information is transmitted to higher levels of image analysis in the brain. It was, for example,

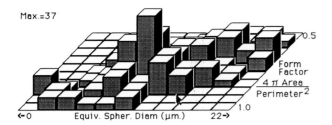

Figure 12.10 Two-way frequency histogram plot. The horizontal axis is the size (equivalent spherical diameter); vertical axis is the shape (form factor) for paint pigment particles.

much more important for early humans to recognize the difference in shape between a gazelle and a lion than to be able to estimate the absolute size of the gazelle. Shape recognition is very important in image analysis as well, and it has been used as an important parameter in expert systems that recognize and classify objects (Russ and Rovner, 1988).

The types of shape descriptors that are available include dimensionless ratios of size parameters (for instance, *form factor* is defined as 4π times area divided by perimeter squared, which is 1.0 for a circle and smaller for more irregular shapes, while *aspect ratio* is the ratio of the maximum caliper dimension to the minimum caliper dimension). There are a variety of these descriptors that tend to emphasize some particular aspect of shape, convexity, and so forth. Correlation of shape with feature size or position is often diagnostic of the object's history. For instance, Figure 12.10 shows such a plot for paint pigment particles. The trend of decreasing form factor with size is often associated with agglomeration, in which the larger particles are made up from many smaller ones, each of which is approximately spherical.

The number of internal holes or external points would also fit into the shape category, as would the radius of curvature, although it is not a dimensionless quantity. The fractal dimension of a feature outline can be measured and is particularly useful for many natural objects, which have this particular type of self-similar irregularity (Mandelbrot, 1977; Kaye, 1984; Mecholsky and Passoja, 1985; Underwood, 1987). Finally, there is harmonic analysis, in which the feature outline is "unrolled" as a plot of radius versus angle, and then subjected to Fourier analysis (Schwartz and Shane, 1969; Ehrlich and Weinberg, 1970; Beddow et al., 1977). The coefficients of the various frequency terms are often difficult to relate to the shape as perceived by human observers, but statistical analysis of chaotic populations of objects (e.g., sand grains from different sources or produced by different weathering histories) often shows the ability to distinguish the classes of objects by these coefficients.

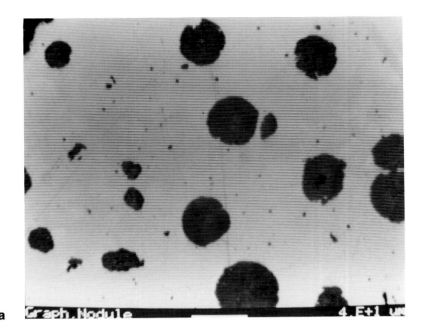

a Graph.Nodule   4.E+1 um

SIZE

Figure 12.11 shows a light microscope image of a cast iron containing carbon nodules. The nodules appear as circular features, and we have no difficulty in convincing ourselves that this reflects a three-dimensional shape that must be a sphere. Measurement of the size of the circles is straightforward, and, by scanning the stage of the microscope to multiple locations on the sample, it is possible to accumulate enough observations to plot a histogram of the size distribution of the features. There are several ways that these data can be plotted. The use of a linear axis for the equivalent circular diameter (the diameter of a circle having the same area as each measured feature, to compensate for slight irregularities in shape) shows a large number of small features and a small number of features spread over a range of larger sizes [Figure 12.11(b)]. This distribution is not easily characterized or interpreted.

If a logarithmic axis is used instead, a clearer pattern emerges, as shown in Figure 12.11(c). The sample contains two distinct populations of features, one much larger than the other (presumably arising from formation at different stages in the metal's solidification and treatment). The use of logarithmic scales for these types of plots is common in materials' description since many of the processes of phase transformation and growth, as well as grinding and other treatment histories, produce distribu-

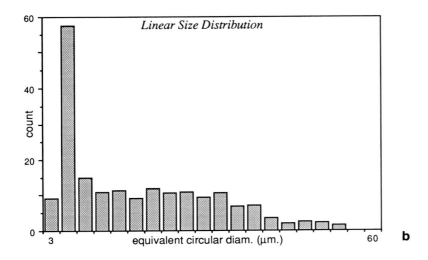

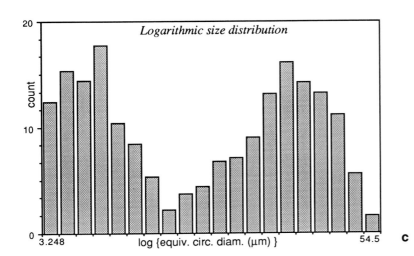

Figure 12.11 (a) Light microscope image of a cast iron specimen containing spherical graphite nodules. (b) Distribution showing the number of features (sections through carbon nodules) as a function of size for multiple fields of the sample shown in (a). (c) The same data as in (b), but with the distribution based on the logarithm of the equivalent circular diameter. The presence of two populations is now evident.

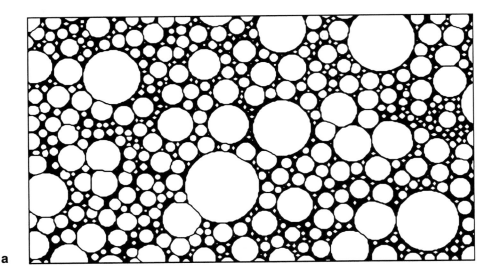

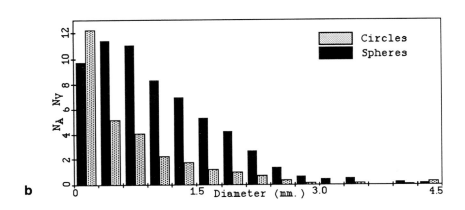

Figure 12.12 (a) Pores or "cells" in bread showing circular features visible on a planar section. (b) Frequency distributions for N_A (number of circles per unit image area, based on circle diameter) and N_V (number of spheres per unit volume, based on the sphere diameter) for the pores shown in (a).

360

tions that are at least approximately log-normal (i.e., give rise to a normal or Gaussian plot of frequency on a logarithmic size scale).

However, it is important to remember that the (geometric) mean circle diameter is not the same as the size of the sectioned sphere. On the average, if a large number of identical spheres are sectioned at random, the mean circle diameter will be $\pi/4$ times the equatorial diameter of the sphere. Furthermore, a complete histogram of circle sizes expected from sectioning uniform size spheres can be generated. For a sphere this can be done analytically, but for more complex shapes a Monte Carlo sampling program is usually the most efficient way to generate the distribution. From this distribution a matrix of probability values of finding a circle of size class i resulting from sectioning a sphere of size class j can be computed. Inversion of this matrix and its multiplication by the measured circle size distribution histogram produces a histogram for the three-dimensional sphere sizes. This operation is referred to as "unfolding" and can also be performed for other shapes such as ellipsoids, polyhedra, cylinders, etc. (Wicksell, 1925; DeHoff, 1962; Cruz-Orive, 1976; Weibel, 1980).

Spheres are the simplest shape, and arise fairly often in real applications. Figure 12.12 shows a macroscopic binary image of bread, in which the "cells" (bubbles in the dough) are distinguished. The distribution of circle sizes, and the calculated distribution of sphere sizes that results from the unfolding, are shown in Figure 12.12(b). Notice that this provides a value for the number of spheres per unit volume N_V, which cannot be determined otherwise from a two-dimensional plane image.

A few comments should be added, however. First, the unfolding works only in the aggregate—we still know nothing about the size of any individual object. Second, the matrix of values generated from geometric probability depends critically on our assumption of object shape. If this is incorrect, or if the real objects have varying shapes (and especially if those shapes vary systematically with size), then the results of the unfolding can be quite wrong. It is in principle possible to unfold both shape and size at the same time, but this requires measurement of many more features for adequate statistical precision, and still demands a proper interpretation of the shapes that are present.

In order to measure such a simple parameter as area to characterize feature size and hence through the principles of stereology reach a description of three-dimensional objects, it is sometimes necessary to carry out prior image processing or editing steps. The common processing operations shown before in the context of global measurements still apply, of course, but it is now also necessary to separate any individual features that touch, so they can be measured separately. Sometimes this touching is present in the image because of the finite pixel size, and sometimes it arises because the features really are in contact. Figure 12.13 shows an example of sintered particles in which the latter is true. The first step in the process is to produce a binary image as shown in Figure 12.13(b). In order to measure the individual feature sizes (which would then be converted to particle sizes using a sphere or similar model for unfolding), it is necessary to disconnect them.

An automatic method for accomplishing this is known as "watershed segmenta-

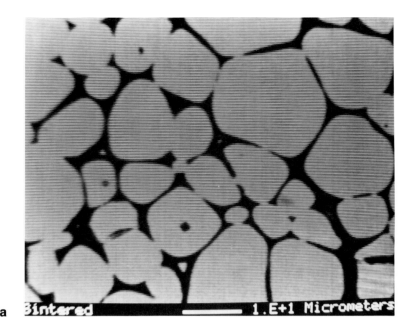

a

b

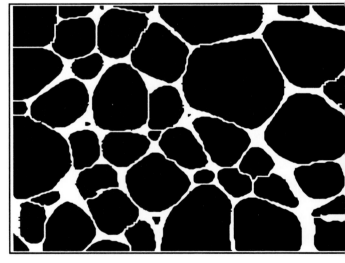

c

Figure 12.13 (a) Section through sintered compact of irregular particles of varying sizes. (b) Binary image produced by thresholding (a). (c) Segmentation of features in (b) using the convex watershed method.

362

tion" (Beucher and Lantejoul, 1979; Coster and Chermant, 1985; Russ and Russ, 1988b). It proceeds by eroding the binary image until each feature disappears (having separated from its neighbors along the way). Then each ultimate eroded point is dilated again until the original feature boundary is reached, but with rules that forbid new connections to form between different features. The result is the lines of separation shown in the image in Figure 12.13(c). Note that this method is very different from a superficially similar but inadequate technique in which erosion is used to separate the features, and they are then dilated back to their original size. The latter method works only if all of the features are similar in size; in the general case of particles covering a range of sizes, the small features may be completely erased before the larger ones have been separated.

The examples for feature-specific size measurement have thus far illustrated applications in which the images are of sections through the material structure. Projected images, especially of particulates, are also commonly encountered and call for a different analytical approach. Figure 12.14 shows an SEM image of alumina powder dispersed on a substrate using camphor–napthalene wax (Thaulow and White, 1971). Measurement of feature size gives an external caliper dimension in this case, but surface irregularities that reduce the volume and increase the surface area compared to that predicted by a smooth and convex model cannot be distinguished. However, models to estimate the volume (usually expressed as an equivalent spherical diameter) and surface area, usually in terms of the feature's projected area, maximum and minimum caliper diameter, and perimeter, can be based on an assumed shape of an ellipsoid of revolution.

As for many particulates, this powder has a log-normal size distribution. The plots in Figure 12.14(c) show the histograms of frequency (number of particles) versus diameter on linear and log axes; the logarithmic plot has a generally symmetrical and "normal" shape. In many particulate processes it is more meaningful to deal with the total volume of particles that exceed a certain size (comparable to the results of a sieve test). When the cumulative volume is plotted, the graph in Figure 12.14(c) shows that the small number of large particles dominates the sample. If this plot is made on a probability scale, the result is a straight line for a log-normal material. In the example, the few percent deficit in small features (causing the line to fall below the straight overall trend) results from the systematic under-counting of small features, either because they have hidden under or adhered to larger ones, because they had too little contrast to show in the SEM image, or because they had a greater tendency to fall off the slide and escape measurement.

POSITION

The location of individual features is usually given in terms of the X, Y coordinates of the centroid. This can be used to measure gradients or neighbor distances. Figure 12.15 shows an example of the type of situation that can arise when surface treatments produce variations in the size or number of second-phase particles. Depend-

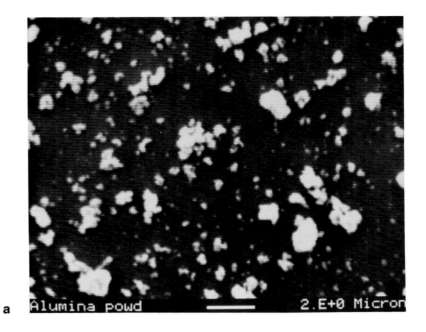

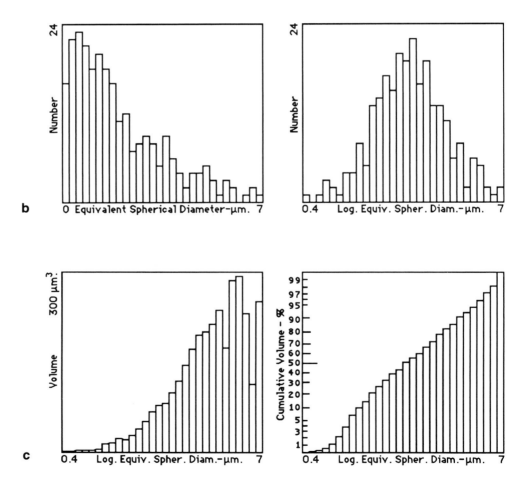

Figure 12.14 (a) Alumina particles viewed in an SEM. (b) Distribution plots for the particles in (a) plotted on linear and log scales. (c) Data from (b) showing volume plots with linear and cumulative probability scales. Note the deficit at small sizes.

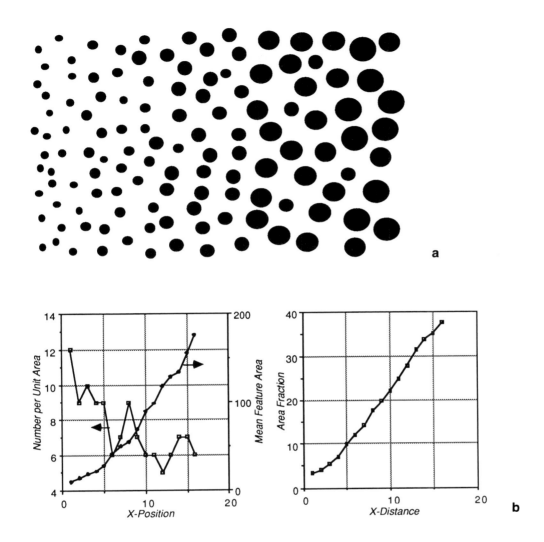

Figure 12.15 (a) Schematic diagram showing a gradient of feature size and density. (b) Distribution plots of feature count and area, and total area fraction versus *x* position, for (a).

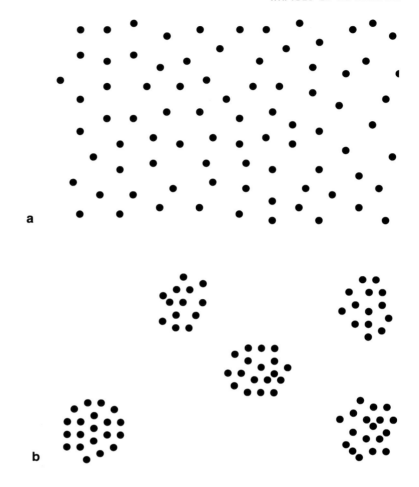

ing on whether that treatment alters the composition or simply applies heat treatment to the material, this may produce a variation in the number of features, their individual size, and/or their total volume fraction. Plots of these parameters are easily obtained once the feature-specific data have been collected. From the plots, meaningful characterization of the thickness of the treated layer can be determined.

Another important property of second-phase particles is their spatial distribution, which can be either random, ordered, or clustered (Schwarz and Exner, 1983; Hare et al., 1988). This can be determined from a histogram of nearest-neighbor distances, which can in turn be derived from the table of feature locations. Computers are well suited to sorting through lengthy tables of coordinates to find and rank the nearest-neighbor distances. As shown in Figure 12.16, departures from the Poisson distribution of nearest-neighbor distances expected for a random distribution [Figure

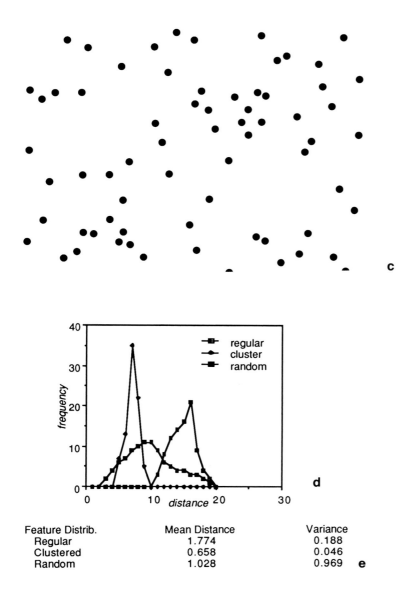

Feature Distrib.	Mean Distance	Variance
Regular	1.774	0.188
Clustered	0.658	0.046
Random	1.028	0.969

Figure 12.16 Binary images with (a) regular, (b) clustered, (c) random feature distributions, (d) their nearest-neighbor distance distributions, and (e) descriptive statistics (mean and standard deviation) ratioed to those for an ideal random Poisson distribution.

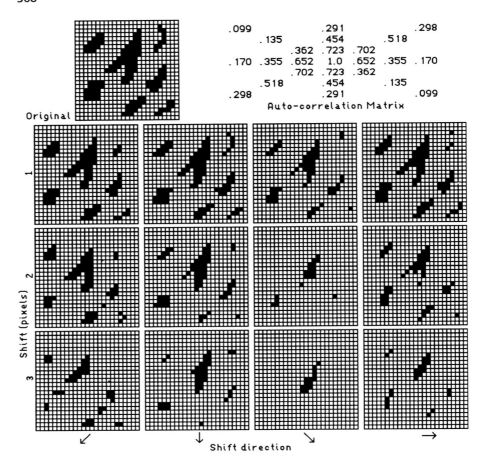

.099					.291				.298
	.135				.454			.518	
		.362		.723	.702				
.170	.355	.652	1.0	.652	.355	.170			
		.702	.723	.362					
	.518				.454			.135	
.298					.291				.099

Auto-correlation Matrix

Original

Shift (pixels)   1   2   3

a

↙   ↓   ↘   →

Shift direction

12.16(c)] of features in either two or three dimensions sensitively identify the presence of regular ordering [Figure 12.16(a)] by a reduction in the standard deviation and increase in the mean value of neighbor distances, or of clustering [Figure 12.16(b)] by a reduction in the mean value. Because of the impracticality of obtaining this type of data by manual measurement, it has only been with the advent of computerized image-analysis systems that such techniques have become widely available. Figure 12.16(d) shows the measurement results.

Preferred orientation can be studied either by manual or computerized methods. The manual technique is to count the number of intercepts per unit length of test line in various directions. This can also be automated in several different ways, including measuring the orientation of the boundary line around features or measurement of feature projected dimensions in several directions. It is also possible to use image-processing methods to shift the image in various directions and AND it with the original. In this autocorrelation method illustrated in Figure 12.17(a), the extent of feature overlap is a measure of the feature dimensions in each direction. This can

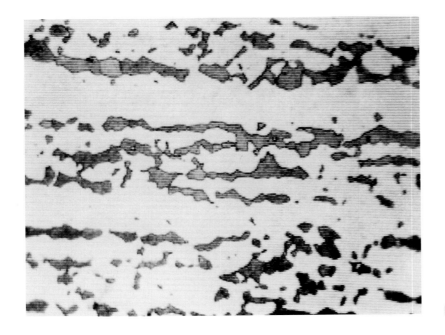

b

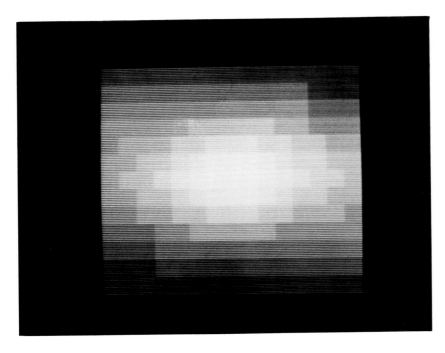

c

Figure 12.17 (a) Autocorrelation performed by shifting a binary image and counting the overlaps. (b) Light metallographic image of a copper alloy showing banding (preferred orientation). (c) Magnitude of autocorrelation results from (b) plotted as a grey-encoded image.

369

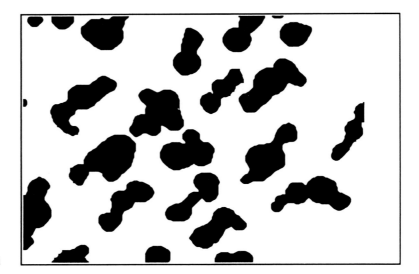

a

be displayed in graphical form, as shown in Figure 12.17(c) for the oriented structure in Figure 12.17(b).

The intercept length data can be used to calculate numerical indices of preferred orientation but for ease of human understanding are most often presented graphically, as shown in Figure 12.18 for the case of pores [Figure 12.18(a)]. Notice that the plot is much more meaningful when presented as a rose of orientation [Figure 12.18(c)], which directly shows both the direction and degree of the orientation of the pores.

SUMMARY

Computer-assisted image analysis is extremely useful for characterizing materials' microstructures. While manual methods are very efficient and powerful for determining simple global properties, they rely on human consistency in interpreting the images and are limited in the number of samples for which data can realistically be accumulated. Computer-based methods are capable of great flexibility to handle many different kinds of images, with processing or editing functions available to deal with specific problems of separating the phase or features of interest from their surroundings. The computer minimizes human measurement errors and can collect large amounts of data for meaningful analysis. In addition, the use of an interactive and heavily graphic computer display to examine data relationships encourages an interactive search for meaningful correlations, which would be impractical by hand.

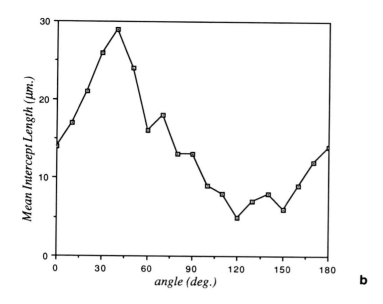

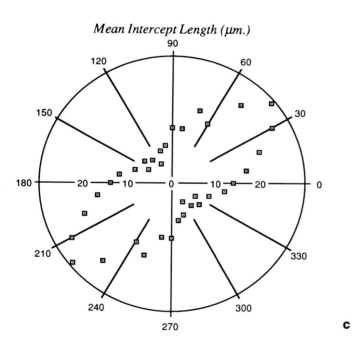

Figure 12.18 (a) Binary image of sections through pores in green compact, produced by brightness thresholding. (b) Histogram of mean intercept length as a function of angle, determined on multiple fields similar to (a). (c) Data from (b), plotted in polar coordinates as a rose of orientation for the features. The degree of preferred orientation and its angle are now evident.

A comprehensive presentation of techniques for computerized image analysis is available in Russ (1990).

REFERENCES

American Society for Testing and Materials, Committee E-4 (Metallography), Standard E112, Philadelphia, Penn.

Ball, M. D., and McCartney, D. G. (1981). *J. Microscopy* **124**, 57.

Beddow, J. K., Philip, G. C., and Vetter, A. F. (1977). *Powder Tehnol.* **18**, 15.

Beucher, S., and Lantejoul, C. (1979). *Proc. Intl. Workshop on Image Processing*, CCETT, Rennes, France.

Castleman, K. R. (1979). *Digital Image Processing*. Prentice-Hall, Englewood Cliffs, N.J.

Coster, M., and Chermant, J-L. (1985). *Précis D'Analyse D'Images*, CNRS, Paris.

Cruz-Orive, L-M. (1976). *J. Microscopy* **107**, 235.

DeHoff, R. T. (1962). *Trans. AIME* **224**, 474.

Ehrlich, R., and Weinberg, B. (1970). *J. Sediment. Petrol.* **40**, 205.

Exner, H. E. (1986). *Proc. 1st Intl. Conf. on Microstructology*, Gainesville, Fla.

Haralick, R. M., Shanmugam, K., and Dinstein, I. (1973). *IEEE Trans. Syst. Man Cybern.* **SMC-3**, 610.

Hare, T., Russ, J., C., and Lane, J. (1988). *J. Electron Microscopy Tech.* **10**, 1.

Heyn, E. (1903). *Metallographist* **6**, 37.

Kaye, B. H. (1984). *Particle Characterization* **1**, 14.

Kendall, M. G., and Moran, P. A. (1963). *Geometrical Probability*. No. 10, Griffith's Statistical Monographs, Charles Griffith, London.

Mandelbrot, B. B. (1977). *The Fractal Geometry of Nature*. Freeman, N.Y.

Matheron, G. (1975). *Random Sets and Integral Geometry*, Wiley, N.Y.

Mecholsky, J. J., and Passoja, D. E. (1985). *Fractals and Brittle Fracture*, in *Fractal Aspects of Materials*, Materials Research Society Abstracts, MRS, Pittsburgh.

Miles, R., and Serra, J. (eds.). (1978). *Geometrical Probability and Biological Structure*. Springer-Verlag, Berlin.

Nagao, M., and Matsuyama, T. (1979). *Computer Graphics, Vision and Image Processing* **9**, 394.

Pentland, A. P. (1983). *Proc. IEEE Comput. Soc. Conf. Comput. Vision Patt. Recog.*, Washington D.C., p. 201.

Pratt, W. K. (1978). *Digital Image Processing*. Wiley, New York.

Rhines, F. N. (1986). *Microstructology: Behavior and Microstructure of Materials*. Dr. Riederer-Verlag, Stuttgart.

Rigaut, J. P. (1988). *J. Microscopy* **150**, 21.

Rosenfeld, A., and Kak, A. C. (1982). *Digital Picture Processing*. Academic Press, N.Y.

Russ, J. C. (1986). *Practical Stereology*. Plenum Press, N.Y.

————. (1990). *Computer Assisted Microscopy: The Analysis and Measurement of Images*. Plenum Press, N.Y.

————, and Rovner, I. (1989). *American Antiquities* **54**, 784.

————, and Russ, J. C. (1987). *Particle Characterization* **4**, 22.

————, and Russ, J. C. (1988a). *J. Microscopy* **148**, 263.

————, and Russ, J. C. (1988b). *Acta Sterologica* **7(1)**, 33.

Schwartz, H. P., and Shane, K. C. (1969). *Sedimentology* **13**, 213.

Schwarz, H. and Exner, H. E. (1983). *J. Microscopy* **129**, 155.

Serra, J. (1982). *Image Analysis and Mathematical Morphology.* Academic Press, London.

Thaulow, N., and White, E. W. (1971). *Powder Technol.* **5**, 377.

Underwood, E. E. (1970). *Quantitative Stereology.* Addison Wesley, Reading, Mass.

———. (1987). *Acta Stereologica* **6**, Suppl. II, 169.

Weibel, E. R. (1980). *Stereological Methods,* Vols. 1 and 2. Academic Press, London.

Weszka, J. (1978). *Computer Graphics, Vision and Image Processing* **7**, 259.

Wicksell, S. D. (1925). *Biometrica* **17**, 84.

Index